T0400901

ADVANCES IN FOOD SAFETY AND FOOD MICROBIOLOGY

BACTERIOPHAGES IN DAIRY PROCESSING

ADVANCES IN FOOD SAFETY AND FOOD MICROBIOLOGY

Dr. Anderson de Souza Sant'Ana
And Dr. Bernadette D. G. M. Franco (Series Editors)
Department of Food and Experimental Nutrition,
Faculty of Pharmaceutical Sciences, University of Sao Paulo, San Paulo, Brazil

Food safety is a specific area of food science focused in the study of microbiological, chemical and physical hazards in foods and beverages.Several tools, systems and approaches have been developed last years to systematize the management of food production chain aiming at protecting public health.Food Microbiology is an interesting and exciting field of microbiology concerned with the study of foodborne microorganisms, their occurrence, interactions and responses to the environmental found in foods and beverages. Depending mainly on the intrinsic or extrinsic properties of foods and beverages, one or more of microorganisms such as protozoan, viruses, yeasts, moulds and bacteria or toxins will be of significance.

Food safety is a specific area of food science focused in the study of microbiological, chemical and physical hazards in foods and beverages. Several tools, systems and approaches have been developed in recent years to systematize the management of food production chain aiming at protecting public health.

Food Microbiology is an interesting and exciting field of microbiology concerned with the study of foodborne microorganisms, their occurrence, interactions and responses to the environmental found in foods and beverages. Depending mainly on the intrinsic or extrinsic properties of foods and beverages, one or more of microorganisms such as protozoan, viruses, yeasts, moulds and bacteria or toxins will be of significance.

In fact, Food Microbiology and Food Safety can be considered a "boiling and changing" field of microbiology and food science, respectively. These fields have risen to that status mainly in the last 30 years. Data from epidemiological studies of foodborne disease outbreaks or from investigations of spoilage episodes have highlighted relevant changes in the pillars supporting the microbial ecology of foods: consumer habits, practices and consumers health status, food preservation techniques, responses of microorganisms to preservation methods, microbial expression of virulence factors and food production chain systems.

In addition, a strong trend in food microbiology and safety is the application of computational, statistical, molecular biology and chemical approaches, resulting in a deeper approach to study and understand the microbial interactions with foods and among them. Trends indicate that the insertion of these four sciences into food safety and microbiology will make an important difference in the evolving of studies of microbial ecology and food safety. Based on this scenario, it is the main aim of the Series "Advances in Food Safety and Food Microbiology" to cover "hot topics" in the field of Food Microbiology and Food Safety. Coverage will be given to applied food microbiology and safety with a collection of texts focused on the study of spoilage, pathogens and industrial applications of foodborne microorganisms. Focus will be given to responses of microorganisms to preservation methods, consumer food safety practices, advances in microbiological methods and on the applications of statistical, molecular biology, computational and the "omics" in food microbiology and safety. Risk analysis, integration between epidemiology and food microbiology, chemical and physical contaminants of foods are also given consideration.

We hope that the books in this Series will be widely used by food microbiologists and those concerned with food safety in their studies or as references for new approaches to be considered in an effort to continuously evolve food microbiology and food safety from farm to fork, from tradition to technology.

Probiotic and Prebiotic Foods: Technology, Stability and Benefits to Human Health
Nagendra P. Shah, Adriano Gomes da Cruz, and Jose de Assis Fonseca Faria (Editors)
2011. ISBN: 978-1-61668-842-4 (Hardcover)
2010. ISBN: 978-1-61728-825-8 (ebook)

Stress Response of Foodborne Microorganisms
Hin-chung Wong (Editor)
2011. ISBN: 978-1-61122-810-6 (Hardcover)
2011. ISBN: 978-1-61324-410-4 (ebook)

Advances in Post-Harvest Treatments and Fruit Quality and Safety
Manuel Vázquez and José A. Ramírez de Leon (Editors)
2011. ISBN: 978-1-61122-973-8 (Hardcover)
2011. ISBN: 978-1-61470-700-4 (ebook)

New Trends in Marine and Freshwater Toxins: Food and Safety Concerns
Ana G. Cabado and Juan Manuel Vieites (Editors)
2011. ISBN: 978-1-61470-324-2 (Hardcover)
2011. ISBN: 978-1-61470-398-3 (ebook)

Clostridium Botulinum: A Spore Forming Organism and a Challenge to Food Safety
Christine Rasetti-Escargueil and Susanne Surman-Lee (Editors)
2011. ISBN: 978-1-61470-575-8 (Hardcover)
2011. ISBN: 978-1-61470-653-3 (ebook)

Enterococcus and Safety
Teresa Semedo-Lemsaddek, Maria Teresa Barreto-Crespo, and Rogério Tenreiro (Editors)
2011. ISBN: 978-1-61470-569-7 (Hardcover)
2011. ISBN: 978-1-61470-664-9 (ebook)

Essential Oils as Natural Food Additives: Composition, Applications, Antioxidant and Antimicrobial Properties
Luca Valgimigli (Editors)
2011. ISBN: 978-1-62100-241-3 (Hardcover)
2012 ISBN: 978-1-62100-282-6 (ebook)

On-Farm Strategies to Control Foodborne Pathogens
Todd R. Callaway and Tom S. Edrington (Editors)
2011.ISBN: 978-1-62100-411-0 (Hardcover)
2012. ISBN: 978-1-62100-480-6 (ebook)

Molecular Typing Methods for Tracking Foodborne Micoorganisms
Steven L. Foley and Rajesh Nayak (Editors)
2011. ISBN: 978-1-62100-643-5 (Hardcover)
2012. ISBN: 978-1-62100-728-9(ebook)

Progress on Quantitative Approaches of Thermal Food Processing
Vasilis P. Valdramidis and Jan F. M. Van Impe (Editors)
2011.ISBN: 978-1-62100-842-2 (Hardcover)
2012. ISBN: 978-1-62100-899-6 (ebook)

Pathogenic Vibrios and Food Safety
Yi-Cheng Su
(Editor)
2011. ISBN: 978-1-62100-866-8
(Hardcover)
2012. ISBN: 978-1-62100-903-0 (ebook)

Foodborne Protozoan Parasites
Lucy J. Robertson and Huw V. Smith
(Editors)
2011. ISBN: 978-1-61470-008-1 (Hardcover)
2012. ISBN: 978-1-62100-903-0 (ebook)

Predictive Mycology
Philippe Dantigny and Efstathios Z. Panagou
(Editors)
2011. ISBN: 978-1-61942-675-7 (Hardcover)
2012. ISBN: 978-1-61942-684-9 (ebook)

Bacteriophages in Dairy Processing
Andrea del Luján Quiberoni and Jorge Alberto Reinheimer
(Editors)
2012. ISBN: 978-1-61324-517-0 (Hardcover)

ADVANCES IN FOOD SAFETY AND FOOD MICROBIOLOGY

BACTERIOPHAGES IN DAIRY PROCESSING

ANDREA DEL LUJÁN QUIBERONI
AND
JORGE ALBERTO REINHEIMER
EDITORS

Nova Science Publishers, Inc.
New York

Library of Congress Cataloging-in-Publication Data

Bacteriophages in dairy processing / editors, Andrea del Lujan Quiberoni and Jorge Alberto Reinheimer.
p. ; cm.
Includes bibliographical references and index.
ISBN 978-1-61324-517-0 (hardcover)
1. Bacteriophages. 2. Dairy microbiology. I. Lujan Quiberoni, Andrea del. II. Reinheimer, Jorge Alberto.
[DNLM: 1. Bacteriophages. 2. Dairy Products--microbiology. 3. Dairying. QW 161]
QR342.B3383 2011
579.2'6--dc23

2011013651

Published by Nova Science Publishers, Inc. † New York

CONTENTS

PREFACE

Fermented dairy products represent one of the oldest forms of biotechnology practiced by mankind. Essential to this process is the growth and development of a highly specialized and specifically selected group of microbes collectively known as Lactic Acid Bacteria (LAB) that bring about the bioconversion of milk to one of a myriad of unique fermented dairy foods. As the original purpose of dairy fermentations was that of preservation, one can only expect this oft overlooked functionality will continue to be of importance as the ever increasing global population continues to stress the world food supply. Beyond the basic production of lactic acid, microbial metabolism is largely responsible for the infinite varieties of organoleptic properties that distinguish the fermented dairy products. Today, the economic value of the global dairy industry is estimated in the billions and is dependent on the consistent and reliable activity of LAB.

As within all microbial ecosystems, dairy fermentations are susceptible to bacteriophage (phage) attack. The omnipresence of these microbial viruses and the very nature of the industrial manufacture of fermented dairy products preclude their practical elimination. Not long after the biological discovery of bacteriophage, Whitehead and Cox established in 1935 their involvement as a causative agent of starter culture inhibition, which can occasionally lead to significant economical losses. Since this finding, a lifetime of research has revealed a rich and complex interaction between the LAB hosts and their respective phages. Beyond the obvious intent to control and mitigate their impact, the cumulative studies of LAB phages have contributed greatly towards the basic processes of these biological entities which are the most abundant and ubiquitous on this planet.

Through the chapters of this book, experts in the field describe the basic biology and genetics of LAB phage providing a basis of understanding for their impact on the dairy fermentation process. While much of the research is modeled upon specific strains of LAB and their lytic or lysogenic phages, one chapter delves into the dynamics of traditional, natural complex starter systems which form a fascinating ecology within itself with regards to phage-host interactions. From this foundation of understanding, the reader moves onto research concerning the practical detection and control of phage. In addition to "traditional" approaches that employ chemical sanitizers and thermal treatments, one chapter focuses on novel technologies that are being implemented in the food industry for food safety purposes as an alternative means of phage control. In parallel, starter bacteria themselves have not been defenseless in addressing predation by virulent phage. The chapter dealing with LAB phage resistance is of particular interest in describing the microbial adaptive process in the face of lethal pressures and the practical exploitation of this knowledge by the dairy industry.

Notably, the description of CRISPR/Cas introduces to the reader a novel adaptive immune system for combating phage that has offered new insights into basic microbiological evolutionary processes to the general scientific community. Coming full circle, the study of LAB bacteriophage has not been exclusively focused upon negative consequences. A resurgent interest in research performed in the early 1900's to employ phage as biocontrol agents is described in the final chapter of this book. As food safety concerns continue to be foremost in the industry as a whole, and considering LAB fermentation was originally a form of preservation, it seems only fitting to close with a positive application of phage for the dairy industry.

Upon completing, this book should provide a comprehensive view of the current knowledge on bacteriophages that infect LAB, and the various means that are available to control their spread in dairy plants.

Dennis A. Romero
Danisco USA Inc.
Philippe Horvath
Danisco France SAS

CONTRIBUTORS

EDITORS

Andrea del Luján Quiberoni. Instituto de Lactología Industrial (INLAIN), Universidad Nacional del Litoral – Consejo Nacional del Investigaciones Científicas y Técnicas, 3000 Santa Fe, Argentina. aquibe@fiq.unl.edu.ar

Jorge A. Reinheimer. Instituto de Lactología Industrial (INLAIN), Universidad Nacional del Litoral – Consejo Nacional del Investigaciones Científicas y Técnicas, 3000 Santa Fe, Argentina. jorreinh@fiq.unl.edu.ar

CONTRIBUTORS

Stephen T. Abedon*. Associate Professor of Microbiology, The Ohio State University, Mansfield, OH 44906, United States. abedon.1@osu.edu

Hans Wolfgang Ackermann*. Department of Microbiology, Immunology and Infectiology, Faculty of Medicine, Laval University, Quebec, G1K 7P4, Canada. ackermann@mcb.ulaval.ca

María Cecilia Aristimuño Ficoseco. Centro de Referencia para Lactobacilos (CERELA) – CONICET, T4000ILC San Miguel de Tucumán, Argentina. ceciaristi@hotmail.com

Zeynep Atamer*. University of Hohenheim, Institute of Food Science and Biotechnology, D-70599 Stuttgart, Germany. zeynep.atamer@uni-hohenheim.de

Ana G. Binetti*. Instituto de Lactología Industrial (INLAIN), Universidad Nacional del Litoral – Consejo Nacional del Investigaciones Científicas y Técnicas, 3000 Santa Fe, Argentina. anabinetti@fiq.unl.edu.ar

Mariángeles Briggiler Marcó*. Instituto de Lactología Industrial (INLAIN), Universidad Nacional del Litoral – Consejo Nacional del Investigaciones Científicas y Técnicas, 3000 Santa Fe, Argentina. mbriggi@fiq.unl.edu.ar

María Luján Capra*. Instituto de Lactología Industrial (INLAIN), Universidad Nacional del Litoral – Consejo Nacional del Investigaciones Científicas y Técnicas, 3000 Santa Fe, Argentina. mcapra@fbcb.unl.edu.ar

---

* **Corresponding author**

Domenico Carminati*. Agriculture Research Council, Fodder and Dairy Productions Research Centre (CRA-FLC), 26900 Lodi, Italy. domenico.carminati@entecra.it

Benjamin K. Chan. Department of Biology, University of Utah, Salt Lake City, UT 84112, United States. bkc5@utah.edu

Hélène Deveau. Département de biochimie, de microbiologie et de bio-informatique, Faculté des sciences et de génie, Université Laval, Québec, G1V 0A6 Canada. Helene.Deveau@bcm.ulaval.ca

Pilar García*. Instituto de Productos Lácteos de Asturias (CSIC), Carretera de Infiesto s/n, 33300 - Villaviciosa, Asturias, Spain. pgarcia@ipla.csic.es

Giorgio Giraffa. Agriculture Research Council, Fodder and Dairy Productions Research Centre (CRA-FLC), 26900 Lodi, Italy. giorgio.giraffa@entecra.it

Elisabetta Guerzoni. University of Bologna, Università di Bologna, 47023 Cesena, Italy. elisabetta.guerzoni@unibo.it

Daniela Guglielmotti. Instituto de Lactología Industrial (INLAIN), Universidad Nacional del Litoral – Consejo Nacional del Investigaciones Científicas y Técnicas, 3000 Santa Fe, Argentina. dgugliel@fiq.unl.edu.ar

Diana Gutierrez. Instituto de Productos Lácteos de Asturias (CSIC), Carretera de Infiesto s/n, 33300 - Villaviciosa, Asturias, Spain. dianagufer@ipla.csic.es

Elvira M. Hébert. Centro de Referencia para Lactobacilos (CERELA) - CONICET, T4000ILC San Miguel de Tucumán, Argentina. ehebert@cerela.org.ar

Knut J. Heller. Max Rubner-Institut, Federal Research Institute of Nutrition and Food, Department of Microbiology and Biotechnology, D-24103 Kiel, Germany. knut.heller@mri.bund.de

Jörg Hinrichs. University of Hohenheim, Institute of Food Science and Biotechnology, D-70599 Stuttgart, Germany. eidner@uni-hohenheim.de

Philippe Horvath. Danisco France SAS, BP 10, F-86220 Dangé-Saint-Romain, France. philippe.horvath@danisco.com

Rosalba Lanciotti. University of Bologna, Università di Bologna, 47023 Cesena, Italy. rosalba.lanciotti@unibo.it

Beatriz Martínez. Instituto de Productos Lácteos de Asturias (CSIC), Carretera de Infiesto s/n, 33300 - Villaviciosa, Asturias, Spain. bmf1@ipla.csic.es

Diego Mercanti*. Instituto de Lactología Industrial (INLAIN), Universidad Nacional del Litoral – Consejo Nacional del Investigaciones Científicas y Técnicas, 3000 Santa Fe, Argentina. diegojav78@gmail.com

Sylvain Moineau*. Département de biochimie, de microbiologie et de bio-informatique, Faculté des sciences et de génie, Université Laval, Québec, G1V 0A6 Canada. sylvain.moineau@bcm.ulaval.ca

Horst Neve. Max Rubner-Institut, Federal Research Institute of Nutrition and Food, Department of Microbiology and Biotechnology, D-24103 Kiel, Germany. horst.neve@mri.bund.de

María J. Olaya Passarell. Centro de Referencia para Lactobacilos (CERELA) - CONICET, T4000ILC San Miguel de Tucumán, Argentina. mjolaya@cerela.org.ar

Francesca Patrignani. Dipartimento di Scienze degli Alimenti, Università di Bologna, 47023 Cesena, Italy. francesca.patrignani@unibo.it

Andrea del Luján Quiberoni. Instituto de Lactología Industrial (INLAIN), Universidad Nacional del Litoral – Consejo Nacional del Investigaciones Científicas y Técnicas, 3000 Santa Fe, Argentina. aquibe@fiq.unl.edu.ar

Raúl R. Raya*. Centro de Referencia para Lactobacilos (CERELA) - CONICET, T4000ILC San Miguel de Tucumán, Argentina. rraya@cerela.org.ar

Jorge A. Reinheimer. Instituto de Lactología Industrial (INLAIN), Universidad Nacional del Litoral – Consejo Nacional del Investigaciones Científicas y Técnicas, 3000 Santa Fe, Argentina. jorreinh@fiq.unl.edu.ar

Ana Rodríguez. Instituto de Productos Lácteos de Asturias (CSIC), Carretera de Infiesto s/n, 33300 - Villaviciosa, Asturias, Spain. anarguez@ipla.csic.es

Lorena Rodríguez. Instituto de Productos Lácteos de Asturias (CSIC), Carretera de Infiesto s/n, 33300 - Villaviciosa, Asturias, Spain. lorenarguez@ipla.csic.es

Dennis Romero. Danisco USA Inc., Madison, WI 53716, United States. dennis.romero@danisco.com

Geneviève Rousseau. Département de biochimie, de microbiologie et de bio-informatique, Faculté des sciences et de génie, Université Laval, Québec, G1V 0A6 Canada. Genevieve.Rousseau@greb.ulaval.ca

Viviana B. Suárez*. Instituto de Lactología Industrial (INLAIN), Universidad Nacional del Litoral – Consejo Nacional del Investigaciones Científicas y Técnicas, 3000 Santa Fe, Argentina. vivisuar@fiq.unl.edu.ar

Miriam Zago. Agriculture Research Council, Fodder and Dairy Productions Research Centre (CRA-FLC), 26900 Lodi, Italy. miriam.zago@entecra.it

In: Bacteriophages in Dairy Processing
Editor: Andrea del Lujan Quiberoni et al.

ISBN 978-1-61324-517-0

Chapter 1

PHAGES OF LACTIC ACID BACTERIA: DISCOVERY AND CLASSIFICATION

Hans-Wolfgang Ackermann*
Department of Microbiology, Immunology and Infectiology,
Faculty of Medicine, Vandry Pavilion,
Avenue de la Médecine, Laval University,
Quebec, Canada

ABSTRACT

Lactic acid bacteria (LAB) produce lactic acid from sugars and comprise lactobacilli, *Lactococcus lactis, Leuconostoc, Pediococcus, Streptococcus thermophilus,* and *Tetragenococcus*. These bacteria are implicated into the production of many fermented foods, notably cheeses and yoghurt. They may be attacked by nuisance bacteriophages, able to annihilate whole fermentation batches. The nefarious role of LAB phages was recognized in 1935.

Over 1,250 LAB phages were studied by electron microscopy, making LAB phages the largest group of all bacteriophages and indeed all viruses. They belong to the *Myoviridae, Siphoviridae,* and *Podoviridae* families of tailed phages. *Lactobacillus* phages comprise many myoviruses and are very varied. About 92% of LAB phages are siphoviruses.

Phages of *L. lactis* belong to 10 species. Those of *Strep. thermophilus* are a very homogenous group and seem to belong to a single species. Electron microscopic data are plentiful, but often of low quality and confusing. Classification into genera and species is not very advanced. Genome sequencing is very useful for the establishment of species and phage genera that transcend host genus boundaries. Further progress is linked to genomics and improved electron microscopy.

* E-mail address: ackermann@mcb.ulaval.ca. Tel.: +1 418 656 2131 # 2558; fax: +1 418 656 7898

INTRODUCTION

1. Lactic Acid Bacteria

Lactic acid bacteria (LAB) ferment carbohydrates to produce lactic acid. They are Gram-positive cocci or rods of low G+C content, mesophilic, nonmotile, nonsporulating, and microaerophilic. Lactic acid bacteria belong to the *Firmicutes* order *Lactobacillales* and comprise the genera *Lactobacillus, Leuconostoc, Pediococcus, Tetragenococcus,* and the species *Streptococcus thermophilus* and *Lactococcus lactis*. The latter was formerly classified as a member of the genus *Streptococcus*. The genus *Lactococcus* was created in the 1990s and has only one species, *L. lactis* (Hardie and Whiley 1997).

These bacteria are generally apathogenic to man, although some infections by lactobacilli (Cannon et al. 2005) and leuconostocs (Cappelli et al. 1999) have been reported. The species *Leuc. garviae* is fish-pathogenic (NCBI Taxonomy Browser). On the other hand, lactic acid bacteria are certainly among the most useful microorganisms to mankind. They are used world-wide and for eons in the preparation of fermented foods as a means to preserve food and to add flavor. Lactic acid bacteria are used in spontaneous and industrial fermentations for the production of cheeses such as Swiss cheese and cheddar, and also of such diverse foods as yoghurt, silage, sauerkraut, sour dough, sour milk and other fermented milks, Korean *kimchi* or fermented spicy cabbage, soy sauce, fermented sausages, and the malolactic fermentation of wine (Teuber and Lembke 1983).

The most important lactic acid bacteria are *L. lactis* and *Strep. thermophilus* for their role in cheese-making. Lactobacilli are widespread in nature and occur in the intestines and vaginas of many animals including humans. Their principal use is yoghurt production. The genus *Lactobacillus* has over 150 species (NCBI Taxonomy Browser) and is extremely diversified. By contrast, *L. lactis* and *Strep. thermophilus* are single species and biologically rather homogeneous. This diversity, or lack of, is reflected in the bacteriophages that infect these bacteria. Leuconostocs are associated with sauerkraut and pickle fermentation. Pediococci are used in the manufacture of fermented sausage and *Tetragenococcus halophilus*, formerly classified as a *Pediococcus*, is employed in soy sauce fermentation (Uchida and Kanbe 1993, Caldwell et al. 1999). Lastly, *Propionibacterium freudenreichii* produces the characteristic flavor and holes of Swiss cheese, but is not considered here because it is not a lactic acid bacterium.

Cheese production, as a method of food preservation, may go back to the domestication of sheep and goats 10,000 years ago in the Middle East. It is mentioned in the Odyssey and, as I observed, is evident archeologically in clay milk strainers from the -4th century of classical Greece. According to its distribution, cheese-making seems to have been primarily a habit of early Indo-Europeans. Cheese-making is now a giant industry practized on an enormous scale in industrialized countries. In the 19th century, cheese-making became increasingly industrialized and rationalized, in particular in developed countries with populations of European descent. Around 1890, defined starter cultures were introduced into cheese-making to ensure a stable quality and predictable yields (Heller 2007). Typically, large milk vats were inoculated with a mixture of a few bacterial "starter" strains (Moineau and Lévesque 2005) and then left to ferment. These were classical monocultures, waiting for accidents to happen. In the 1930s, microbiologists began to worry about faulty fermentations that resulted in low yields and unpalatable products.

2. Bacteriophage Discovery and Harmful Phages

Bacteriophages or "phages" were discovered almost simultaneously and described in 1915 and 1917, respectively, by the English bacteriologist Frederick William Twort (1915) and the Canadian-born Felix d'Herelle (1917), then working at the Pasteur Institute of Paris. d'Herelle coined the term "bacteriophage" and postulated that his agent was involved in the cure of infectious diseases. Contrary to Twort, d'Herelle understood immediately the significance of his discovery and asserted that phages were viruses, analogous to the already discovered tobacco mosaic virus and the agent of foot-and-mouth disease. This was not accepted by all the scientific community and a long controversy on the nature of phages, viral or enzymatic, ensued (Summers 1999).

Dairy microbiologists adopted the viral hypothesis and explained faulty fermentations in cheese-making by phage infestation. As early as 1926, Hadley and Dabney described phage lysis of *Strep. lactis* (1926). In 1935, Whitehead and Cox, working at the New Zealand Dairy Research Institute, observed indeed phage infection in *Strep. lactis* cultures (Whitehead and Cox 1935). The nefarious activity of bacteriophages in cheese-making was soon confirmed (Whitehead and Hunter 1937). In 1950, it was clear that destruction of lactic acid bacteria by bacteriophages "represents one of most important causes of insufficient acid development in manufactured dairy products" (Elliker 1950). Similar faulty fermentations were caused by *Strep. thermophilus* phages present in Swiss cheese whey (Deane et al. 1953). Although initial attempts to find *Lactobacillus* phages in nature and human stools had been negative (Schmager 1951), a phage for *Lactobacillus plantarum* was found in 1954 (White et al. 1954). *Leuconostoc citrovorum* phages were blamed in 1946 for low-quality sour milk production (Mosimann and Ritter 1946). Nuisance phages destroying valuable bacteria, in particular starter cultures, could be found in the factory environment and milk and cheese whey, entering fermentation vats through airborne or waterborne routes. It was thus a milestone to discover that *Strep. lactis* was lysogenic and that viable temperate phages could be induced from it (Reiter 1949). This raised the possibility that nuisance phages in the dairy industry could be of endogenous origin. Indeed, 1-10% of *Strep. thermophilus* starter strains produced phages after induction with mitomycin C (Brüssow 1999) and high proportions of lysogenic strains were observed in lactococci (Cappelli et al. 1999).

3. Further Reading

Phages of dairy or lactic acid bacteria were reviewed many times in articles and book chapters, starting in 1950 (Elliker 1950). Reviews are devoted to all LAB phages (Teuber and Lembke 1983, Mata and Ritzenthaler 1988, Jarvis 1989, Brüssow 2001, Desiere et al. 2002, Emond and Moineau 2007) or phages of *Lactobacillus* (Sozzi et al. 1981, Brüssow and Suárez 2006, Villion and Moineau 2009), *Lactococccus lactis* (Brondsted and Hammerhammer 2006), and *Strep. thermophilus* (Brüssow et al. 1998, Brüssow 1999,

Quiberoni et al. 2010). Dimensions and physicochemical properties of numerous LAB phages may be found in a general book on viruses of prokaryotes (Ackermann and DuBow 1987). In addition, there are excellent reviews on LAB phage genomics and molecular genetics, lysogeny, phage-host interactions, and phage resistance against inactivation, but they are outside the scope of this chapter.

BACTERIOPHAGE OBSERVATIONS

Early Studies

Electron microscopes were introduced in the 1930s. The first prototype was produced in 1938 in Berlin by the Siemens and Halske AG (Haguenau et al. 2003) and the first electron microscopic images of phages, in fact coliphages, were published simultaneously in 1940 (Pfankuch and Kausche 1940, Ruska 1940). This was followed by a period of improvements in electron microscope construction and preparation techniques. The first electron micrographs of LAB phages, namely of phages of *Lactococcus lactis*, were published in 1949 (Parmelee 1949) and 1953 (Deane et al. 1953). The phages had long, noncontractile tails and appeared to be members of the *Siphoviridae* family (below). The phages had been shadowed for contrast. In 1959, Brenner and Horne (Brenner and Horne 1959) introduced negative staining by salts of phosphotungstic acid. This novel technique soon replaced shadowing and allowed the visualization of viruses with unprecedented clarity and detail. The principle of negative staining was to mix viruses (or any particles) with an electron-dense salt solution, causing them to stand out white against a dark background like particles in Chinese ink. Indeed, negative staining revolutionized virology. Uranyl acetate, already used in the staining of tissue sections, proved to be another useful negative stain. Other compounds were tested without results by many researchers including the author of this chapter. Ammonium molybdate, while providing excellent negative staining, is no improvement over phosphotungstates. The first LAB phage described after the introduction of negative staining, as early as 1960, was *Lactococcus* phage 3ML (synonym Φv3ML), supplied by Reiter (1949). The phage had an elongated, apparently oval head of 55 x 40 nm and a long, noncontractile tail of 100 x 9 nm (Bradley and Kay 1960). It is ascribed to the c2 species of *Lactococcus* phages (Jarvis et al. 1991). The sequence of discoveries is described in Table 1.

A Bacteriophage Explosion

Electron microscopes evolved from curiosities to mainstream research instruments and became widely available in the 1960s. This and the powerful technique of negative staining prompted the investigation of phages on a grand scale. The number of phages examined by electron microscopy grew exponentially (Ackermann 1992). The first phage survey, published in 1969, listed 111 viruses, 99 of which were tailed and 12 isometric or filamentous (Eisenstark 1967). Only eight of them were LAB phages. By 1992, at least 4,007 phages, belonging to 13 families of bacterial and archaeal viruses, had been examined in the electron microscope. About 96% of these phages were tailed and only 4% were cubic, filamentous, or pleomorphic.

Table 1. Discovery of LAB phages

Phage host	First report	First electron microscopical observation	
		Shadowing	Negative staining
Lactobacillus	Schmager 1951 White et al. 1954		De Klerk et al. 1965
Lactococcus lactis	Whitehead and Cox 1935	Parmelee et al. 1949	Bradley and Kay 1960
Leuconostoc	Mosimann and Ritter 1946		Sozzi et al. 1978
Pediococcus	Caldwell et al. 1999		Caldwell et al. 1999
Streptococcus thermophilus	Deane et al. 1953	Deane et al. 1953	Bauer et al. 1970
Tetragenococcus	Uchida and Kanbe 1993		Uchida and Kanbe 1993

This included 724 LAB phages (Ackermann 1992). These surveys are now conducted periodically. In 2007, the number of phages examined in the electron microscope had grown to over 5,500, making viruses of prokaryotes by far the largest division in the viral world (Ackermann 2007). Again, over 96% of these phages were tailed. Presently, the total amount of LAB phages examined in the electron microscope is an astonishing 1,267 viruses. The numbers of LAB phages known today and in 1992 are compared in Table 2. LAB phages, numerically, certainly constitute the largest group in the bacteriophage world and continue to generate great interest because of their economic importance. This interest is reflected, for example, in the ever-growing numbers of *Lactobacillus* phages observed in the electron microscope. In 1981, Sozzi et al. reported only 58 phages, several of which were tailless heads and certain to be defective (Sozzi et al. 1981). By contrast, the present tally is 212 *Lactobacillus* phages, reflecting an increase of about 400% in 30 years.

Most LAB phages were isolated in countries with large dairy industries and a lively research tradition, namely Australia, Canada, Denmark, France, Germany, Ireland, New Zealand, the Netherlands, Switzerland, the U.S.A., and now Argentina. The New Zealand Dairy Research Institute was particularly active in this field and large numbers of *Lactococcus* phages were isolated there under the leadership of E. A. and B. E. Terzaghi and A. W. Jarvis. Research centered on phages of *L. lactis* and *Strep. thermophilus*. Phages were generally characterized by electron microscopy, a series of biological criteria (host range, latent period, burst size, cofactor requirements, adsorption velocity), and resistance against physical and chemical agents such as heating and chloroform. The investigation of restriction endonuclease patterns and DNA-DNA hybridization came later, but played crucial roles in LAB phage classification. The first complete DNA sequences of LAB phages were published in 1994 and 1995. They belonged to two small siphoviruses with elongated heads, named c2 and biL67 (Schouler et al. 1994, Jarvis et al. 1995). Belatedly, research was extended to phages of the "minor" lactic acid bacteria, *Pediococcus* and *Tetragenococcus* (Table 1). The impressive number of over 1,260 LAB phages presently examined in the electron microscope obscures the fact that their dimensions are often unreliable (see below).

According to gross morphology, phages may be divided into morphotypes (Ackermann and Eisenstark 1974). They are derived from "basic phage groups" proposed in 1967 by Bradley in a seminal paper (Bradley 1967). These morphotypes were devised for rapid identification by electron microscopy.

Table 2. Number of LAB phage observations by morphotype

Host	Morphotype								References
	A1	B1	B2	B3	C1	C2	C3	Total	
Lactobacillus	72	131	1	7	1			212	Ackermann 2007, Villion and Moineau 2009, U
	56	*46*	*1*	*2*				*105*	Ackermann 1992
Lactococcus lactis	3	448	200		1			661	Ackermann 2007, U
	2	*419*	*160*		*1*			*588*	Ackermann 1992
Leuconostoc	8	33						41	Ackermann 2007, U
		31						*31*	Ackermann 1992
Pediococcus		4						4	Ackermann 2007, U
Streptococcus thermophilus		345			1			346	Ackermann 2007, Quiberoni et al. 2010
		116			*1*			*117*	Ackermann 1992
Tetragenococcus	1	2						3	Ackermann 2007
	84	963	201	7	3			1267	
Total	*58*	*612*	*161*	*2*	*2*			*841*	Ackermann 1992

U, updated on November 30, 2010. Data of 1992 are in italics.

However useful, they are artificial creations and do not constitute taxonomical groups. Bradley's groups A, B, and C correspond to the *Myoviridae, Siphoviridae*, and *Podoviridae*, respectively. Each of them may be subdivided according to head shape into phages with isometric, moderately elongated, or very long heads (types 1, 2, and 3, respectively). One encounters seven types among LAB phages, namely A1, B1, B2, B2, B3, C1, C2, and C3 (Figure 1).

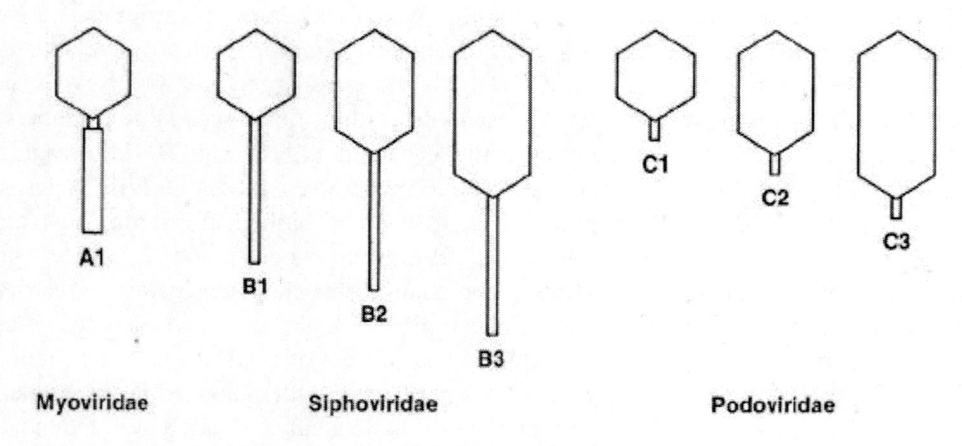

Figure 1. Morphotypes of LAB phages. From Ackermann et al. (1984); with permission of S. Karger AG, Basel.

Type A2, exemplified by coliphage T4, is not present in LAB phages. Type A3 is very rare and found in a few *Salmonella* phages only. Table 1 shows the incidence of LAB phage morphotypes (below) in 1992 and today. Table 2 reports the dimensions of selected LAB phages and Table 3 illustrates the enormous variability observed in the dimensions of these viruses. The great variability on LAB phages is illustrated in Figures 2-5.

Table 3. Dimensions of selected phages

Host	Morphotype	Phage	Head, nm	Tail length, nm	Comments	References
Lactobacillus	A1	LgcI	63	178		Ismail et al. 2009
		LP65	80	193	SPO1-like	Chibani-Chennoufi et al. 2004
	B1	ϕadh	62	398		Quiberoni et al. 2010
		lb6	56-62	200-210	Tail bars	Sozzi et al. 1981
	B2	223	107 x 53	176		Tohyama et al. 1972
	B3	0235	125 x 50	230		Séchaud et al. 1988
Lactococcus lactis	A1	RZh	90	200-210	SPO1-like	Tikhonenko 1968, 1970
	B1	936	53-58	137-159		Jarvis 1977, Jarvis et al. 1991
		P335	54	145		Jarvis et al. 1991
		1358	54	110		Jarvis 1977, Jarvis et al. 1991
		P087	65	200		Jarvis et al. 1991, Lavigne et al. 2008
		1706	58	276		Deveau et al. 2006
		949	79-88	499-527		Jarvis 1977, Jarvis et al. 1991
	B2	c2	50 x 36	94		Pillidge and Jarvis 1988
		Q54	56 x 43 *	109 *		Deveau et al. 2006
	C2	P034	65 x 44	24	ϕ29-like	Lembke et al. 1980
	C3	KSY1	230 x 50	35		Saxelin et al. 1979
Leuconostoc	A1	ϕR105	78	188	SPO1-like	Barrangou et al. 2002
	B1	ggg	50	125		Greer et al. 2006
Pediococcus	B1	pa97	63	300		Caldwell et al. 1999
Streptococcus thermophilus	B1	ALQ1.3	61	299		Quiberoni et al. 2006
		Sfi21	65	230-260		Brüssow et al. 1994
Tetragenococcus	A1	ϕ7116	87-98	200	SPO1-like	Uchida and Kanbe 1993

* Dimensions are probably too high.

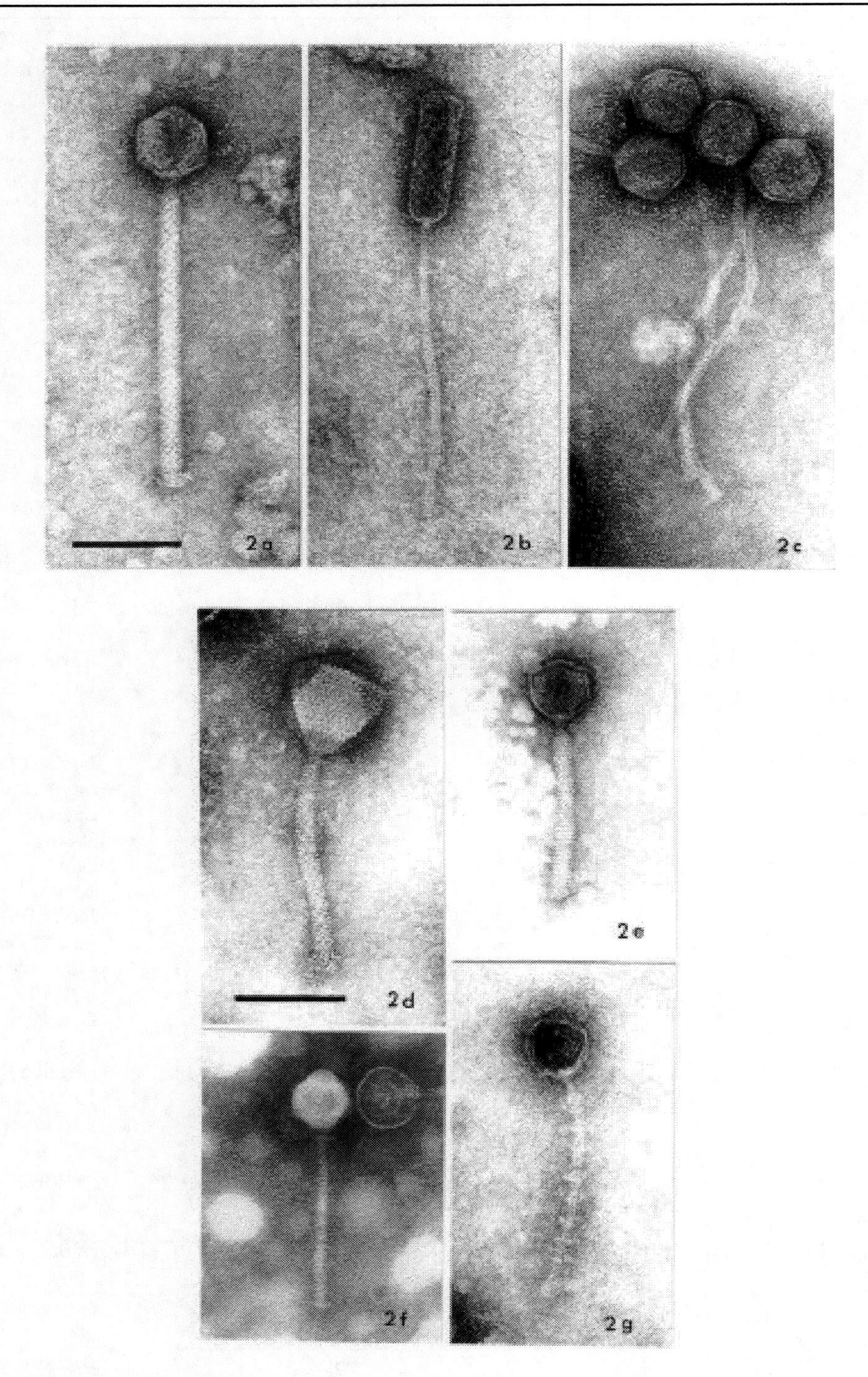

Figure 2. Lactobacillus phages. (a). Phage hv of Lb. helveticus. (b). Phage 0235 of Lb. delbrueckii subsp. lactis. (c). Three particles of Lb. casei phage PL-1. (d). Phage fri of Lb. plantarum; note the capsomers and facets on the phage head. (e). Phage b2 of Lb. helveticus. (f). Phage MLC-A of Lb. paracasei showing a tiny collar below the head. (g). Phage lb6 of Lb. bulgaricus showing transverse tail bars.

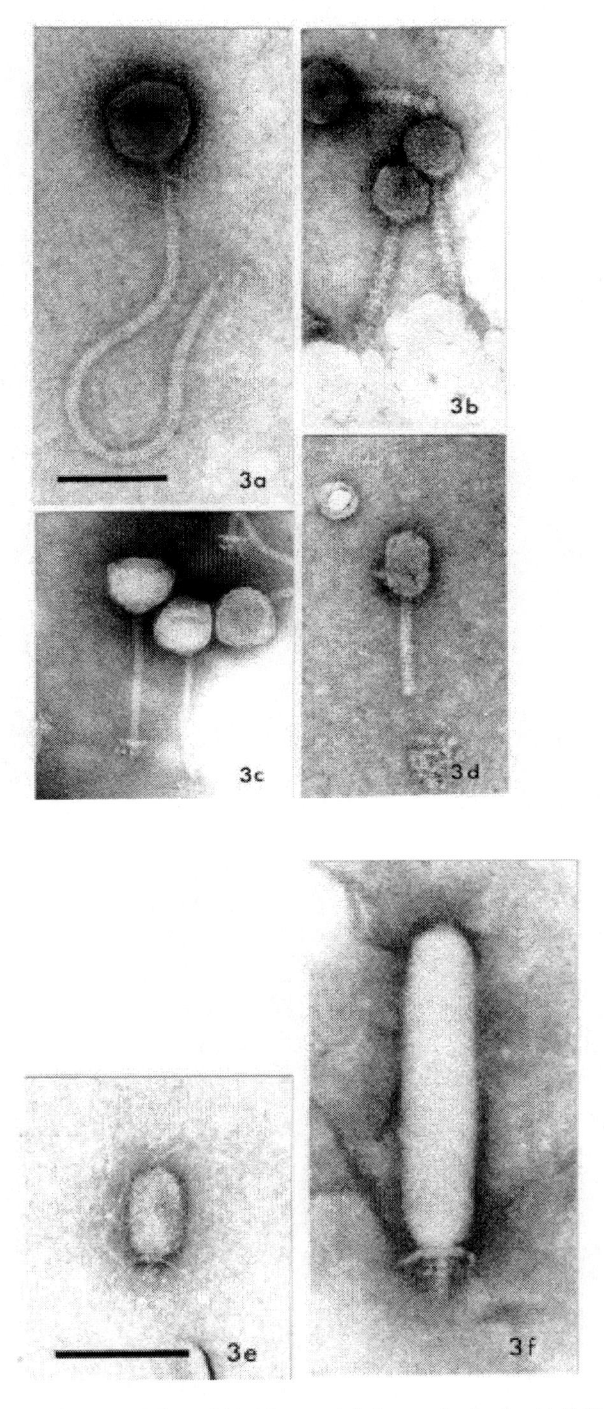

Figure 3. Lactococcus lactis phages. (a). Phage 949. The particle has a single short tail fiber. (b). Phage 1358. The transverse tail bars give the tail a fuzzy appearance and make it appear thicker than it is. (c). Phage UL37.1 of species P335 showing a base plate with bulbous tail spikes. (d). Phage P004. Contrary to other phages of the c2 species, no collar is visible here. (e). Phage P034 displaying short fibers. (f). Phage KSY1.

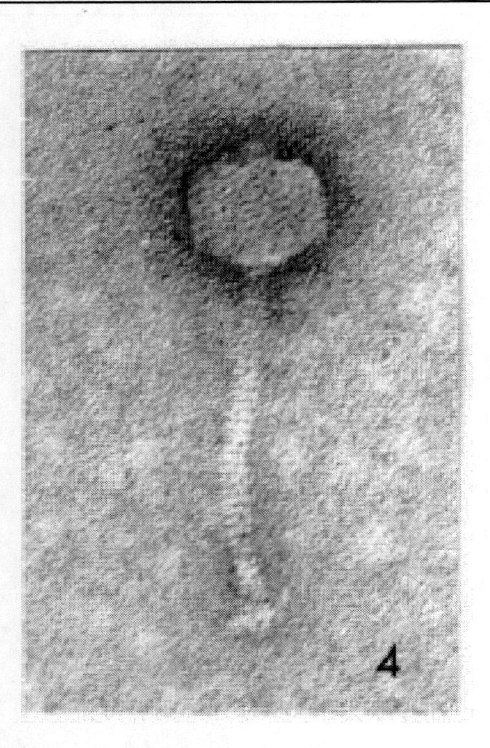

Figure 4. Phage Ln-6 of *Leuconostoc mesenteroides*.

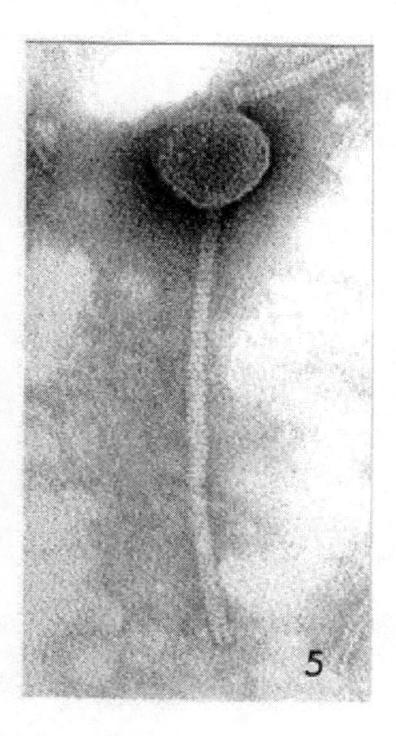

Figure 5. Phage DT4 of Streptococcus thermophilus.

CLASSIFICATION

High-level classification addresses categorization into orders and families. In viruses, these categories are defined by the nature of nucleic acid, gross morphology, and presence or absence of an envelope. Classification into families is generally easy and virus families are the most stable part of the edifice of virus classification of the ICTV (International Committee on Taxonomy of Viruses). Low-level classification concerns subfamilies, genera, and species. It is emphasized that no biologist can certify what a species is. There are some 22 species definitions in the literature (Fauquet et al. 2005). The ICTV has adopted the "polythetic species concept" (Van Regenmortel 1998), meaning that a species is defined by a set of common properties that may or may not be present in all members. This definition is clearly elastic and does not provide practical or precise guidance. Genus and species definition rely on many if not all detectable "low-level" properties, none of which has pre-eminence. They include, for example, dimensions, G+C%, DNA-DNA hybridization, resistance against physical and chemical agents, host range, and now increasingly genomic data. Determination of these properties is often labor-intensive and low-level virus classification is much of an art.

1. Macro-Classification

Bacterial viruses are classified into one order and 10 families (Fauquet et al. 2005, Ackermann 2007). The tailed phages constitute the order *Caudovirales* with three families, namely the *Myoviridae* (24.5% % of phages) with long, contractile tails, the *Siphoviridae* (61%) with long, noncontractile tails, and the *Podoviridae* (14%) with short tails. These three families comprise over 96% of phages. All members contain a single molecule of linear dsDNA. The remaining seven families, all very small and sometimes having a single member, comprise (a) isometric phages with cubic symmetry and ssDNA (*Microviridae*), dsDNA (*Corticoviridae, Tectiviridae*), ssRNA (*Leviviridae*) or dsRNA (*Cystoviridae*), (b) filamentous phages with ssDNA *(Inoviridae)*, and (c) pleomorphic phages with dsDNA *(Plasmaviridae)*. The phages of these seven families are thus very varied with respect to basic morphology and nucleic acid nature, but do not occur in lactic acid bacteria and shall not be discussed here.

2. Micro-Classification

2.1. Sequence-Based and Transcending Host Genus Boundaries

Over 150 fully sequenced phage genomes of the Podoviridae and Myoviridae families from the NCBI and EBI databases were analyzed for protein sequence similarity. Using BLAST-P based tools, called CoreExtractor and CoreGenes, phage genera were defined by measuring genomic relationships by the numbers of shared homologous/orthologous proteins (Lavigne et al. 2008, 2009). There was generally an excellent correlation between genomics and morphology; mainly, phages of different morphology always differed by their genomes. However, not surprisingly, phages of simple, uncharacteristic morphology sometimes were found in different genomic groups. It is clear that this genomic classification is only a beginning and that new taxa will emerge once more phage genomes are sequenced.

Among 55 phages of the Podoviridae family (Mahony et al. 2006), the T7- and □29-related phages were given subfamily rank. The latter, characterized by their small size and prolate heads, were named Picovirinae. Although the sequence of L. lactis phage P034 is unknown, this phage can confidently be attributed to the Picovirinae by its shape and dimensions. The other Picovirinae are famous Bacillus phage □29 and its immediate relatives, Streptococcus pneumoniae phage Cp-1 and its relatives, and some Staphylococcus and Kurthia phages. All host bacteria have low G+C contents and are evolutionary related. L. lactis phage KSY1 was recognized as a completely independent virus and might represent a genus of its own (Lavigne et al. 2008).

This approach was extended to 102 Myoviridae genomes (Lavigne et al. 2009). Three subfamilies and eight genera were individualized. The subfamilies corresponded to the T4-, P2-, and SPO1-like phages. Phage SPO1 is a very well-characterized Bacillus virus. The SPO1-like phages were named Spounavirinae and comprise phages of Lactobacillus paracasei and plantarum, bacilli, and staphylococci. Spounavirinae are specific to Gram-positive bacteria with low G+C content and have large heads of about 90 nm in diameter, conspicuous capsomers and contractile tails with double base plates. According to tail length, 130-140 or 180-200 nm, they fall into to two groups. Spounavirinae have large genomes of 140-150 kb in size.

Once defined, the Spounavirinae subfamily was extended by using the criteria of characteristic morphologic and genome properties, for example the presence of 5-hydroxymethyluracil instead of thymine (Klumpp et al. 2010). It appears that Spounavirinae, besides occurring preferentially in bacilli, staphylococci and lactobacilli, also occur in L. lactis, Leuconostoc and Tetragenococcus (Table 3). The spounavirus of L. lactis was already described in 1968 (Tikhonenko 1968, 1970), but was not mentioned since and may be lost. The Spounavirinae probably includes six phages of Lactobacillus fermentum and Lactobacillus casei found in 1965 (De Klerk et al. 1965).

Among the Siphoviridae,

α. Two genera, centered on Streptocccus thermophilus phages Sfi11 and Sfi21 (Brüssow et al. 1994) and based on similar genome organization, were proposed by the Brüssow group in Switzerland. Sfi-like viruses have cos (cohesive) sites and Sfi21-like phages have pac sites (mediating headful packaging of DNA). They comprise phages not only of Strep. thermophilus, but also of Bacillus, Lactobacillus, Lactococcus, Listeria, and Staphylococcus. It is noteworthy that the analyzed phages genomes show gradients of relatedness, reflecting the taxonomic position of their bacterial hosts and suggesting some degree of vertical evolution (Brüssow et al. 1994, Desiere et al. 2002).

β. Λ. lactis phage 1358 and Listeria phages P35 and P40 share similar genome organization, constitutive proteins and relatively short tails (Dupuis and Moineau 2010).

2.2. Phages by Host Group

a. *Lactobacillus* phages of the myovirus family are very heterogeneous. Some are SPO1-like and belong to the *Spounavirinae* (Table 3), but most of them have not been classified. One remarkable type, exemplified by *Lb. helveticus* phage hv (Figure 2a), stands out by its size (head diameter 54 nm, tail length 235 nm) (Accolas and

Spillmann 1979). Clearly, many more phage species are to be discovered if genome sequencing is extended to *Lactobacillus* myoviruses. The siphoviruses of *Lactobacillus* fall into three morphotypes. Those with elongated heads (types B2 and B3) are easily individualized. The phages with isometric heads (type B1) are more problematic. Not only are they numerous and morphologically heterogeneous, with tails ranging from 116 to 500 nm in length, but most electron microscopic data in the literature are dubious. A proposal has been made to classify these phages by DNA homology (Mata and Ritzenthaler 1988). The B3 type phages, e.g., *Lb. delbrueckii* subsp. *lactis* phage 0235, resemble very closely *Bacillus licheniformis* phage BLE (synonym θc) which has a head of 130 x 50 nm and a 240 nm long tail (Ludvik et al. 1977).

b. *L. lactis* phages were classified in 1991 into 12 species defined by DNA homology and morphology (Jarvis et al. 1991). This classification proved to be extremely useful for research on dairy phage ecology and identification of contaminating phages. After revision, the number of species was reduced to eight (Brondsted and Hammer 2006) and finally enlarged by the addition of two likely novel species (Deveau et al. 2006) (Table 5). The P335 species is fairly heterogeneous by genome structure (Deveau et al. 2006). Phage species 936, P335 and c2 are ubiquitous and account for 98% of faulty fermentations by *L. lactis* (Moineau 1999, Desiere et al. 2002). Species 949 is interesting because its only member, phage 949, resembles giant *Bacillus mycoides* phage N5, described as having a head of 91-96 nm and a tail of 545 nm in length (Tikhonenko 1968). The novel species Q54 appears as a hybrid that retained c2 species morphology and acquired a module from 936-like phages via recombination (Fortier et al. 2006). The 936 species has an inconstant collar (Mahony et al. 2006). Heteroduplex analysis revealed that its presence depends on a nonessential gene that may be deleted (Loof and Teuber 1986). Similarly, I observed that the collar of c2-like phages is inconstant. The presence of a collar is therefore not a taxomical property in *Lactococcus* phages. Most interesting is the observation that a phage of the P335 species, due to pressure exerted by an abortive phage resistance mechanism, acquired a slightly longer tail and a different base plate from a resident prophage (Moineau et al. 1994). Morphology is therefore not immutable and somewhat plastic.

c. *Leuconostoc* myoviruses are clearly members of the *Spounavirinae* subfamily (Barrangou et al. 2002, Yoon et al. 2002). *Leuconostoc* siphoviruses were briefly reviewed by Jarvis (1989). *Leuc. mesenteroides* phages were divided into two DNA homology groups (Boizet et al 1992). A glance at the dimensions of *Leuconostoc* phages suggests that they belong to three or more species according to tail length.

d. *Pediococcus* phages are siphoviruses which, according to tail length, may belong to two species (Table 4). This remains to be investigated by DNA-DNA hybridization or genome sequencing.

e. *Strep. thermophilus* phages, despite their great number, appear to be a single polythetic species with two large groups based on the mode of DNA packaging and the number of major structural proteins, namely *cos*-type phages with two major proteins and *pac*-type phages with three major proteins, respectively (Quiberoni et al. 2010). Size variants have been reported, but may be explained by the observation of freak or damaged particles or by faulty electron microscopy.

f. *Tetragenococcus* phages are either SPO1-like myoviruses or siphoviruses of uncharacteristic morphology (Tables 2 and 3).

Table 4. Size variation in frequent morphotypes

Host	Morphotype	Head, nm	Tail length, nm	Comments
Lactobacillus	A1	49-99	123-234	Many SPO1-like phages
	B1	40-110	116-500	Very heterogeneous
Lactococcus	B1	17-98	37-330	Several species
	B2	46-66 x 32-57	74-120	Homogeneous group
Leuconostoc	A1	81-110	174-203	SPO1-like
	B1	49-78	117-370	Three types of tail length?
Pediococcus	B1	51-63	130-300	Heterogeneous
Streptococcus thermophilus	B1	40-71	175?-330	Two species?

Table 5. *Lactococcus lactis* phage species

Family	*Siphoviridae*		*Podoviridae*		References
Morphotype	B1	B2	C2	C3	
Phages	936, P335, P107, P087, 949, 1483, 1358, BK5-T	c2	P034	KSY1	Jarvis 1989
	936, P335, P107, P087, 949	c2	P034	KSY1	Brondsted and Hammer 2006
	936, P335, P107, P087, 949, 1706	c2, Q54	P034	KSY1	Fauquet et al. 2005, Deveau et al. 2006

ELECTRON MICROSCOPY

1. General

Based on the sheer number of phages examined, electron microscopy (EM) is the single most important technique in the study of LAB phages. In this case, it is a relatively simple science. Indeed, LAB phage electron microscopy is almost exclusively transmission EM and examination of negatively stained viruses with only two stains, without more complicated techniques such as shadowing, sectioning, and cryo- or scanning electron microscopy. DNA spreading for in order to find deletions by heteroduplex mapping and immuno-electron microscopy have been applied to a very few LAB phages. For added simplicity, all LAB phages are tailed and members of only three families. The verdict on the results is ambivalent.

On the positive side is an enormous expansion of knowledge making LAB phages, numerically, the largest of all virus groups. As their diagnosis is simple, many phage *Lactococcus* species have been identified by EM and many phage infections in the industry have been elucidated. The imprint of EM is visible everywhere, even in genomics. Since EM is suitable for examining preparations made from phage plaques or crude lysates, EM allows for instant diagnosis and is the fastest technique in microbiology after slide agglutination and Gram staining. On the negative side is the fact that many EM data, primarily dimensions, are of questionable value or downright worthless. It is as if they characterize less the phages than electron microscopists.

Things were complicated by the advent of digital electron microscopes and CCD (charged-couple device) cameras in the 1990s. Hitherto, electron microscopes were "manual" in the sense that users could adjust all parts of the columns of their instruments. Images were recorded on films or plaques and developed and printed in a darkroom. Contrast was essentially adjusted by means of graded filters and papers. Magnification was easily corrected in the darkroom by adjusting a photographic enlarger. All this was tedious and labor-intensive, but, again, users controlled every step of the procedure.

"Digital" electron microscopes may still have film or plate cameras, but most are operated jointly with CCD cameras. There is no longer a darkroom. Imaging is thus greatly simplified and recorded images are easy to store and exchange. On the negative side, one notes a general deterioration of image quality, particularly poor contrast and thus grey images. This deterioration seems to have started before the introduction of digital electron microscopes and is partly attributable to inadequate purification of viruses before examination. It is certainly not due to the negative stains used, phosphoungstate and uranyl acetate. Both stains, in use since the emergence of electron microscopy, are time-tested and give excellent staining. Uranyl acetate often gives excellent contrast, but is acidic and prone to stain dsDNA, tends to crystalize on the grid, and gives rise to many artifacts (Ackermann and DuBow 1987). Attempts to replace these compounds by others, whether related or not, have been essentially unsuccessful. One of the biggest problems of digital electron microscopes seems to be magnification control. The manufacturer's specifications cannot be trusted in any way, as they are adjusted by company technicians at the time of installation and have no absolute value. As the author has seen, the adjustment of magnification of a digital EM may be long and difficult. The author also encountered several poorly adjusted digital electron microscopes that produced wildly improbable dimensions.

2. The Disease

The general name of the disease is poor electron microscopy. It is multifactorial and widespread. The following causes may be listed in their increasing order of gravity:

a. Misalignment of the electron microscope, evident in unsharp images and astigmatism.

b. Errors due to the presence of adventitious lysogenic phages or contaminants, deformed phage heads, broken phage tails, or abnormally long tails. The latter represent assembly errors and are common in siphoviruses.
c. Staining artifacts. Uranyl acetate has a strong affinity to dsDNA and often stains phage heads in deep black (positive staining). Such heads are always shrunken by about 15% and their dimensions are useless (Ackermann and DuBow 1987). Unfortunately, these phage heads are highly contrasted and therefore popular with inexperienced researchers.
d. Dirty preparations. Some investigators do not purify their phages and examine crude lysates or phage suspensions made directly from plaques. Another problem seems to be linked to the MRS and M17 media, so frequently used in LAB phage research. For reasons unknown to this author, the lysates so prepared tend to give low-contrast preparations.
e. Poor contrast.
f. Lump descriptions. For example, it happened that 5 or 10 manifestly different *Strep. thermophilus* phages were described in one sentence as having "tails of 150-300 nm in length" without indicating the dimensions of individual phages. Such descriptions are meaningless.
g. Poor or no magnification control, resulting in unlikely dimensions. For example, head diameters inferior to 50 nm are almost certain to be wrong. In a general way, the specifications of the manufacturers are just indications. Magnification depends on the installing engineer and lens current, thus on local current supply, meaning that it can vary within the same day. The possible consequences are considerable variations in virus dimensions, evident in manifestly identical viruses such as *L. lactis* phages of the c2 species (Table 4). Indeed, the dimensions of these phages are distributed in a broad Gaussian bell and do not form a needle-sharp peak as they should. Further, we are unable to sort siphoviruses by their dimensions. This hinders their identification and classification. The problem is old. With respect to *Lactococcus* phages, Teuber and Lembke (1983) stated that "comparison of the data is almost impossible because of the different qualities of the staining and enlargement procedures used." This need not be. An electron microscope is a costly high-precision instrument and there is no excuse for mishandling it.
h. "Farming out" to other laboratories. The latter may be incompetent and disinterested, cumulating all the above problems while the original laboratory has no control and is even unable to complain.

3. The Cure

Misalignment is, in principle, easily correctable and should not even occur. Sources of errors such as contaminating phages and abnormal particles may require some experience, prolonged observation and, generally, easy access to an electron microscope. Positively stained heads are easily detected; all the electron microscopist has to do, is not to photograph them. Lump descriptions should not exist or be rejected by journals.

Lysates must be sterilized by filtration. They can be purified by CsCl or sucrose density gradient centrifugation. However, it is much simpler to clean them by centrifugation and simple washing. This author sediments phages in an Eppendorf vial of 1.5 ml and by centrifugation at 25,000 g for 1 hour, using a medium-sized centrifuge and a fixed-angle rotor. This is followed by two washes in buffer, e.g., ammonium acetate (0.1 M, pH 7). Even tap water will do, this author found it necessary to wash some LAB phages thrice.

In "manual" electron microscopes, contrast is classically controlled by apertures and choice of the right films, developers and papers. Considerable improvements can be obtained in the darkroom by means of graded filters for printing. In digital electron microscopes, the contrast can be greatly improved by manipulation of the "histogram" that comes with camera software (Tiekotter and Ackermann 2009). It is regrettable that EM companies have so far neglected to provide detailed instructions, leaving users to themselves. Ultimately, contrast can be improved with Photoshop technology.

In all electron microscopes, magnification is controlled by means of test specimens, namely beef liver catalase crystals or, which is particularly interesting for a phage laboratory, bacteriophage T4 tails. Catalase crystals can be obtained commercially. They have parallel lines with a periodicity of 8.8 nm (Luftig 1967), but are somewhat difficult to use. T4 tails are more convenient and have, as compared to catalase crystals, a length of 114 nm. Latex spheres and diffraction grating replicas are for low magnification only (10-30,000 times) and should not be used in phage studies.

In conventional electron microscopes, the adjustment is very easy and done within 1-3 minutes by means of a photographical enlarger. Magnification control in digital electron microscopes may be achieved by photographing test specimens and calculating correction factors. Sadly, this seems to be done rarely or not at all.

4. Twelve Practical Commandments

1. Wash your phages. Proteins and bacterial slime in a lysate are your enemies.
2. Stain with phosphotungstate and uranyl acetate. They are not equivalent and may give different results.
3. Be wary of damaged, deformed, or aberrant particles. Remember that particles of the same virus are supposed to be identical. Variations indicate malformations, contamination, presence of lysogenic phages, or just poor work.
4. Be wary of phages in UA crystals. They are invariably swollen and deformed.
5. Do not measure positively stained phages.
6. Describe different phages individually.
7. Select a sufficiently high final magnification (200,000 or better on paper) and stay with it.
8. Photograph sufficient numbers of phages (10-20) at the same magnification.
9. Prepare a calibration specimen. A grid with UA-stained T4 phages may last for two years.
10. Control magnification regularly (e.g., with every film).

11. Measure phage heads between opposite apices (it is easier than between opposite sides). Measure all three diameters of isometric heads as they are complementary.
12. State the number of particles measured and your calibration technique in Materials and Methods.

Conclusion

LAB phages continue to generate interest because of the economic importance of their host bacteria, but the study of LAB phages has increasingly shifted to the investigation of phage resistance mechanisms and genome sequencing. This does not mean that no novel phages will be found and described in the future. Many LAB phages have relatives in other Firmicutes of the low G+C group. *Lactobacillus* phages are particularly varied and interesting. By contrast, *Leuconostoc* phages may belong to 3-4 species only. There is an urgent need to sequence more phages genomes and to improve phage electron microscopy. Great progress in phage classification, and the understanding of phage relationships and evolution, is to be expected from applying BLAST-P-based programs for comparison of protein sequence similarity (*CoreExtractor, CoreGenes*) to phages of the *Siphoviridae* family. In summary, species classification of LAB phages is still unsatisfactory and in its beginnings.

At the same time, it appears that the investigation of LAB phages is going nowhere, turning in circles and around the same phages. This is largely due to industrialized cheese-making. Indeed, in the quest for uniform and exclusive products and high-yield fermentations, the same starters are used all over the world (Heller 2007, Quiberoni et al. 2006). It is no accident that the strange and unique *L. lactis* phage KSY1 was found in a fermented milk specialty, *viili,* that is particular to Finland and no other country (Saxelin et al. 1979). Thus, new phages are likely to be found in farms or home-style, nonindustrial cheese factories (Reinheimer et al. 1997), especially if research is extended to sheep, goat, or camel milk in Third-World countries.

Finally, we plead to phase out the Greek letter "phi" in phage names. Normally, virus names are immutable and, indeed, frozen in time for consistency and identification. "Phi", when used once, becomes part of the name; for example, famous microvirus ϕX174 is not ΦX174 or simply X174. The creation of synonyms and variant spellings should be avoided in all circumstances. Some investigators of LAB phages consider the letter ϕ or Φ as an accessory, an abbreviation of "phage". It is not. The ICTV and databases have complicated the problem in replacing Greek letters by their Latin transcripts. Hence, one and the same phage 1205 may be named 1205, ϕ1205, or phi1205. This is quite nonsensical and we suggest the Greek letter ϕ or phi not be used for novel phages.

References

Accolas, J.-P. and Spillmann, H. (1979). Morphology of bacteriophages of *Lactobacillus bulgaricus, L. lactis* and *L. helveticus*. *Journal of Applied Bacteriology, 47,* 309-319.

Ackermann, H.-W. (1992). Frequency of morphological phage descriptions. *Archives of Virology, 124,* 201-209.

Ackermann, H.-W. (2007). 5500 Phages examined in the electron microscope. *Archives of Virology, 152,* 277-243.

Ackermann, H.-W., Cantor, E. D., Jarvis, A. W., Lembke, J. and Mayo, J. A. (1984). New species definitions in phages of gram-positive cocci. *Intervirology, 22,* 181-190.

Ackermann, H.-W. and DuBow, M. S. (1987). *Viruses of prokaryotes. Vol I. General properties of bacteriophages.* Boca Raton, FL, USA: CRC Press.

Ackermann, H.-W. and DuBow, M. S. (1987). *Viruses of prokaryotes. Vol II. Natural groups of bacteriophages.* Boca Raton, FL, USA: CRC Press.

Ackermann, H.-W. and Eisenstark, A. (1974). The present state of phage taxonomy. *Intervirology 3,* 201-219.

Barrangou, R., Yoon, S.-S., Breidt, F., Fleming H. P. and Klaenhammer, T. R. (2002). Characterization of six *Leuconostoc fallax* bacteriophages isolated from an industrial sauerkraut fermentation. *Applied and Environmental Microbiology, 68,* 5452-5458.

Bauer, H., Dentan, E. and Sozzi T. (1970). The morphology of some *Streptococcus* bacteriophages. *Journal of Microscopy (Paris), 9,* 891-898.

Boizet, B., Mata, M., Mignot, O., Ritzenthaler, P. and Sozzi, T. (1992). Taxonomic characterization of *Leuconostoc mesenteroides* and *Leuconostoc oenos* bacteriophage. *FEMS Microbiology Letters, 90,* 211-216.

Bradley, D. E. (1967). Ultrastructure of bacteriophages and bacteriocins. *Bacteriological Reviews, 31,* 230-314.

Bradley, D. E. and Kay, D. (1960). The fine structure of bacteriophages. *Journal of General Microbiology, 23,* 553-563.

Brenner, S. and Horne, R. W. (1959). A negative staining method for high resolution electron microscopy of viruses. *Biochimica et Biophysica Acta 34,* 103-110.

Brondsted, L. and Hammer, K. (2006). Phage of Lactococcus lactis. In R. Calendar (Ed.), *The bacteriophages* (2nd ed., 572-592). New York: Oxford University Press.

Brüssow, H. (1999). Phages of *Streptococcus thermophilus.* In A. Granoff and R. G. Webster (Eds.), *Encyclopedia of virology* (2nd ed., Vol 2, 1253-1262). San Diego, CA, USA: Academic Press.

Brüssow, H. (2001). Phages of dairy bacteria. *Annual Review of Microbiology, 55,* 283-303.

Brüssow, H., Bruttin A., Desiere, F., Lucchini, F. and Foley, S. (1998). Molecular ecology and evolution of *Streptococccus thermophilus* bacteriophages – a review. *Virus Genes, 16,* 95-109.

Brüssow, H. and Desiere F. (2001). Comparative phage genomics and the evolution of *Siphoviridae*: Insights from dairy phages. *Molecular Microbiology, 39,* 213-222.

Brüssow, H., Frémont, M., Bruttin, A., Sidoti, J., Constable, A. and Fryder, V. (1994). Detection and classification of *Streptococcus thermophilus* bacteriophages isolated from industrial milk fermentation. *Applied and Environmental Microbiology 60,* 4537-4543.

Brüssow, H. and Suárez J (2006). *Lactobacillus* phages. In R. Calendar (Ed.), *The bacteriophages* (2nd ed, 653-663). New York, USA: Oxford University Press.

Caldwell, S. L., McMahon, D. J., Oberg, C. J. and Broadbent, J. R. (1999). Induction and characterization of *Pediococcus acidilactici* temperate bacteriophage. *Systematic and Applied Microbiology, 22,* 514-519.

Cannon, P., Lee, T. A., Bolanos, J. T. and Danziger, I. H. (2005). Pathogenic relevance of *Lactobacillus:* a retrospective review of over 200 cases. *European Journal of Clinical Microbiology, 24,* 31-40.

Cappelli, E. A., Barros, R. R., Camello, T. C., Teixeira, L. M., Merquior, V. L. (1999). *Leuconostoc pseudomesenteroides* as a source of nosocomial urinary tract infections. *Journal of Clinical Microbiology, 37,* 4124-4126.

Chibani-Chennoufi, S., Dillmann, M. L., Marvin-Guy, L., Rami-Shojaei, S. and Brüssow, H. (2004). *Lactobacillus plantarum* bacteriophage LP65: a new member of the SPO1-like genus of the family *Myoviridae*. *Journal of Bacteriology, 186*, 7096-7083.

Davidson, B. E., Powell, I. B. and Hillier, A. J. (1990). Temperate bacteriophages and lysogeny in lactic acid bacteria. *FEMS Microbiology Reviews, 87,* 79-90.

Deane, D. D., Nelson, F. E., Ryser, F. C. and Carr, P. H. (1953). *Streptococcus thermophilus* bacteriophage from Swiss cheese whey. *Journal of Dairy Science, 36,* 185-191.

De Klerk, H. C., Coetzee, J. N. and Fourie, J. T. (1965). The fine structure of *Lactobacillus* bacteriophages. *Journal of General Microbiology, 38,* 35-38.

Desiere, F., Lucchini, S., Canchaya, C., Ventura M. and Brüssow, H. (2002). Comparative genomics of phages and prophages in lactic acid bacteria. *Antonie Leeuwenhoek, 82,* 73-91.

Deveau, H., Labrie, S., Chopin, M.-C., Sylvain, S. (2006). Biodiversity and classification of lactococcal phages. *Applied and Environmental Microbiology, 72,* 4338-4346.

d'Herelle, F. (1917). Sur un microbe invisible antagoniste des bacilles dysentériques. *Comptes Rendus Hebdomadaires des Séances de l'Académie des Sciences D, 165,* 373-375.

Dupuis, M. E. and Moineau, S. (2010). Genome organization and characterization of the virulent lactococcal phage 1358 and its similarities to *Listeria* phages. *Applied and Environmental Microbiology, 76,* 1623-1632.

Eisenstark, A. (1967). Bacteriophage techniques. In K. Maramorosch, H. Koprowski (Eds.), *Methods in Virology, Vol. 1* (pp. 449-524). London, UK: Academic Press.

Elliker, P. R. (1950). The problem of bacteriophage in the dairy industry. *Journal of Milk and Food Technology, 14,* 13-14, 44.

Emond, E. and Moineau, S. (2007). Bacteriophages and food fermentations. In S. McGrath and D. van Sinderen (Eds.), *Bacteriophage – genetics and molecular biology* (93-123). Norfolk, UK: Caister Academic Press.

Fauquet, C. M., Mayo, M. A., Maniloff, J., Desselberger, U. and Ball, L. A. (Eds.). (2005). *Virus Taxonomy. VIIIth Report of the International Committee on Taxonomy of Viruses*. London, UK: Elsevier Academic Press.

Fortier, L.-C., Bransi, A. and Moineau, S. (2006). Genome sequence and global gene expression of Q54, a new phage species linking the 936 and c2 phage species of *Lactococcus lactis*. *Journal of Bacteriology, 188,* 6101-6114.

Greer, G. G., Dilts, B. D. and Ackermann, H.-W. (2006). Characterization of a *Leuconostoc gelidum* bacteriophage from pork. *International Journal of Food Microbiology, 114*, 370-375

Hadley, P. and Dabney, E. (1926). The bacteriophage relationship between *B. coli, S. fecalis,* and *S. lacticus*. *Proceedings of the Society for Experimental Biology and Medicine, 13,* 13-18.

Haguenau, F., Hawkes, P. W., Hutchison, J. L., Satiat-Jeunemaître, B., Simon, G. T. and Williams, D. B. (2003). Key events in the history of electron microscopy. *Microscopy and Microanalysis, 9,* 96-138.

Hardie, J. M. and Whiley, R. A. (1997). Classification and overview of the genera *Streptococccus* and *Enterococcus*. *Journal of Applied Microbiology Symposium Supplement, 83,* 1S-11S.

Heller, K. J. (2007). Editorial. Perspectives on milk – a special kind of food. *Biotechnology Journal, 2,* 399-401.

Ismail, E. A., Neve, H., Geis, A. and Heller, K. J. (2009). Characterization of temperate *Lactobacillus gasseri* phage LgaI and its impact as prophage on autolysis of its lysogenic host strains. *Current Microbiology, 58,* 648-653.

Jarvis, A. W. (1977). The serological differentiation of lactic streptococcal bacteriophage. *New Zealand Journal of Dairy Science Technology, 12,* 176-181.

Jarvis, A. W. (1984). Differentiation of lactic streptococcal phages into phage species by DNA-DNA homology. *Applied and Environmental Microbiology, 47,* 343–349.

Jarvis, A. W. (1989). Bacteriophages of lactic acid bacteria. *Journal of Dairy Science, 72,* 3406–3428.

Jarvis, A. W., Fitzgerald, G. F., Mata, M., Mercenier, A., Neve, H., Powell, I. B., Ronda, C., Saxelin, M. and Teuber, M. (1991). Species and type phages of lactococcal bacteriophages. *Intervirology, 32,* 2-9.

Jarvis, A. W., Lubbers, M. W., Waterfield, N. R., Collins, L. J. and Polzin, K. M. (1995). Sequencing and analysis of the genome of lactococcal phage c2. *International Dairy Journal, 5,* 963-976.

Klumpp J., Lavigne R., Loessner, M.J. and Ackermann, H.-W. (2010). The SPO1-related bacteriophages. *Archives of Virology, 155,* 1547-1561.

Lavigne, R., Darius, P., Summer, E. J., Seto, D., Mahadevan, P., Nilsson, A. S., Ackermann, H.-W. and Kropinski, A. M. (2009). Classification of *Myoviridae* bacteriophages using protein sequence similarity. *BMC Microbiology, 9,* 224.

Lavigne, R., Seto, D., Mahadeva, P., Ackermann, H.-W. and Kropinski, A. M. (2008). Unifying classical and molecular taxonomic classification: analysis of the *Podoviridae* using BLASTP-based tools. *Research in Microbiology, 159,* 406-414.

Lembke, J., Krusch, U., Lompe, A. and Teuber, M. (1980). Isolation and ultrastructure of bacteriophages of group N (lactic) streptococci. *Zentralblatt für Bakteriologie, Mikrobiologie und Hygiene 1. Abt. Originale B, Hygiene, 1,* 79-91.

Loof, M. and Teuber, M. (1986). Heteroduplex analysis of the genomes of *Streptococcus lactis* subsp. *diacetylactis* bacteriophages of the P008-type isolated from German cheese factories. *Systematic and Applied Microbiology, 8,* 226-229.

Ludvik, J., Erbenova L. and Lipavska, H. (1977). Ultrastructure of *Bacillus licheniformis* bacteriophage BLE and its DNA. *Virology, 77,* 872-875.

Luftig, R. B. (1967). An accurate measurement of the catalase crystal period and its use as an internal marker for electron microscopy. *Journal of Ultrastructure Research, 20,* 91-102.

Mahony, J., Deveau, H., McGrath, S., Ventura, M., Canchaya, C., Moineau, S., Fitzgerald, G. F. and van Sinderen, D. (2006). Sequence and comparative genomic analysis of lactococcal bacteriophages jj50, 712 and P008: evolutionary insights into the 936 phage species. *FEMS Microbiology Letters, 261,* 253-261.

Mata, M. and Ritzenthaler, P. (1988). Present state of lactic acid bacteria phage taxonomy. *Biochimie, 70,* 395-399.

Moineau, S. (1999). Applications of phage resistance in lactic acid bacteria. *Antonie Leeuwenhoek, 76,* 377-382.

Moineau, S. and Lévesque, C. (2005). Control of bacteriophages in industrial fermentations. In E. Kutter, A. Sulakvelidze (Eds.), *Bacteriophages: biology and applications* (285-295). Boca Raton, FL, USA: CRC Press.

Moineau, S., Pandian, S. and Klaenhammer, T. R. (1994). Evolution of a lytic bacteriophage via DNA acquisition from the *Lactococcus lactis* chromosome. *Applied and Environmental Microbiology, 60,* 1832-1841.

Mosimann, W. and Ritter, W. (1946). Bakteriophagen als Ursache von Aromaschwund in Rahmsäuerungskulturen. *Schweizerische Milchzeitung, 72,* 211-212.

NCBI Taxonomy Browser. http://www.nlm.nih.gov/Tax.

Parmelee, C. E., Carr, P. H. and Nelson F. E. (1949). Electron microscope studies of bacteriophage active against *Streptococcus lactis. Journal of Bacteriology, 57,* 391-397.

Pfankuch, E. and Kausche, G. A. (1940). Isolierung und übermikroskopische Abbildung eines Bakteriophagen. *Naturwissenschaften, 28,* 46.

Pillidge, C. J. and Jarvis, A. W. (1988). DNA restriction maps and classification of the lactococcal bacteriophages c2 and sk1. *New Zealand Dairy Science and Technology, 23,* 411-416.

Quiberoni, A., Moineau, S., Rousseau, G. M., Reinheimer, J. and Ackermann H.-W. (2010). *Streptococcus thermophilus* bacteriophages. *International Dairy Journal, 30,* 657-664.

Quiberoni, A., Tremblay, D., Ackermann, H.-W., Moineau, S. and Reinheimer, J. A. (2006). Diversity of *Streptococcus thermophilus* phages in a large-production cheese factory in Argentina. *Journal of Dairy Science, 89,* 3791-3799.

Raya, R. R., Kleeman, E. G., Luchansky, J. B. and Klaenhammer, T. R. (1989). Characterization of the temperate bacteriophage ϕadh and plasmid transduction in *Lactobacillus acidophilus* ADH. *Applied and Environmental Microbiology, 55,* 2206-2213.

Reinheimer, J. A., Binetti, A. G., Quiberoni, A., Bailo, N. B., Rubiolo, A., Giraffa, G. (1997). Natural milk cultures for the production of Argentinian cheeses. *Journal of Food Protectection, 60,* 59–63.

Reiter, B. (1949). Lysogenic strains of lactic streptococci. *Nature, 164,* 667-668.

Ruska, H. (1940). Über die Sichtbarmachung der bakteriophagen Lyse im Übermikroskop. *Naturwissenschaften, 28,* 45-46.

Saxelin, M.-L., Nurmiaho, E. L., Korhola, M. P. and Sundman, V. (1979). Partial characterization of a new C3-type capsule-dissolving phage of *Streptococcus cremoris. Canadian Journal of Microbiology, 25,* 1182-1187.

Schmager, A. (1951). Über Züchtungsversuche von Phagen, die aktiv gegen grampositive Milchsäurestäbchen sind. *Zeitschrift für Hygiene und Infektionskrankheiten, 132,* 536-539.

Schouler, C., Ehrlich, S. D. and Chopin M.-C. (1994). Sequence and organization of the lactococcal prolate-headed bIL67 phage genome. *Microbiology (UK), 140,* 3061-3069.

Séchaud, L., Cluzel, P.-J., Rousseau, M., Baumgartner, A. and Accolas, J.-P. (1988). Bacteriophages of lactobacilli. *Biochimie, 70,* 401–410.

Sozzi, T., Poulin, J. M., Maret, R. and Pousaz R. (1978). Isolation of a bacteriophage of *Leuconostoc mesenteroides* from dairy produce. *Journal of Applied Bacteriology, 44,* 159-161.

Sozzi, T., Watanabe, K., Stetter, K. and Smiley, M. (1981). Bacteriophages of the genus *Lactobacillus. Intervirology, 16,* 129-135.

Summers, W. C. (1999). *Felix d'Herelle and the origins of molecular biology*. New Haven, CT, Yale, UK: University Press.

Teuber, M. and Lembke, J. (1983). The bacteriophages of lactic acid bacteria with emphasis on genetic aspects of group N lactic streptococci. *Antonie Leeuwenhoek, 49,* 471-478.

Tiekotter, K. L. and Ackermann, H.-W. (2009). High-quality virus images obtained by TEM and CCD technology. *Journal of Virological Methods, 159,* 87-92.

Tikhonenko, A. S. (1968). *Ultrastructure of bacterial viruses*. Moscow, Izdadelstvo Nauka (Russian); New York: Plenum Press (English).

Tikhonenko, A. S. (1970). *Ultrastructure of bacterial viruses*. Moscow, Izdadelstvo Nauka (Russian); New York: Plenum Press (English).

Tohyama, K., Sakurai, T., Arai, H. and Oda, A. (1972). Studies on temperate phages of *Lactobacillus salivarius*. I. Morphological, biological, and serological properties of newly isolated temperate phages of *Lactobacillus salivarius*. *Japanese Journal of Microbiology, 16,* 385-395.

Twort, F. W. (1915). An investigation on the nature of ultra-microscopic viruses. *Lancet, ii,* 1241-1243.

Uchida, K. and Kanbe, C. (1993). Occurrence of bacteriophages lytic for *Pediococcus halophilus*, a halophilic lactic-acid bacterium, in soy sauce fermentation. *Journal of General and Applied Microbiology, 39,* 429-437.

Van Regenmortel, M. H. V. (1998). Virus species, a much overlooked but essential concept in virus classification. *Intervirology, 31,* 241-154.

Villion, M. and Moineau, S. (2009). Bacteriophages of *Lactobacillus*. *Frontiers in Bioscience, 14,* 1661-1683.

White, R., Georgi, C. E. and Militzer, W. (1954). Heat studies on a thermophilic bacteriophage. *Proceedings of the Society for Experimental Biology and Medicine, 85,* 137-139.

Whitehead, H. R. and Cox, G. A. (1935). The occurrence of bacteriophage in cultures of lactic streptococci. A preliminary note. *New Zealand Journal of Science Technology, 16,* 319-320.

Whitehead, H. R. and Hunter, G. J. E (1937). Observations on the activity of bacteriophage in the group of lactic streptococci. *Journal of Pathology and Bacteriology, 44,* 337-347.

Yoon, S.-S., Barrangou-Poueys, R., Breidt, F., Klaenhammer, T. R. and Fleming, H. P. (2002). Isolation and characterization of bacteriophages from fermenting sauerkraut. *Applied and Environmental Microbiology, 68,* 973-976.

In: Bacteriophages in Dairy Processing
Editor: Andrea del Lujan Quiberoni et al.
ISBN 978-1-61324-517-0

Chapter 2

BACTERIOPHAGE ADAPTATION, WITH PARTICULAR ATTENTION TO ISSUES OF PHAGE HOST RANGE

***Benjamin K. Chan*[1] *and Stephen T. Abedon*[2*]**

[1]Department of Biology, University of Utah, 257 South 1400 East, Salt Lake City, Utah, US

[2] Department of Microbiology, Ohio State University, 1680 University Dr., Mansfield, Ohio, US

ABSTRACT

Bacteriophages (phages), the viruses of bacteria, are ecological entities that evolutionarily adapt to their hosts along with the extracellular environments that they inhabit. Adaptations can be observed in terms of phage survival characteristics, rates of bacterial acquisition (adsorption rate constant), infection durations (a component of generation time), and overall infection fecundities (burst size). These adaptations can impact rates of phage population growth as that occurs over the broader environment, that is, change in phage densities as a function of time (P'). Shorter generation times, greater infection fecundities, and greater survival characteristics all can give rise to a greater P' and, in terms of dairy processing, the greater P' then the sooner the impact of phage contaminants on fermentation properties. A second and related measure of phage adaptation, that also combines multiple elements of phage biology, is what we call an effective burst size (EBS). These are those phages derived from an infection's *absolute* burst size (ABS) which, given exposure to susceptible hosts, succeed in initiating successful new infections. Another important measure of phage adaptation is a culture's total phage yield (TPY). Here we discuss phage evolution particularly in terms of phage adaptations that enhance P', EBS, or TPY, and as informed, in part, by classical (macro) ecological concepts. We consider in particular issues pertaining to the evolution of phage host range.

[*] E-mail address: abedon.1@osu.edu. Tel.: +1 419 755 4343; fax: +1 419 755 4327

INTRODUCTION

Bacteriophages – the viruses of bacteria and "phages" for short – are quantitatively the most numerous organisms on Earth, with numbers ranging up to perhaps 10^{32} in total (Wommack and Colwell 2000, Abedon 2008c). Phages directly impact bacteria by killing them via lytic as well as abortive infections (Hyman and Abedon 2010), contribute to the phenotype of persisting bacterial hosts via lysogenic conversion (Wagner and Waldor 2002, Abedon and LeJeune 2005, Los et al. 2010), effect 'kill the winner' which is a frequency-dependent selection against more-common bacterial types (Levin and Lenski 1985, Levin 1988, Thingstad and Lignell 1997, Thingstad 2000, Weinbauer and Rassoulzadegan 2004, Thingstad et al. 2008, Abedon 2009d, 2009f), and, in the course of transduction, serve as major contributors to gene exchange among bacteria (Miller 2001, Paul and Jiang 2001, Miller 2004, Brabban et al. 2005, Breitbart et al. 2005, Abedon 2009f). Phages similarly can substantially impact bacteria contributing to industrial-scale dairy (Whitehead and Cox 1935, Marth and Steele 2001, Moineau and Lévesque 2005) and non-dairy fermentations (Bogosian 2006), especially by killing these bacteria. In the course of their various interactions with bacteria and other aspects of their environments, phages evolve. The consequence of natural selection in the course of that evolution is adaptation: "a surplus of beneficial over deleterious changes" (Mustonen and Lassig 2010) that result in improvement in phage survival and reproductive characteristics. Here we consider issues of phage adaptation, with partial emphasis on host range adaptation.

1. INTRODUCTION TO PHAGE BIOLOGY AND ADAPTATION

Phage adaptations can be viewed organismally (as opposed to molecularly or biochemically) as impacting either a phage's adsorption, latent period, or what we describe as an effective burst size (EBS) (Abedon 2008a, 2008b, 2009d, 2009f, Abedon and Thomas-Abedon 2010, Abedon 2011a). See Figure 1 for summary. Adsorption involves a combination of virion movement such as via diffusion (the extracellular search for new bacteria to infect), virion encounter with a bacterium, reversible and especially irreversible virion attachment to that bacterium, and subsequent phage-genome translocation from the phage capsid into the bacterial cytoplasm. A phage's ability to adsorb bacteria, as well as post-adsorption abilities, determine a phage's host range (Hyman and Abedon 2010). Phages with narrow host ranges can be described as specialists, that is, specialists on certain bacterial types, while phages with wide host ranges can be considered more as generalists (Weinbauer 2004).

Latent period follows adsorption and is a measure of the duration of the phage infection period. It begins with phage-genome translocation into the cytoplasm and, especially for lytic infections (Hyman and Abedon 2009), ends with phage-progeny release from the infected bacterium. Translocation is followed by an eclipse. Maturation of the first phage virion within cells defines the end of the eclipse and initiation of what we will call the post-eclipse. For lytic infections the post-eclipse ends with a combination of phage-progeny release, infection death, and bacterial destruction. Release of phages from infected bacteria connects the infection aspect of phage life cycles with the adsorption aspect. Shortening either a phage's extracellular search or latent period will have the effect of reducing a phage's generation time and thus the rate of phage population growth.

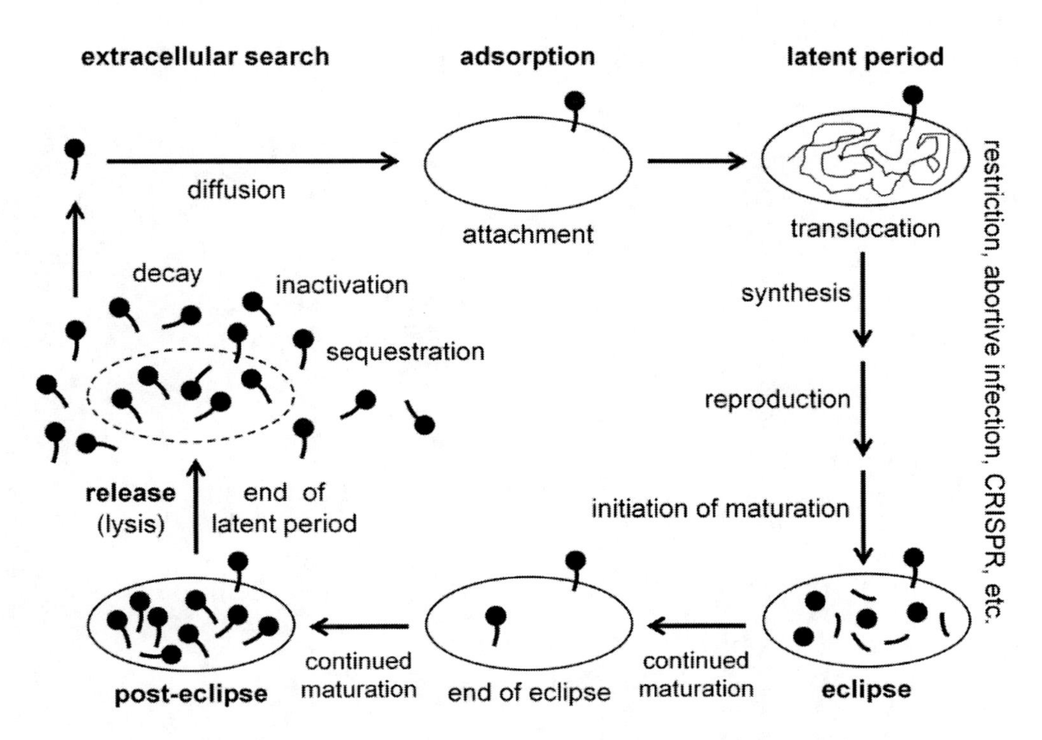

Figure 1. Lytic phage life cycle. Starting with a diffusion-driven extracellular search, free phages encounter and then attach to bacteria in the course of phage adsorption. Transitioning to the phage infection period, or latent period, phage nucleic acid is translocated from the phage particle into the bacterial cytoplasm. This initiates the eclipse portion of the phage latent period during which macromolecule synthesis, genome replication (reproduction), and phage particle maturation is initiated. By definition the eclipse ends when the first mature phage particle has been produced, initiating the portion of the latent period that we are referring to as the post-eclipse. During the post-eclipse phage progeny continue to accumulate at a rate that is characteristic of the phage that is infecting in combination with the host that it is infecting along with environmental conditions. Phage release occurs at what typically is a fairly well defined time during the infection and for lytic phages this release, via lysis, results in termination of the phage infection as well as destruction of host bacterium. Released phages are subject to decay/inactivation or sequestration such as by biofilm extracellular polymeric substances in the course of their extracellular search. Note that phages are also subject to inactivation following adsorption to bacteria as indicated on the right with the words restriction, abortive infection, CRISPR, etc. Note also that not-yet fully mature virion components, for the sake of clarity, are not shown over "end of eclipse" or "post-eclipse". See Breitbart et al. (2005) and Abedon (2006) for further discussion of the phage life cycle from ecological or evolutionary ecological perspectives.

Free phages, those that have been released from bacteria, are subject to decay (a.k.a., inactivation). The number of phages that are inactivated following this release is equal to ABS minus EBS, where ABS is the actual or absolute phage burst size, that is, the burst size as traditionally defined. Towards minimizing the difference between ABS and EBS, a phage population can increase its inherent resistance to inactivation and, in addition, can increase its rate of adsorption since the longer a phage takes until it infects a bacterium then the more likely that phage will become inactivated while in the extracellular state (Abedon 2006). See

Figure 2 for illustration. EBS as a term alternatively has been used to describe phage decay that occurs due to lethal mutagenesis prior to bacterial lysis (Bull and Wilke 2008), the number of phage progeny produced per phage adsorbed in a high-multiplicity adsorption context (Gadagkar and Gopinathan 1980, Patel and Rao 1984), or the number of viruses released during chronic virus infection (Spouge and Layne 1999).

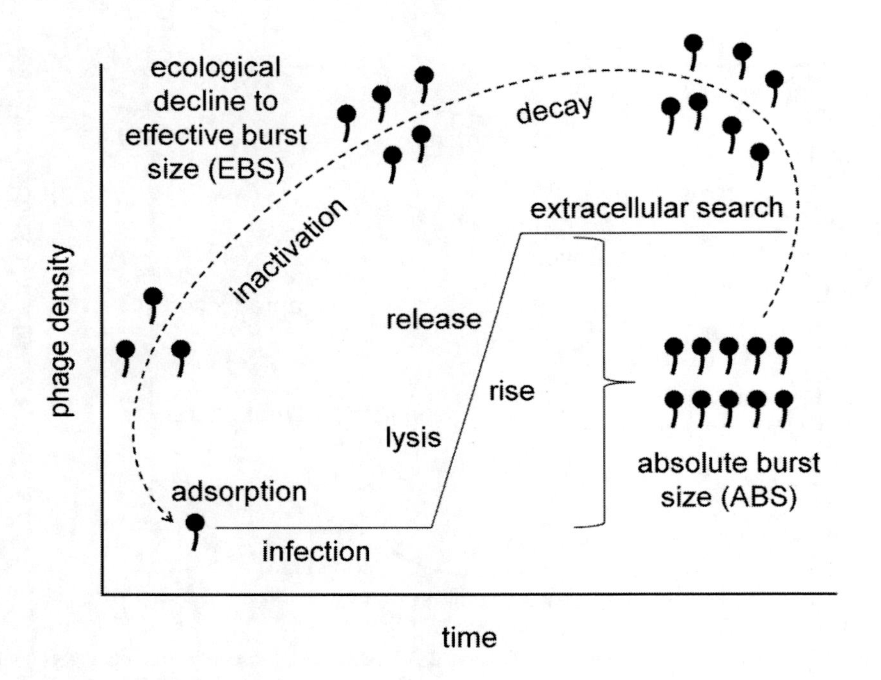

Figure 2. Relationship between effective burst size (EBS) and absolute burst size (ABS) as shown within the context of a phage single-step growth experiment (Hyman and Abedon 2009). Phage adsorption is followed by a latent period during which phage progeny are produced. For lytic phages, release via lysis tends to not be simultaneous across the phage population, resulting in what was known as a rise, for rise in phage titer, that spans some duration. During the rise a burst size of phages is released, which is a per-infection measure of phage fecundity. During the phage extracellular search this absolute burst size (ABS) is reduced as phages decay/are inactivated. Further declines from the ABS occur upon phage adsorption to bacteria that are unable to support a phage infection. The number of phages remaining from the initial ABS that are able to acquire and successfully infect bacteria we describe as an effective burst size (EBS).

Though phages can adapt to their environments in these multiple ways, evolutionary fixation of new alleles within populations, which can be used to measure of the occurrence of adaptation (Wichman et al. 1999), does not always occur (Rousseau and Moineau 2009). Failures of such fixation are suggestive of limitations on phage adaption that can stem from low likelihoods of generation of sufficiently beneficial alleles. Such limitations include too low mutation rates, genotypic constraints on beneficial allele generation (a.k.a., historical contingencies), and too small population sizes (resulting in both low absolute numbers of mutations and weaker natural selection due to genetic drift; Lenski and Levin 1985, Burch and Chao 2000, Bull 2008, Guyader and Burch 2008). The basic steps of adaptation thus include (1) genetic variation within populations generated by mutation (or recombination), (2) rise in the frequency of relatively beneficial alleles as mediated especially by natural selection (resulting in creation of polymorphisms), and, in many instances, (3) fixation of the corresponding adaptation within a population.

2. Phage Evolutionary Fitness

Changes in phage density as a function of time we abbreviate as the first derivative of phage density, *P*' using Lagrange's notation, which is one measure of phage evolutionary fitness. So long as sufficient densities of phage-susceptible bacteria are present within environments – defining the so-called phage proliferation threshold (Abedon and Thomas-Abedon 2010) – then phage populations will not decline in numbers, implying an absolute phage fitness of one or greater. In general we expect that greater EBSs or shorter generation times will have the effect of increasing *P*'. Thus, mechanisms that can lead to shorter phage latent periods, faster phage adsorption, larger bursts (ABS), or slower phage inactivation all should result in increases in *P*' and therefore greater phage evolutionary fitness (Abedon 2009d, 2009f). These different fitness-enhancing mechanisms, however, will not necessarily vary independently.

Reductions in *P*' in some cases also can have the effect of increasing the final yields of phages within cultures (TPY or total phage yield). TPY is another measure of phage evolutionary fitness as well as a key determinant of the phage potential to impact bacterial populations (Abedon and Thomas-Abedon 2010, Abedon 2011a). Increases in TPY can occur most notably in association with increases in EBS, but also if bacterial populations can grow to larger densities, such as can occur given delays in phage population growth (that is, reduced *P*'). The result can be greater numbers of phage infections and thereby more phage bursts, where TPY varies as a function of the product of the total number of bacteria infected and phage EBS (Abedon 2009d). Such a result is seen in Abedon *et al.* (2003) where a wild-type phage enjoyed a yield advantage over a mutant phage in part because of wild-type's slower population growth. A similar result is presented by Storms *et al.* (2010) who observed increased TPY when employing phages that were less effective adsorbers. Such delays can also be seen in phage-bacterial chemostat experiments (Abedon 2006). For example, delays in phage growth can be mediated by initiating phage populations at lower densities, even if bacterial densities are also started at lower densities (Abedon 2009b, 2009d), or by extending phage latent periods (Abedon et al. 2003). The consequence of such delays in phage population growth is reduced chemostat stability, and reduced chemostat stability also can be mediated by increasing phage bursts sizes (Kerr et al. 2008) or by increasing rates of bacterial population growth (Bohannan and Lenski 1997, Abedon 2006). Less effective at reducing chemostat stability, through still reducing *P*', are reductions in phage adsorption constants (Abedon 2006) or greater phage decay rates (Chao et al. 1977, Abedon 2006, 2009d) since these will also negatively impact TPY.

Mechanisms that can increase phage evolutionary fitness by one measure, *P*', thus will not necessarily increase phage fitness by other measures such as TPY, e.g., (Abedon et al. 2003, Bull 2006). Which measure is the more appropriate determinant of phage fitness will depend on various factors. For example, *P*' can be more important when phages are competing head-to-head within well-mixed cultures. Alternatively, TPY can be more important when phage populations are competing between cultures (e.g., different vats or batches), particularly if sampling occurs prior to population-wide lysis of a culture's bacteria. TPY also can be more important within spatially structured environments such that direct competition between phage populations for individual bacteria is lacking (Abedon and Culler 2007b, Abedon 2008b). In terms of phage impact on bacteria, the greater *P'* then the *sooner* that impact, the greater TPY then the *greater* that impact, and the greater the ability of phages to penetrate to target bacteria, including in terms of phage host range, then also the greater the phage impact (Abedon 2006, 2009d).

3. Adaptation: General Principles

Ecology, as defined by Haeckel (1866), is the study of organism interactions with environments. Those environments consist of both biotic and abiotic components, organisms and non-organisms, respectively, together defining what are known as ecological niches (Hutchinson 1957). The most important biotic component of niches, to phages, typically will be phage-susceptible bacteria. Abiotic components we can differentiate into those that can contribute to phage success, such as phage adsorption cofactors; nutrients utilized by host bacteria during infection; or a general benignness of the physical environment (not too hot, not too cold, not too acidic, not too basic); etc. (Abedon 2009d). Alternatively, abiotic as well as biotic components of environments can be antagonistic to phages (Suttle 2000, Weinbauer 2004), such as UV radiation (Suttle and Chen 1992, Sinton et al. 1999), physical extremes (Müller-Merbach et al. 2005), phagotrophic protists (González and Suttle 1993, Bettarel et al. 2005), animal immune systems (Geller et al. 1998, Merril 2008), or irreversible absorption to abiotic components of environments (Rossi and Aragno 1999). Even bacteria can be antagonistic to phages, particularly ones that can adsorb phages but are unable to support subsequent phage infection (Labrie et al. 2010, Hyman and Abedon 2010). Phage adaptations increase phage abilities to effectively persist within and otherwise exploit the niches in which they are found.

These adaptive traits can be any aspect of an organism's biology, though can be categorized into various loosely defined categories such as behavioral, structural, and physiological adaptations. Importantly, each of these types of adaptation in many cases can be viewed in at least two ways: (1) as evolutionary adaptations, that is, products of evolution and especially of natural selection, and which is our primary emphasis here, or (2) as non-genetic changes in the current states of organisms. Thus, organisms can change non-genetically over the course of their lives, such as one sees over the course of maturation. Importantly, many of those changes follow patterns that have been set down by evolution. Thus, even non-genetic changes in organism states can be guided by evolutionary adaptations.

3.1. Behavioral Adaptations

Behavior, traditionally, involves the activity of sensory organs in combination with musculoskeletal responses. Analogs to sensory organs as well as motion-mediating functions also exist in non-animal organisms, including plants opening their leaves in response to light, or bacterial taxis. In translating these ideas into the characteristics of phages, behaviors might be considered to involve especially the actions of virions prior to their adsorption. For example, phages change their adsorption behaviors in response to adsorption cofactors (Kutter et al. 1994). The result of this sensing can involve a sequestration of tail fibers against the virion body, a musculoskeletal-like action (Kellenberger et al. 1996, Letarov et al. 2005). More loosely, behaviors can be defined as phase shifts that can occur within evolutionary algorithms, i.e., adaptations that can take on more than on state in response to environmental input. Lytic-lysogenic decisions by temperate phages (Little 2006) thus can potentially be described as phage behaviors.

3.2. Structural Adaptations

Structurally adaptive traits are morphological features such as shape, size, and associated function. Virion stability, for example, can be a structural adaptation (de Paepe and Taddei 2006) as too can tape measure proteins, which control phage tail lengths. The latter can have additional phenotypic effects such as enabling infection of *Mycobacterium smegmatis* hosts when the latter are in stationary phase (Piuri and Hatfull 2006). Indeed, phage adsorption characteristics in general are dependent on virion structural adaptations, such as phage T4's wac (whisker antigen control) protein which, as described above, is involved in tail fiber retraction in response to adverse conditions (Conley and Wood 1975, Letarov et al. 2005) – a behavioral adaptation that is dependent on a structural adaptation.

3.3. Physiological Adaptations

Changes in gene expression or the activity of other molecules, especially enzymes, can be described as physiological adaptations and are relevant to phages mainly during infection. The timing of phage latent periods is a particularly well studied physiological adaptation as it involves both the extent of gene expression, such as production of holin proteins, and the activity of the individual holin proteins once produced (Young and Wang 2006). Latent period thus can be modified as a consequence of the host physiological state (Hadas et al. 1997).

Interestingly, the actual action of these holin proteins, initiating the lysis of a phage-infected bacterium, can be seen as a phase shift and therefore roughly as a behavior. Therefore, while delineating adaptations into behavioral, physiological, and structural categories can be useful, particularly when connecting phage ecological characteristics with those phenomena studied in classical ecology, in practice such distinctions are not necessarily absolute. The distinction between genetic (i.e., evolutionary) and non-genetic changes in organism properties (i.e., behavioral, structural, or physiological), along with the tendency of the latter to be dependent on the former, nevertheless is a crucial point to keep in mind as one considers phage organismal adaptations. This distinction can also be described in terms of ultimate causation (evolutionary changes) versus proximal causation (particularly mechanistic underpinnings of adaptations), where our emphasis here is on ultimate causation. That is, why do particular phage features exist rather than alternative features?

3.4. Tradeoffs

One ultimate causation perspective on adaptation places an emphasis on understanding tradeoffs associated with particular traits. Such tradeoffs, for example, might exist between eclipse length and the post-eclipse productivity of infections, with longer gearing-up times potentially allowing for subsequently faster virion-production rates while a shorter eclipse directly results in shorter infections and therefore shorter generation times. Conflicts also likely exist in terms of phage host-range breadth: Narrower host-range phages (specialists) may have extended extracellular search times as susceptible hosts could be rarer, while a broader host range could interfere with phage adaptation to particular hosts. Perhaps the classic tradeoff for phages is found between latent period length and burst size where increases in the former result in increases in the latter, but at the expense of lengthening phage generation time. Each of these possible tradeoffs is discussed more fully in subsequent sections.

4. Adaptation and Community Ecology

Communities consist of two or more different species found within a single environment. The potentially resulting inter-specific interactions considerably complicate ideas of phage adaptation, in particular leading to what can be described variously as antagonistic coevolution (Brockhurst et al. 2007, Paterson et al. 2010), arms races (Lenski 1984, Lenski and Levin 1985, Weitz et al. 2005) or the "Red Queen's Hypothesis" (van Valen 1973, O'Malley 2009), all as occurring between phages and their hosts. Since our emphasis here is more on the phage- rather than bacterial-adaptation side of the coevolution equation, we will bypass the question of whether or not antagonistic coevolution strictly requires "exclusive antagonism" (Lenski and Levin 1985), that is, one bacterial species interacting within only a single phage species. Indeed, one of the issues we will consider in depth is the potential for phages to interact with multiple bacterial strains or species.

4.1. Predator-Prey Dynamics

Phages, whether viewed as predators or as parasites (Kerr et al. 2008, Abedon 2011b), are constantly under selective pressure to become better hunters of bacteria as are bacteria to become better evaders of phages (Vos et al. 2009). When contrasted with more classically studied predator-prey interactions, however, phages display a very large conversion efficiency of prey into predators, i.e., tens, hundreds, and even thousands of phages may be produced per bacterial prey obtained. Therefore, we might predict that phages would quite frequently overwhelm any bacterial potential to evade phages, resulting in regional bacterial extinction. The potential for individual phages to acquire bacteria, though, may be relatively small in comparison with equivalent animal predators of animal prey, since phages are unable to move on their own accord such as directionally towards their bacterial prey. On the other hand, the prey in many cases are sessile, or at least unable to effectively run away from phages. Thus, we can envisage phage-bacterial acquisition dynamics as the equivalent of a drunkard's walk by phages towards bacteria that are invisible but nonetheless "sitting" ducks.

Phage omnipotence stemming from their high conversion efficiencies, a.k.a., large absolute burst sizes (ABS), can be further constrained by phage susceptibility to environment-mediated degradation (i.e., virion decay); *limitations* on the phage ability to physically penetrate to otherwise susceptible bacteria as can occur given environments that are spatially structured such as by agar, biofilm extracellular matrix, or simply lack of broth-culture mixing (as seen with static microcosms; Schrag and Mittler 1996, Abedon 2006); limitations on phage ability to respond to bacterial resistance mechanisms (Lenski 1988, Labrie et al. 2010, Hyman and Abedon 2010); and what can be described as numerical refuges for bacteria (Chao et al. 1977), an anti-kill the winner-type situation that could be dubbed 'spare the loser'. An interesting juxtaposition of some of these ideas is seen in two studies of phage-bacterial coevolution by Lopez-Pascua, Buckling, and colleague (Lopez-Pascua and Buckling 2008, Lopez-Pascua et al. 2010): In restraining phage omnipotence, the spatial structure of a static microcosm allowed extensive coevolution to occur (Buckling and Rainey 2002), but with even greater rates of coevolution observed when phage infection potential was enhanced. This greater phage infection potential was achieved by supplying bacteria with greater nutrient densities; see also (Bohannan and Lenski 1997, Mizoguchi et al. 2003). Consistently, Ratner and Rodin (1984) noted "that the most preferable ecological situation for coevolution of genes responsible for parasitism of bacteriophages to occur is the complete absence of flow and resource limits on bacterial growth" (p. 408) where static microcosms by definition display an "absence of flow" as they lack both fluid mixing and outflow.

The greater potential for phage-bacterium antagonistic coevolution in these experiments was probably manifest through greater phage-bacterium encounter rates and/or greater phage burst sizes. The result would be greater selective pressure on bacteria due to greater TPY through a combination of more phage infections and greater productivity per infection, e.g., (Hadas et al. 1997). At the same time, greater phage and bacterium numbers could supply greater mutation opportunities for both bacteria and phages, though perhaps more crucially for the latter (Lopez-Pascua and Buckling 2008). This greater phage infection potential can also be achieved via periodic (but not continuous) disruption of spatial structure (Brockhurst et al. 2003). It is important, though, that phage infection rates not be so high that phages drive bacteria to extinction, thereby obviating any potential for phage-bacterium antagonistic coevolution, nor for phage population size or diversity to be so low that phage host-range mutations are unavailable for overcoming bacterial resistance (Lenski 1988, Buckling and Rainey 2002, Brockhurst et al. 2007). By culturing bacteria within well mixed environments; by protecting phages from physical, chemical, or biological antagonists; by growing bacteria and thereby potentially also phages to relatively high densities; and by limiting bacterial variability, i.e., as one often observes in fluid laboratory or industrial ferments, then the phage predatory impact on bacteria can come to dominate their interactions rather than antagonistic coevolution. See Abedon (2009d) for additional consideration of how the phage ability to impact bacterial populations changes with environmental complexity.

4.2. Bacterial Resistance to Phage Attack

Bacterial populations can respond to substantial phage attack by evolving adaptations of phage-resistance (Kerr et al. 2008, Hyman and Abedon 2010, Labrie et al. 2010). Macro-ecologically, we can consider this resistance from the perspective of what has been described as a "life-dinner principle", the idea that rabbits run faster than foxes because the rabbit is running for its life (resistance analog) while the fox is running only for its dinner (infection analog) (Dawkins and Krebs 1979). The cost of lytic infection thus can be considered to be much greater for the bacterium than it is beneficial to the phage since the phage may be able to find a different host at a later time. As Dawkins and Krebs put it, "A fox may reproduce after losing a race against a rabbit. No rabbit has ever reproduced after losing a race against a fox." Therefore, we might predict that selective pressure would act much more strongly on bacterial avoidance of infection than phage ability to infect.

Missing from this metaphor, as applied to phages and bacteria, is that while the genome sizes of foxes and rabbits are roughly equivalent (both ~ 2,800 MBp; Sandritter et al. 1960, Wurster-Hill et al. 1988), those of bacteria and phages are not: 0.5 to 10 MBp for bacteria (Cole and Saint-Girons 1999) *versus* 0.005 to 0.5 MBp for phages (Abedon 2009f, 2011c). Bacteria thus potentially have an edge in evolving phage avoidance, in terms of direct locus-by-locus coevolutionary pairing, than rabbits may have relative to foxes. This edge may be particularly seen with bacterial mutation to phage resistance in terms of loss of bacterial surface molecules, a.k.a., phage receptors (Hyman and Abedon 2010). Here, however, it is the bacteria that are potentially giving up dinner, by giving up access to resources that lost receptors may have been involved in acquiring, while a phage may lose its life through decay given a lack of availability of susceptible bacteria. Another way of making the same point is to note that bacteria can possess greater lifestyle redundancy than do phages. This redundancy is associated with their larger genome sizes, multiple routes toward obtaining dinner such that some aspects of nutrient acquisition can be deemphasized through receptor loss or modification without total loss of bacterial fitness (Lenski and Levin 1985, Lenski 1988). In fact, presumed removal of some of this redundancy, though UV mutagenesis of bacteria, reportedly impedes the bacterial potential to evolve resistance to phages (Buckling et al. 2006).

As Lenski (1984) put it (p. 321, emphasis his), "...structural constraints on the highly site-specific phage adsorption process are probably more severe than the physiological constraints on resource assimilation by bacteria, so that *there is a fundamental asymmetry in the coevolutionary potential of bacteria and phage*." Thus, bacteria may eliminate molecules required for phage infection simply through loss of function (e.g., mutational deletion of a receptor; Kerr et al. 2008) while countering mutations by phages may be expected to be less frequent because they must retain function, i.e., as adsorption organelles. These mutational constraints, which potentially lower the phage coevolutionary potential, have been described as a mechanism by which bacteria and phages are able to coexist in a population (Lenski and Levin 1985). Alternatively, phages can potentially display higher per-gene mutation rates, larger population sizes, and greater access to horizontal gene transfer mechanisms (Abedon 2009f, 2011c), all of which could give phages an edge in these arms races in terms of genomic variability (Wichman et al. 2005).

Notwithstanding these issues, that coevolution between phages and bacteria can occur has been documented extensively in the laboratory, particularly given phage-bacterial coexistence in static microcosms. Furthermore, changes in phage genes in one study were found, as expected, to code predominantly for virion structural proteins associated with phage-host attachment (Paterson et al. 2010). Our suspicion, therefore, is that phage-bacterial antagonistic coevolution, in many or most natural ecosystems, overall favors neither party, but such balances can be completely lost in artificial ecosystems, and particularly so if cultures are fluid – and well mixed – rather than solid or semi-solid (Moineau and Lévesque 2005).

5. Phage Host-Range Adaptation

The phage life cycle involves adsorption and infection, with the latter generally giving rise to multiple phage progeny. Selection should favor phage adaptations that serve to either optimize or maximize these characteristics under specific ecological conditions. Classical ecology also deals with similar issues of phenotype optimization (Bull 2006). In this section we explore how phage growth characteristics can fit into these more general models of ecological utility, focusing on phage adsorption and associated host range (Abedon 2009f); see also Dennehy (2009) who reviewed phage host range issues from the perspective of virus "emergence" as well as Hyman and Abedon (2010) who review issues pertaining to phage host range from an intentionally non-ecological point of view.

5.1. Adsorption and Host Range

From an ecological standpoint, adsorption – and particularly the extracellular search – can be viewed as the dissemination phase of phages as virions. This extracellular search also can be viewed in terms of optimal foraging as well as dietary breadth (Wang et al. 1996, Bull 2006, Kerr et al. 2008, Heineman et al. 2008). Foraging costs for phages include exposure to dangers such as DNA damaging agents or capsid-degrading extracellular proteases, where the likelihood of virion damage should go up the longer foraging continues. Additional concerns include the following: The potential for adsorption to such entities as starved or stationary phase bacteria (Miller and Ripp 2002, Miller and Day 2008); the potential for adsorption to non-bacterial components of environments (Daniels and Wais 1998) including materials in that phages can become entangled in such as the extracellular matrix of biofilms (which can serve as general virus traps; Abedon 2011a); and the potential for adsorption to genetically suboptimal bacterial hosts, e.g., (Ferris et al. 2007, Heineman et al. 2008).

Foraging occurs in part over a pre-bacterial-encounter interval and the lengh of this interval is relevant also because of its impact on the overall phage generation time. One way by which phages can decrease their generation times, therefore, can be through reducing their adsorption times, which can be accomplished by three different mechanisms: (1) increasing phage affinity for bacteria so that more phage-bacterial encounters result in phage attachment, (2) increasing the number of bacteria that are phage susceptible (particularly by broadening phage host range), or (3) increasing the rate at which phages move through aqueous environments. The latter is seen especially when environments are not well mixed since under such conditions it is phage diffusion rates that limit phage movement, and phage diffusion rates, being a function of virion size and shape, are partially under phage genetic control. Faster diffusion, for example, is expected to contribute to larger plaque sizes, where plaques form within unmixed environments (Abedon 2006, Abedon and Culler 2007a).

Within genetically diverse cultures of bacteria, as typically is the case in natural environments, the probability of encountering a bacterium that differs in phenotype from a phage's parental host can be quite high. Should a phage infect this different bacterial type (broader host range) if infecting a bacterium displaying a new phenotype comes with the potential cost of suboptimal reproduction, or should a phage instead move on to the next bacterium (narrower host range)? The alternative cost, for a phage that displays too narrow a host range, is increased exposure to extracellularly acting phage-decay agents and, indeed, doubt that a suitable host will ever be found. The predominant costs of foraging thus are different between specialists and generalists where the former can be subject for longer periods to the dangers of foraging while the latter is subject to greater risks of wasting time (and/or yield) on suboptimal forage. Which strategy should prevail thus should be dependent on which of these costs, as well as associated rewards, is greater.

Phage optimal foraging questions, with regard to host quality and quantity, have been addressed experimentally by Heineman et al. (2008) and Guyander and Burch (2008), and theoretically by Bull (2006). Heineman et al. (2008) were able to demonstrate a phage ability to discriminate between hosts of differing quality, with a utility to specialization apparent when higher quality hosts were more abundant or the costs of infecting lower quality hosts greater (Heineman et al. 2008). Consistently, Gyander and Burch (2008) found that generalist phages were favored during times of lower host density. As considered especially by Bull (2006) and Heineman et al. (2008), decisions by phages of whether or not to specialize on higher quality hosts can be considered, ecologically, using what is known as the marginal value theorem (Wang et al. 1996, Bull et al. 2004, Wang 2006, Kerr et al. 2008).

5.2. The Marginal Value Theorem

Foraging animals are assumed to employ strategies that maximize their benefits (forage obtained) while minimizing their costs, such as energy output or their exposure to predators. Within the field of behavioral ecology, these issues concerning optimal foraging – such as the utility of greater specialization, for example, see Kawecki (1998) – can be addressed by what is known as the marginal value theorem (Charnov 1976). The marginal value theorem predicts that when forage density is low, such as bacteria, which are the forage for phages, then a forager should choose to consume rather than ignore suboptimal forage (Heineman et al. 2008). However, if forage density is high, but still variable in quality, then specialization on higher quality forage should occur due to the opportunity costs of dining on inferior food.

These predictions, for phages, seem to be contradicted in static microcosms where higher host availability in fact can result in reduced phage specialization (Lopez-Pascua and Buckling 2008, Lopez-Pascua et al. 2010). Our interpretation of this seeming inconsistency, though, is that it may be a consequence of a combination of relatively low host availability imposed by environmental spatial structure and a greater degree of host diversity due to ongoing antagonistic coevolution. Thus, very speculatively, opportunities for phage specialization may be rare due to a lack of extensive pockets of a single host type to specialize on while simultaneously the utility of greater generalization may be enhanced due to greater host diversity, i.e., as can result from greater phage-bacterial coevolution given greater environmental nutrient densities. Thus, while bacteria may be available in reasonably high densities, specific host susceptibility types may not.

In the just-discussed experiments we can predict that relatively little phage decay was observed. It therefore is interesting that given an absence of phage decay during phage extracellular searches we have *less* of an expectation that lower densities of high quality hosts would drive a broadening of adsorption specificity, since foraging costs, particularly the likelihood of virion inactivation, would be relatively low. However, an equivalent to virion decay exists if phage access to susceptible bacteria is sufficiently low, that is, as may be observed given an imposition of environmental spatial structure such as may be observed given phage-bacterial propagation in static microcosms. Thus, a phage that is unable to easily exit a local environment containing predominantly phage resistant bacteria presumably could benefit from being less choosy about what bacteria it infects (Lopez-Pascua et al. 2010), and this could be true even if susceptible bacteria overall are reasonably available in the environment. Alternatively, extremely low host quality could modify these predictions since if a phage were to become inactivated *upon* adsorption to low-quality hosts, then that would be expected to strongly select against such adsorption (Kutter et al. 1994, Daniels and Wais 1998, Bull 2006, Heineman et al. 2008). Regardless of what complications might be involved, selection nevertheless will favor a phage host-range strategy that tends to maximize P', TPY, of both.

5.3. Optimizing Receptor Specificity

Phages are limited in how they can modify host ranges through adsorption characteristics alone. In particular, bacterial acquisition ability is highly dependent upon whether a virion is able to recognize specific receptors found on the surface of bacteria: Either a phage uses a receptor that is common to highly susceptible bacteria but not found on less susceptible bacteria, with such choosiness potentially resulting in an adsorption-rate costs (Heineman et al. 2008), or a phage uses a receptor that simply is commonly found on bacteria, though not necessarily useful towards distinguishing among bacteria in terms of suitability for infection. It is also possible for phages to display broader host ranges not so much because of receptor choice but instead because of a lower affinity threshold for receptors before adsorption commences. As a consequence of such host-range broadening, however, virions can be more prone to inactivation (Lenski and Levin 1985, Lenski 1988, Hyman and Abedon 2010). Regardless of the molecular mechanisms involved, it is likely that selection favors phages that use receptors that strike some balance between the costs of adsorbing to especially highly unsuitable entities given too-broad an adsorptive range and the costs of displaying too-long extracellular searches given too narrow an adsorption range.

So far not considered is that adaptations besides host range can be associated with specialization. For instance, greater affinity for a given receptor might evolve if a phage is exposed only to a single receptor variant, resulting in greater adsorption rates, or, alternatively, a phage might switch to a more prevalent receptor molecule associated with the same bacterial target (again potentially increasing adsorption rates). The result, potentially, is an increase in the availability of bacteria to a given phage since the rate of phage adsorption to bacteria is not just a function of bacterial densities but also of the likelihood of successful adsorption given bacterial encounter (Stent 1963). Such adaptation in the form of more-rapid phage adsorption to a specific host has been observed in laboratory culture (Daniels and Wais 1998).

Burst size and infection duration may also improve as phages adapt to a given host (increasing and decreasing, respectively), that is, once adsorption specialization to that host has occurred (Crill et al. 2000, Bull 2006). This improvement can result in additional reductions in ability to effectively infect hosts upon which a phage has *not* specialized, a consequence known as antagonistic pleiotropy (Heineman et al. 2008). Meanwhile, a further cost to generalization – beyond the potential that latent periods may be extended or burst sizes reduced upon infection of low quality hosts – is the potential that adsorption to individual bacteria will be less likely due to deficits in adsorption affinities, thus reducing bacterial availability to some degree. More generally (Bull 2006), "The evolution of host range in viruses and other organisms has sometimes been considered from the perspective that a broad range is better than a narrow one, except that breadth impairs the ability to infect any one host well" (p. 934).

Notwithstanding the potential for phage specialization resulting in faster adsorption, there exist at least two arguments against such adaptation bestowing substantial utility. The first is that faster phage adsorption in fact will not necessarily give rise to faster phage population. growth. This caveat is particularly so given infection of bacteria that are somewhat immobilized, such as one sees during phage plaque formation (Yin and McCaskill 1992, You and Yin 1999, Abedon 2006, Abedon and Culler 2007a, 2007b, Abedon and Yin 2008, Krone and Abedon 2008, Abedon and Yin 2009, Gallet et al. 2009, Abedon and Thomas-Abedon 2010, Abedon 2010, 2011a). The second point is that the strength of selection acting on phage adsorption rates even in well-mixed broth is expected to decline at higher bacterial densities (Abedon 2009e); see also Guyader and Burch (2008) for consistent experimentally observation. An additional consideration is that modification of phage adsorption rates can potentially have pleiotropic impacts on phage fitness, as Pepin et al. (2006) found with phage ϕX174 adsorption-rate mutants, for which increases in phage adsorption rates did not always lead to increases in phage broth-growth fitness. The factors contributing to what receptors a phage uses for adsorption along with their affinity for those receptors – thereby impacting the rapidity of phage adsorption – thus could be both complex and numerous. Phage host-range evolution generally is less complex, though, if a phage population is adapting to only a single rather than multiple bacterial strains (Bull 2006).

5.4. Distinguishing Hosts in Terms of Health

Differences in host genotype are but one factor that defines a host's quality since, even within a genotypic monoculture of bacteria, differences in health status may be present. In particular, a phage may produce more offspring in a host that is in prime health rather than a host of sub-prime condition (Abedon 2011a), where "prime" possibly could be defined more subtly than just a bacterium's ability to grow fast (Bull 2006). Should a phage encounter a susceptible, genetically high quality host that is in peak health within well-mixed fluid culture, then from an optimal foraging perspective it presumably should always infect that host. However, should a phage encounter an otherwise identical host that is in sub-peak health, then according to the marginal value theorem it should only infect that host if the reproductive benefits of infection are greater than those that it would receive, on average, if it were to reject this host to search for another, potentially higher quality host.

Phages could potentially limit their adsorptions to only well-nourished bacteria by employing receptors that are not found on physiologically compromised bacteria such as stationary phase bacteria (Heineman et al. 2008). This supposition ignores a number of important considerations, however. First, by choosing a receptor based on bacteria-health-based criteria, a phage basically is less flexible in terms of what receptors it targets. Second, if one bacterium within an environment is starving, then presumably other bacteria that a phage could adsorb to also may be starving, so a phage would be left choosing between adsorbing to a suboptimal host and not adsorbing at all. The latter option, as discussed above, may or may not work out for a phage given some potential for virion decay. Third, for many bacteria and ecosystems, starvation is the normal state (Morita 1997) suggesting that a phage that holds out for physiological optimality may end up holding out for an awfully long time. Notwithstanding these objections, it appears that some phages do distinguish among hosts in a manner that may be *indirectly* associated with host physiological suitability for infection, particularly by avoiding hosts based on distinctions between different environments, i.e., as mediated in T4 phages by protein wac (Conley and Wood 1975, Kutter et al. 1994). Perhaps similarly, certain cyanophages appear to avoid infecting their cyanobacterial hosts under low-light conditions (Clokie and Mann 2006, Jia et al. 2010).

There may be a way around this concern of adsorbing to low-health bacteria that has nothing to do with adsorption. That is, by displaying pseudolysogeny (Miller and Ripp 2002, Miller and Day 2008, Abedon 2009c) – by which we mean post-adsorption delays in the initiation of phage infections until, for example, host access to nutrients improves – a phage can absorb even to physiologically suboptimal bacteria and simply wait until those bacteria physiology improve before fully initiating its infection. This trades off generation time for burst size (Abedon 2009f), but at least the phage, now inside a bacterium, is less vulnerable to virion decay.

6. Latent Period

The latent period of a phage can be differentiated into two components, one of which is pre-reproductive and the other which coincides with the production of phage progeny (Abedon 2006, 2009f). The pre-reproductive phase more-or-less overlaps with the phage eclipse, though strictly the process of new-phage production, such as genome replication, begins prior to the end of the eclipse (that is, prior to the maturation of the first progeny virion). For phages, the pre-reproductive phase consists of a combination of the extracellular search (discussed above) and this earlier portion of the phage eclipse. Progeny accumulation occurs during the post-eclipse period, and the longer that interval lasts then the more phage progeny that are produced. Optimization of latent period likely relies heavily on current host density with shorter periods being selected for when hosts are common. Before addressing that issue, though, we briefly consider phage eclipse period evolution.

6.1. Optimizing the Phage Eclipse

The phage eclipse is a highly complex interval during which phages divert the metabolism of host bacteria towards what ultimately will be phage progeny production. In perhaps simplistic terms, the shorter the eclipse then the shorter the phage latent period and therefore the shorter the phage generation time. However, little is known about how a phage might reduce its eclipse period nor whether noticeable reductions can be manifest through the mutation of individual genes. A default assumption therefore might be that natural selection has already reduced the eclipse to a reasonably short interval and that reducing that interval further could be difficult due to a combination of inefficiencies in natural selection (essentially a diminishing returns argument where each subsequent reduction could be sufficiently small that natural selection may have difficulty distinguishing it from neutral variation), antagonistic pleiotropy (Williams 1957, Crill et al. 2000, Duffy et al. 2006, Duffy and Turner 2008, Heineman et al. 2008) meaning that improvement during the infection of one host or under particular circumstances might lead to costs while infecting different hosts or under different circumstances, or tradeoffs (also a kind of antagonistic pleiotropy) where reductions in the length of eclipse periods could produce costs during other stages of the phage life cycle.

Reductions in the phage eclipse, for example, might come at a cost in terms of the rate of phage production post-eclipse. This latter possibility is equivalent to the idea that extended maturation, as seen in many organisms, can allow for increased fecundity, nutrient acquisition, strength, survival, experience, etc. (Schaffer and Rosenzweig 1977, Bell 1980, Stearns and Crandall 1981). Particularly, numerous plant and animal species can enhance their reproductive output through delayed reproduction in response to environmental cues (Ituarte and Spivak 2007, Blanck and Lamourous 2007, Wallace and Sabe 2009). A shift in life-history strategy toward delayed reproduction also is an example of a '*K* strategy' i.e., emphasis on competitive ability (where *K* stands for environmental carrying capacity), whereas a shift toward earlier or more-rapid reproduction is an example of '*r* strategy' (MacArthur and Wilson 1967, Suttle 2007) where *r* is an organism's intrinsic growth-rate constant. See Breitbart et al. (2005) and Abedon (2009d) for additional musings on how selection might impact the duration of the phage eclipse.

6.2. Optimizing the Post-Eclipse

All else held constant, the faster infections produce phage progeny, intracellularly, then the more progeny that will be produced by indivdiual phage-infected bacteria. Limitations on these rates could come as a consequence of factors implemented during the eclipse, as just speculated, or instead could be borne from conflicts between rates of phage production and subsequent virion robustness (de Paepe and Taddei 2006, Abedon 2006). A likely example of the latter is seen with ssRNA phages. These phages are very simple and rapidly produced, with burst sizes measured in the thousands, but also are more susceptible to genome damage due to a lack of repair enzymes plus are subject to high mutations rates in the course of their extremely rapid replication (Drake and Holland 1999). Phages likely also are able to improve their burst size in the course of adaptation to new hosts, such as in terms of adapting to new-host codon biases (Carbone 2008). An alternative take on this idea of maximizing or at least optimizing rates of phage maturation comes from the idea of pseudolysogeny as discussed above: Phages infecting starving bacteria may delay the initiation of infections until environmental conditions improve such that likely rates of phage progeny production improve (Miller and Day 2008).

The most studied of phage tradeoffs also is associated with the post-eclipse, and this is the impact of phage latent period on burst size. Simply stated, the longer phage infections go on producing phage progeny then the more such progeny will ultimately be produced, that is, burst size (ABS) is a function of how fast and for how long phage progeny are generated within individual infected bacteria. This was first seen in terms of length with the discovery of the eclipse, by Doermann (1952), where decreases in burst size were observed upon artificial and premature lysis. Increases in burst sizes with increased latent periods was seen in earlier work, also by Doermann, who showed that burst size increases can be associated with latent period extensions (Doermann 1948); see also (Hershey 1946). However, extensions in the post-eclipse, just as with possible extensions in the eclipse itself, directly extends the phage latent period and therefore the phage generation time, implying that this means of improving ABS, and thereby EBS as well as TPY, can negatively impact P'. Phage latent periods, post-eclipse durations, and burst sizes thus all may be subject to an optimization that balances generation time and per-infection fecundity, that is, neither too long nor too small, respectively. For lists of pertinent references, see Abedon and Culler (2007b) and Abedon (2009f).

6.3. Complications, Certainty, and Narrow Applicability of the Concept of Latent-Period Optimization

The above scenario is complicated particularly by the phage generation time being defined by a combination of the duration of latent period and the phage extracellular search. The extracellular search, in turn, changes in duration as a function of the phage adsorption rate, which is a function of a combination of the phage adsorption rate constant (rate of adsorption of a single phage to a single bacterium) and the number of bacteria that are present in the environment. What latent period is optimum for a given phage thus will vary depending on the length of the phage eclipse, the number of phage-susceptible bacteria present in an environment, the magnitude of the phage adsorption rate constant, and even, to some degree, the rate at which phage progeny mature post eclipse. These issues become further complicated if one introduces spatial structure into the mix, such as one sees during phage plaque formation (Abedon and Culler 2007b), along with genetical or physiological diversities of possible hosts (Bull 2006).

What is reasonably certain, however, is that when phage adsorption rates are high, such as when bacteria are common in the environment, then the latent period will make up a larger proportion of the overall generation time. This means that selection should be stronger for shorter latent periods when host bacteria are more readily attained since such shortening is a more meaningful fraction of total generation time under those conditions. Therefore under conditions of higher host density the optimal latent period should be more biased towards shorter latent periods than larger burst sizes. With lower bacterial density, by contrast, the opposite should occur, with weaker selection acting on latent period length since it makes up a smaller proportion of the overall generation time. Consequently, latent period optima should be more readily biased towards larger burst sizes when bacteria are less abundant. Equivalent arguments can be made based on parallels between conversion by phages of bacteria into phage progeny and enzyme conversion of substrate into enzymatic products (Abedon 2009e).

7. Effective Burst Size

The idea of an effective burst size (EBS) can be viewed as a means of making the concept of reproductive number, from epidemiology, more phage specific in its connotations (Abedon and Thomas-Abedon 2010). A reproductive number of one is that parasite transmission rate that results in a parasite's maintenance within a community, whereas a reproductive number of greater than one coincides with an increase in parasite incidence. Equivalently, an effective burst size of one is sufficient to sustain a phage population within a bacterial community whereas a phage population that is increasing in number or density is displaying an effective burst size of greater than one. Specifically, an effective burst size is that number of phages produced by an infection, i.e., the burst size as traditionally defined (or "absolute" burst size; ABS), minus those phage progeny that are lost to decay either as virions or in the course of infection, post adsorption (e.g., such as due to phage restriction). In terms of phage-bacterial community stability, e.g., as seen in chemostats (Abedon 2006), the greater the burst size (measured either as EBS or ABS) then the lower the community stability and thereby the greater likelihood of bacterial and then phage extinction (Gons et al. 2006, Kerr et al. 2008).

7.1. Impact of Decay

Decay can be viewed as impacting phages either while they exist as free virions or while they are infecting bacteria (Abedon 2009a, 2009d). The former occurs independently of potential hosts and can be the result of consumption by protozoa, sunlight, etc. The latter, phage inactivation that occurs while infecting individual bacteria, is associated with various phenomena such as restriction endonucleases, abortive infection systems, CRISPR systems, and even genetic or physiological incompatibilities between phages and the bacteria they are infecting (Hyman and Abedon 2010, Labrie et al. 2010). Ecologically, this host-associated phage inactivation is more or less independent of the duration of the phage extracellular search and in fact can be viewed most productively in terms of limitations on phage host range. The impact of virion degradation, by contrast, should be greater the longer phages are exposed to extracellular conditions, though with the caveat that the exponential nature of phage adsorption means that the largest fraction of phages from any cohort will adsorb sooner and therefore endure less exposure to extracellular virion damaging agents (Abedon et al.

2001). Faster adsorbing phages thus will impact bacteria sooner, will be less susceptible to inactivation, and will display a larger EBS than equivalent slower-adsorbing phages (Abedon 2009d).

Phages may be shielded from extracellularly acting, virion-damaging agents through an extension of the phage latent period. This idea forms the basis of Stewart and Levin's "hard times" hypothesis (Stewart and Levin 1984) where lysogens represent the latent period extension and hard times are a function of low susceptible-host densities; by contrast, perhaps we can invoke a corresponding "easy times" hypothesis to explain the propensity for temperate phages during dairy fermentations to mutate such that they lose their ability to lysogenize bacteria (Brüssow and Kutter 2005). Obligately lytic phages don't have this option of hiding from extracellular degradants as lysogens. Certain phages, though, can nonetheless benefit during hard times by displaying longer non-lysogenic latent periods. One way these extensions can be helpful is by increasing EBS through increases in ABS, that is, longer latent periods producing larger burst sizes. However, if latent periods can be extended until extracellular free phage antagonists are no longer present, then that could supply additional benefits to such a strategy (Abedon 1990, 2009a). Pseudolysogeny, as already discussed, can similarly be viewed not just as a mechanism that could result in burst size increases but also as a means of limiting exposure of future phage progeny to conditions that are not terribly conducive to progeny replication.

Note that the idea of delaying reproduction is not unique to phages, though it can be labeled differently depending on taxonomic group. Several invertebrates and fish species, for example, go through what is known as diapause in response to environmental stimuli/stress such as temperature (Bradford and Roff 1993, Li et al. 2008), water (Kroll 1984, Podrabsky and Hand 1999), food (Gilbert and Schreiber 1998), and predation (Pijanowska and Stolpe 1996). Dauer larvae of nematodes similarly delay reproduction during conditions not favorable for reproduction (Cassada and Russell 1975). Within plants, seed dormancy achieves similar ends (Finch-Savage and Leubner-Metzger 2006). Delayed reproduction strategies tend to be favored only in variable or unpredictable environments since there should exist little incentive to delay reproduction solely for delay's sake given constant conditions, i.e., where conditions, by definition, will never get better.

7.2. Summary of Factors Impacting Effective Burst Size

The phage effective burst size (EBS) is affected by at least four variables: the absolute phage burst size (increasing ABS increases EBS), the existence of virion-impacting environmental degradants (decreases in degradants increases EBS), the duration of the phage extracellular search (decreases in search times increases EBS given some propensity for virion decay during this search), and various issues associated with phage host range. The latter is not as easily quantified as the others since while increases in host range can decrease search times, suboptimal hosts can decrease ABS. An additional issue is that increasing EBS is not always compatible with increasing P' (above). In any case, anything that can be done to reduce phage EBS or P' within industrial ferments should have the effect of reducing or delaying the negative impact of phage on those cultures.

Host Range Yet Revisited (and Conclusion)

In ecological terms, phage host range evolution is essentially an issue of specialization: Either a phage focuses on a specific phenotype within a variable community of potential prey – so as to achieve, evolutionarily, larger absolute burst sizes, shorter latent periods, or faster adsorption – or a phage takes on a generalist strategy of infecting a broader range of phenotypes, but potentially resulting in a cost of smaller absolute burst sizes, etc. As environmental variability can influence the prevalence of potential prey (bacteria), so too can it influence which strategy, specialist *versus* generalist, is preferable for a given phage population. Specifically, optimal foraging theory would predict that stable environments should select for a specialist foraging strategy whereas unstable or unpredictable environments should select for a more generalist strategy (Futuyma and Mereno 1988, Townsend et al. 2003). These expectations are a consequence of specialists, in unpredictable environments, facing the possibility of at least regional extinction whereas generalist may be less able to effectively exploit the bacteria they infect. It is such inefficiency that in stable environments can result in generalists being out competed by specialists.

There are at least three relevant caveats to this scenario. One is that phages may to some degree be able to adapt independently to different bacterial hosts, such as through expression of a diversity of phage accessory genes. The second is that given a stable environment, then there is no reason to expect that a generalist phage might not evolve into a specialist phage. Lastly, it is uncertain how stable even otherwise stable environments can be in terms of bacterial availability, particularly since bacteria that are especially successful will be expected to succumb to phage-mediated reductions in prevalence (that is, kill the winner).

One circumstance in which environmental stability is expected is within the context of industrial ferments, particularly those that are consistently seeded with specific starter strains. Under these circumstances we can predict an ongoing evolution by phages towards enhanced population growth rates and/or total phage yield in the presence of these starter strains. In the absence of successful removal of phages present from prior fermentations, then inoculating a fermentation with the same starter strains would be expected to not just support additional phage proliferation but phage adaptation towards more effective propagation as well.

References

Abedon, S. T. (1990). Selection for lysis inhibition in bacteriophage. *Journal of Theoretical Biology*, *146*, 501-511.

Abedon, S. T. (2006). Phage ecology. In R. Calendar and S. T. Abedon (Eds.), *The Bacteriophages* (pp. 37-46). Oxford: Oxford University Press.

Abedon, S. T. (2008a). Ecology of viruses infecting bacteria. In B. W. J. Mahy and M. H. V. Van Regenmortel (Eds.), *Encyclopedia of Virology, 3rd ed.* (pp. 71-77). Oxford: Elsevier.

Abedon, S. T. (2008b). Phage population growth: constraints, games, adaptation. In S. T. Abedon (Ed.), *Bacteriophage Ecology* (pp. 64-93). Cambridge, UK: Cambridge University Press.

Abedon, S. T. (2008c). Phages, ecology, evolution. In S. T. Abedon (Ed.), *Bacteriophage Ecology* (pp. 1-28). Cambridge, UK: Cambridge University Press.

Abedon, S. T. (2009a). Bacteriophage intraspecific cooperation and defection. In H. T. Adams (Ed.), *Contemporary Trends in Bacteriophage Research* (pp. 191-215). New York: Nova Science Publishers.

Abedon, S. T. (2009b). Deconstructing chemostats towards greater phage-modeling precision. In H. T. Adams (Ed.), *Contemporary Trends in Bacteriophage Research* (pp. 249-283). New York: Nova Science Publishers.

Abedon, S. T. (2009c). Disambiguating bacteriophage pseudolysogeny: an historical analysis of lysogeny, pseudolysogeny, and the phage carrier state. In H. T. Adams (Ed.), *Contemporary Trends in Bacteriophage Research* (pp. 285-307). New York: Nova Science Publishers.

Abedon, S. T. (2009d). Impact of phage properties on bacterial survival. In H. T. Adams (Ed.), *Contemporary Trends in Bacteriophage Research* (pp. 217-235). New York: Nova Science Publishers.

Abedon, S. T. (2009e). Kinetics of phage-mediated biocontrol of bacteria. *Foodborne Pathogens and Disease*, *6*, 807-815.

Abedon, S. T. (2009f). Phage evolution and ecology. *Advances in Applied Microbiology*, *67*, 1-45.

Abedon, S. T. (2010). Bacteriophages and Biofilms. In W. C. Bailey (Ed.), *Biofilms: Formation, Development and Properties* (pp. 1-58). New York: Nova Science Publishers.

Abedon, S. T. (2011a). *Bacteriophages and Biofilms: Ecology, Phage Therapy, Plaques*. New York: Nova Science Publishers.

Abedon, S. T. (2011b). Communication among phages, bacteria, and soil environments. In G. Witzany (Ed.), *Biocommunication of Soil Microorganisms* (pp. 37-65). New York: Springer.

Abedon, S. T. (2011c). Size does matter – distinguishing bacteriophages by genome length (and 'breadth'). *Microbiology Australia* (in press).

Abedon, S. T., and Culler, R. R. (2007a). Bacteriophage evolution given spatial constraint. *Journal of Theoretical Biology*, *248*, 111-119.

Abedon, S. T., and Culler, R. R. (2007b). Optimizing bacteriophage plaque fecundity. *Journal of Theoretical Biology*, *249*, 582-592.

Abedon, S. T., Herschler, T. D., and Stopar, D. (2001). Bacteriophage latent-period evolution as a response to resource availability. *Applied and Environmental Microbiology*, *67*, 4233-4241.

Abedon, S. T., Hyman, P., and Thomas, C. (2003). Experimental examination of bacteriophage latent-period evolution as a response to bacterial availability. *Applied and Environmental Microbiology*, *69*, 7499-7506.

Abedon, S. T., and LeJeune, J. T. (2005). Why bacteriophage encode exotoxins and other virulence factors. *Evolutionary Bioinformatics Online*, *1*, 97-110.

Abedon, S. T., and Thomas-Abedon, C. (2010). Phage therapy pharmacology. *Current Pharmaceutical Biotechnology*, *11*, 28-47.

Abedon, S. T., and Yin, J. (2008). Impact of spatial structure on phage population growth. In S. T. Abedon (Ed.), *Bacteriophage Ecology* (pp. 94-113). Cambridge, UK: Cambridge University Press.

Abedon, S. T., and Yin, J. (2009). Bacteriophage plaques: theory and analysis. M. Clokie and A. Kropinski (Eds.). Bacteriophages: Methods and Protocols. *Methods in Molecular Biology*, *501*, 161-174.

Bell, G. (1980). The costs of reproduction and their consequences. *American Naturalist, 116*, 45-76.

Bettarel, Y., Sime-Ngando, T., Amblard, C., and Bouvy, M. (2005). Low consumption of virus-sized particles by heterotrophic nanoflagellates in two lakes of the French Massif central. *Aquatic Microbial Ecology*, *39*, 205-209.

Blanck, A., and Lamourous, N. (2007). Large-scale intraspecific variation in life-history traits of European freshwater fish. *Journal of Biogeography*, *34*, 862-875.

Bogosian, G. (2006). Control of bacteriophage contamination in commercial microbiology and fermentation facilities. In R. Calendar and S. T. Abedon (Eds.), *The Bacteriophages* (pp. 667-673). Oxford: Oxford University Press.

Bohannan, B. J. M., and Lenski, R. E. (1997). Effect of resource enrichment on a chemostat community of bacteria and bacteriophage. *Ecology*, *78*, 2303-2315.

Brabban, A. D., Hite, E., and Callaway, T. R. (2005). Evolution of foodborne pathogens via temperate bacteriophage-mediated gene transfer. *Foodborne Pathogens and Disease*, *2*, 287-303.

Bradford, M. J., and Roff, D. A. (1993). Bet hedging and the diapauses strategies of the cricket *Allonemobius fasciatus*. *Ecology*, *74*, 1129-1135.

Breitbart, M., Rohwer, F., and Abedon, S. T. (2005). Phage ecology and bacterial pathogenesis. In M. K. Waldor, D. I. Friedman, and S. L. Adhya (Eds.), *Phages: Their Role in Bacterial Pathogenesis and Biotechnology* (pp. 66-91). Washington DC: ASM Press.

Brockhurst, M. A., Morgan, A. D., Rainey, P. B., and Buckling, A. (2003). Population mixing accelerates coevolution. *Ecology Letters*, *6*, 975-979.

Brockhurst, M. A., Morgan, A. D., Fenton, A., and Buckling, A. (2007). Experimental coevolution with bacteria and phage. The *Pseudomonas fluorescens*-F2 model system. *Infection, Genetics and Evolution*, *7*, 547-552.

Brüssow, H., and Kutter, E. (2005). Phage ecology. In E. Kutter and A. Sulakvelidze (Eds.), *Bacteriophages: Biology and Application* (pp. 129-164). Boca Raton, Florida: CRC Press.

Buckling, A., and Rainey, P. B. (2002). Antagonistic coevolution between a bacterium and a bacteriophage. *Proceedings of the Royal Society (London) B: Biological Science*, *269*, 931-936.

Buckling, A., Wei, Y., Massey, R. C., Brockhurst, M. A., and Hochberg, M. E. (2006). Antagonistic coevolution with parasites increases the cost of host deleterious mutations. *Proceedings of the Royal Society (London) B: Biological Science*, *273*, 45-49.

Bull, J. J. (2006). Optimality models of phage life history and parallels in disease evolution. *Journal of Theoretical Biology*, *241*, 928-938.

Bull, J. J. (2008). Patterns in experimental adaptation of phages. In S. T. Abedon (Ed.), *Bacteriophage Ecology* (pp. 217-247). Cambridge, UK: Cambridge University Press.

Bull, J. J., Pfenning, D. W., and Wang, I.-W. (2004). Genetic details, optimization, and phage life histories. *Trends in Ecology and Evolution*, *19*, 76-82.

Bull, J. J., and Wilke, C. O. (2008). Lethal mutagenesis of bacteria. *Genetics*, *180*, 1061-1070.

Burch, C. L., and Chao, L. (2000). Evolvability of an RNA virus is determined by its mutational neighbourhood. *Nature*, *406*, 625-628.

Carbone, A. (2008). Codon bias is a major factor explaining phage evolution in translationally biased hosts. *Journal of Molecular Evolution*, *66*, 210-223.

Cassada, R. C., and Russell, R. L. (1975). The dauerlarva, a post-embryonic developmental variant of the nematode *Caenorhabditis elegans*. *Developmental Biology*, *46*, 326-342.

Chao, L., Levin, B. R., and Stewart, F. M. (1977). A complex community in a simple habitat: An experimental study with bacteria and phage. *Ecology*, *58*, 369-378.

Charnov, E. L. (1976). Optimal foraging: the marginal value theorem. *Theoretical Population Biology*, *9*, 129-136.

Clokie, M. R. J., and Mann, N. H. (2006). Marine cyanophages and light. *Environmental Microbiology*, *8*, 2074-2082.

Cole, S. T., and Saint-Girons, I. (1999). Bacterial genomes—all shapes and sizes. In R. L. Charlebois (Ed.), *Organization of the Prokaryotic Genome* (pp. 35-62). Washington, DC: ASM Press.

Conley, M. P., and Wood, W. B. (1975). Bacteriophage T4 whiskers: A rudimentary environment-sensing device. *Proceedings of the National Academy of Sciences of the United States of America*, *72*, 3701-3705.

Crill, W. D., Wichman, H. A., and Bull, J. J. (2000). Evolutionary reversals during viral adaptation to alternating hosts. *Genetics*, *154*, 27-37.

Daniels, L. L., and Wais, A. C. (1998). Virulence of phage populations infecting *Halobacterium cutirubrum*. *FEMS Microbiology Ecology*, *25*, 129-134.

Dawkins, R., and Krebs, J. R. (1979). Arms races between and within species. *Proceedings of the Society for Experimental Biology and Medicine*, *205*, 489-511.

de Paepe, M., and Taddei, F. (2006). Viruses' life history: Towards a mechanistic basis of a trade-off between survival and reproduction among phages. *Public Library of Science Biology*, *4*, e193.

Dennehy, J. J. (2009). Bacteriophages as model organisms for virus emergence research. *Trends in Microbiology*, *17*, 450-457.

Doermann, A. H. (1948). Lysis and lysis inhibition with Escherichia coli bacteriophage. *Journal of Bacteriology*, *55*, 257-275.

Doermann, A. H. (1952). The intracellular growth of bacteriophages I. liberation of intracellular bacteriophage T4 by premature lysis with another phage or with cyanide. *Journal of General Physiology*, *35*, 645-656.

Drake, J. W., and Holland, J. J. (1999). Mutation rates among RNA viruses. *Proceedings of the National Academy of Sciences of the United States of America*, *96*, 13910-13913.

Duffy, S., and Turner, P. E. (2008). Introduction to phage evolutionary biology. In S. T. Abedon (Ed.), *Bacteriophage Ecology* (pp. 147-176). Cambridge, UK: Cambridge University Press.

Duffy, S., Turner, P. E., and Burch, C. L. (2006). Pleiotropic costs of niche expansion in the RNA bacteriophage f6. *Genetics*, *172*, 751-757.

Ferris, M. T., Joyce, P., and Burch, C. L. (2007). High frequency of mutations that expand the host range of an RNA virus. *Genetics*, *176*, 1013-1022.

Finch-Savage, W. E., and Leubner-Metzger, G. (2006). Seed dormancy and the control of germination. *New Phytologist*, *171*, 501-523.

Futuyma, D. J., and Mereno, G. (1988). The evolution of ecological specialization. *Annual Review of Ecology, Evolution, and Systematics*, *19*, 207-233.

Gadagkar, R., and Gopinathan, K. P. (1980). Bacteriophage burst size during multiple infections. *Journal of Bioscience*, *2*, 253-259.

Gallet, R., Shao, Y., and Wang, I. N. (2009). High adsorption rate is detrimental to bacteriophage fitness in a biofilm-like environment. *BMC Evolutionary Biology*, *9*, 241.

Geller, B. L., Kraus, J., Schell, M. D., Hornsby, M. J., Neal, J. J., and Ruch, F. E. (1998). High titer, phage-neutralizing antibodies in bovine colostrum that prevent lytic infection of *Lactococcus lactis* in fermentations of phage-contaminated milk. *Journal of Dairy Science*, *81*, 895-900.

Gilbert, J. J., and Schreiber, D. K. (1998). Asexual diapauses induced by food limitation in the rotifer *Synchaeta pectinata*. *Ecology*, *79*, 1371-1381.

Gons, H. J., Hoogveld, H. L., Simis, S. G. H., and Tijdens, M. (2006). Dynamic modelling of viral impact on cyanobacterial populations in shallow lakes: implications of burst size. *Journal of the Marine Biological Association of the United Kingdom*, *86*, 537-542.

González, J. M., and Suttle, C. A. (1993). Grazing by marine nanoflagellates on viruses and virus-sized particles: Ingestion and digestion. *Marine Ecology Progress Series*, *94*, 1-10.

Guyader, S., and Burch, C. L. (2008). Optimal foraging predicts the ecology but not the evolution of host specialization in bacteriophages. *Public Library of Science One*, *3*, e1946.

Hadas, H., Einav, M., Fishov, I., and Zaritsky, A. (1997). Bacteriophage T4 development depends on the physiology of its host *Escherichia coli*. *Microbiology*, *143*, 179-185.

Haeckel, E. (1866). *Generelle Morphologie der Organismen*. Berlin: Georg Reimer.

Heineman, R. H., Springman, R., and Bull, J. J. (2008). Optimal foraging by bacteriophages through host avoidance. *American Naturalist*, *171*, E150-E157.

Hershey, A. D. (1946). Mutation of bacteriophage with respect to type of plaque. *Genetics*, *31*, 620-640.

Hutchinson, G. E. (1957). *A Treatise on Limnology*. New York: Wiley and Sons.

Hyman, P., and Abedon, S. T. (2009). Practical methods for determining phage growth parameters. M. Clokie and A. Kropinski (Eds.). Bacteriophages: Methods and Protocols. *Methods in Molecular Biology*, *501*, 175-202.

Hyman, P., and Abedon, S. T. (2010). Bacteriophage host range and bacterial resistance. *Advances in Applied Microbiology*, *70*, 217-248.

Ituarte, R. B., and Spivak, E. D. (2007). Intraspecific variability in life-history traits of a "freshwater shrimp", Palaemonetes argentines. *Annals of Limnology*, *43*, 293-302.

Jia, Y., Shan, J., Millard, A., Clokie, M. R., and Mann, N. H. (2010). Light-dependent adsorption of photosynthetic cyanophages to Synechococcus sp. WH7803. *FEMS Microbiology Letters*, *310*, 120-126.

Kawecki, T. J. (1998). Red queen meets Santa Rosalia: arms races and the evolution of host specialization in organisms with parasitic lifestyles. *American Naturalist*, *152*, 635-651.

Kellenberger, E., Stauffer, E., Haener, M., Lustig, A., and Karamata, D. (1996). Mechanism of the long tail-fiber deployment of bacteriophages T- even and its role in adsorption, infection and sedimentation. *Biophysical Chemistry*, *59*, 41-59.

Kerr, B., West, J., and Bohannan, B. J. M. (2008). Bacteriophage: models for exploring basic principles of ecology. In S. T. Abedon (Ed.), *Bacteriophage Ecology* (pp. 31-63). Cambridge, UK: Cambridge University Press.

Kroll, W. (1984). Morphological and behavioral embryology and spontaneous diapauses in the African killifish, *Aphyosemion gardneri*. *Environmental Biology of Fishes*, *11*, 21-28.

Krone, S. M., and Abedon, S. T. (2008). Modeling phage plaque growth. In S. T. Abedon (Ed.), *Bacteriophage Ecology* (pp. 415-438). Cambridge, UK: Cambridge University Press.

Kutter, E., Kellenberger, E., Carlson, K., Eddy, S., Neitzel, J., Messinger, L., North, J., and Guttman, B. (1994). Effects of bacterial growth conditions and physiology on T4 infection. In J. D. Karam (Ed.), *The Molecular Biology of Bacteriophage T4* (pp. 406-418). Washington, DC: ASM Press.

Labrie, S. J., Samson, J. E., and Moineau, S. (2010). Bacteriophage resistance mechanisms. *Nature Reviews Microbiology*, *8*, 317-327.

Lenski, R. E. (1984). Coevolution of bacteria and phage: Are there endless cycles of bacterial defences and phage counterdefences? *Journal of Theoretical Biology*, *108*, 319-325.

Lenski, R. E. (1988). Dynamics of interactions between bacteria and virulent bacteriophage. *Advances in Microbial Ecology*, *10*, 1-44.

Lenski, R. E., and Levin, B. R. (1985). Constraints on the coevolution of bacteria and virulent phage: A model, some experiments, and predictions for natural communities. *American Naturalist*, *125*, 585-602.

Letarov, A., Manival, X., Desplats, C., and Krisch, H. M. (2005). Gpwac of the T4-type bacteriophages: structure, function, and evolution of a segmented coiled-coil protein that controls viral infectivity. *Journal of Bacteriology*, *187*, 1055-1066.

Levin, B. R. (1988). Frequency-dependent selection in bacterial populations. *Philosophical Transactions of the Royal Society of London B: Biological Sciences*, *319*, 459-472.

Levin, B. R., and Lenski, R. E. (1985). Bacteria and phage: a model system for the study of the ecology and co-evolution of hosts and parasites. In D. Rollinson and R. M. Anderson (Eds.), *Ecology and Genetics of Host-Parasite Interactions* (pp. 227-242). London: Academic Press.

Li, W., Li, J., Coudron, T. A., Lu, Z., Pan, W., Liu, X., and Zhang, Q. (2008). Role of photoperiod and temperature in diapauses induction of endoparasitoid wasp *Microplitis mediator* (Hymenoptera: Braconidae). *Annals of the Entomological Society of America*, *101*, 613-618.

Little, J. (2006). Gene regulatory circuitry of phage l. In R. Calendar and S. T. Abedon (Eds.), *The Bacteriophages* (pp. 74-82). Oxford: Oxford University Press.

Lopez-Pascua, L. C., and Buckling, A. (2008). Increasing productivity accelerates host-parasite coevolution. *Journal of Evolutionary Biology*, *21*, 853-860.

Lopez-Pascua, L. D., Brockhurst, M. A., and Buckling, A. (2010). Antagonistic coevolution across productivity gradients: an experimental test of the effects of dispersal. *Journal of Evolutionary Biology*, *23*, 207-211.

Los, M., Kuzio, J., McConnell, M. R., Kropinski, A. M., Wegrzyn, G., and Christie, G. E. (2010). Lysogenic conversion in bacteria. In P. M. Sabour and M. W. Griffiths (Eds.), *Bacteriophages in the Control of Food- and Waterborne Pathogens* Washington, DC: ASM Press.

MacArthur, R. H., and Wilson, E. O. (1967). *The Theory of Island Biogeography*.

Marth, E. H., and Steele, J. (2001). *Applied Dairy Microbiology*. Boca Raton,FL: CRC Press.

Merril, C. R. (2008). Interaction of bacteriophages with animals. In S. T. Abedon (Ed.), *Bacteriophage Ecology* (pp. 332-352). Cambridge, UK: Cambridge University Press.

Miller, R. V. (2001). Environmental bacteriophage-host interactions: Factors contribution to natural transduction. *Antonie van Leeuwenhoek Journal of Microbiology*, *79*, 141-147.

Miller, R. V. (2004). Bacteriophage-mediated transduction: an engine for change and evolution. In R. V. Miller and M. J. Day (Eds.), *Microbial Evolution: Gene Establishment, Survival, and Exchange* (pp. 144-157). Washington DC: ASM Press.

Miller, R. V., and Day, M. (2008). Contribution of lysogeny, pseudolysogeny, and starvation to phage ecology. In S. T. Abedon (Ed.), *Bacteriophage Ecology* (pp. 114-143). Cambridge, UK: Cambridge University Press.

Miller, R. V., and Ripp, S. A. (2002). Pseudolysogeny: A bacteriophage strategy for increasing longevity *in situ*. In M. Syvanen and C. I. Kado (Eds.), *Horizontal Gene Transfer* (pp. 81-91). San Diego: Academic Press.

Mizoguchi, K., Morita, M., Fischer, C. R., Yoichi, M., Tanji, Y., and Unno, H. (2003). Coevolution of bacteriophage PP01 and *Escherichia coli* O157:H7 in continuous culture. *Applied and Environmental Microbiology*, *69*, 170-176.

Moineau, S., and Lévesque, C. (2005). Control of bacteriophages in industrial ferments. In E. Kutter and A. Sulakvelidze (Eds.), *Bacteriophages: Biology and Application* (pp. 285-296). Boca Raton, Florida: CRC Press.

Morita, R. Y. (1997). *Bacteria in Oligotrophic Environments: Starvation-Survival Lifestyle*. London: Chapman and Hall Ltd.

Müller-Merbach, M., Rauscher, T., and Hinrichs, J. (2005). Inactivation of bacteriophages by thermal and high-pressure treatment. *International Dairy Journal*, *15*, 777–784.

Mustonen, V., and Lassig, M. (2010). Fitness flux and ubiquity of adaptive evolution. *Proceedings of the National Academy of Sciences of the United States of America*, *107*, 4248-4253.

O'Malley, M. A. (2009). What did Darwin say about microbes, and how did microbiology respond? *Trends in Microbiology*, *17*, 341-347.

Patel, I. R., and Rao, K. K. (1984). Bacteriophage burst size as a function of multiplicity of infection. *Current Science*, *53*, 198-200.

Paterson, S., Vogwill, T., Buckling, A., Benmayor, R., Spiers, A. J., Thomson, N. R., Quail, M., Smith, F., Walker, D., Libberton, B., Fenton, A., Hall, N., and Brockhurst, M. A. (2010). Antagonistic coevolution accelerates molecular evolution. *Nature*, *464*, 275-278.

Paul, J. H., and Jiang, S. C. (2001). Lysogeny and transduction. *Methods in Microbiology*, *30*, 105-125.

Pepin, K. M., Samuel, M. A., and Wichman, H. A. (2006). Variable pleiotropic effects from mutations at the same locus hamper prediction of fitness from a fitness component. *Genetics*, *172*, 2047-2056.

Pijanowska, J., and Stolpe, G. (1996). Summer diapauses in *Daphnia* as a reaction to the presence of fish. *Journal of Plankton Research*, *18*, 1407-1412.

Piuri, M., and Hatfull, G. F. (2006). A peptidoglycan hydrolase motif within the mycobacteriophage TM4 tape measure protein promotes efficient infection of stationary phase cells. *Molecular Microbiology*, *62*, 1569-1585.

Podrabsky, J. E., and Hand, S. C. (1999). The bioenergetics of embryonic diapause in an annual killifish, austrofundulus limnaeus. *Journal of Experimental Biology*, *202*, 2567-2580.

Ratner, V. A., and Rodin, S. N. (1984). Coevolution of directly contacting proteins in phage-bacterium ecosystem: Possibility or fiction? *Journal of Theoretical Biology*, *108*, 405-411.

Rossi, P., and Aragno, M. (1999). Analysis of bacteriophage inactivation and its attenuation by adsorption onto colloidal particles by batch agitation techniques. *Canadian Journal of Microbiology*, *45*, 9-17.

Rousseau, G. v. M., and Moineau, S. (2009). Evolution of Lactococcus lactis phages within a cheese factory. *Applied and Environmental Microbiology*, *75*, 5336-5344.

Sandritter, W. D., Müller, D., and Gensecke, O. (1960). Ultraviolettmikro spektr ophotometrische Messungen des Nukleinsäuregehaltes von Spermien und diploiden Zellen. *Acta Histochemica*, *10*, 139-154.

Schaffer, W. M., and Rosenzweig, M. L. (1977). Selection for optimal life histories. II. Multiple equilibria and the evolution of alternative reproductive strategies. *Ecology*, *58*, 60-72.

Schrag, S., and Mittler, J. E. (1996). Host-parasite persistence: the role of spatial refuges in stabilizing bacteria-phage interactions. *American Naturalist*, *148*, 348-377.

Sinton, L. W., Finlay, R. K., and Lynch, P. A. (1999). Sunlight inactivation of fecal bacteriophages and bacteria in sewage-polluted seawater. *Applied and Environmental Microbiology*, *65*, 3605-3613.

Spouge, J. L., and Layne, S. P. (1999). A practical method for simultaneously determining the effective burst sizes and cycle times of viruses. *Proceedings of the National Academy of Sciences of the United States of America*, *96*, 7017-7022.

Stearns, S. C., and Crandall, R. E. (1981). Quantitative predictions of delayed maturity. *Evolution*, *35*, 455-463.

Stent, G. S. (1963). *Molecular Biology of Bacterial Viruses*. San Francisco, CA: WH Freeman and Co.

Stewart, F. M., and Levin, B. R. (1984). The population biology of bacterial viruses: Why be temperate. *Theoretical Population Biology*, *26*, 93-117.

Storms, Z. J., Arsenault, E., Sauvageau, D., and Cooper, D. G. (2010). Bacteriophage adsorption efficiency and its effect on amplification. *Bioprocess and Biosystems Engineering*, *33*, 823-831.

Suttle, C. A. (2000). The ecology, evolutionary and geochemical consequences of viral infection of cyanobacteria and eukaryotic algae. In C. J. Hurst (Ed.), *Viral Ecology* (pp. 248-286). New York: Academic Press.

Suttle, C. A. (2007). Marine viruses - major players in the global ecosystem. *Nature Reviews Microbiology*, *5*, 801-812.

Suttle, C. A., and Chen, F. (1992). Mechanisms and rates of decay of marine viruses in seawater. *Applied and Environmental Microbiology*, *58*, 3721-3729.

Thingstad, T. F. (2000). Elements of a theory for the mechanisms controlling abundance, diversity, and biogeochemical role of lytic bacterial viruses in aquatic systems. *Limnology and Oceanography*, *45*, 1320-1328.

Thingstad, T. F., Bratbak, G., and Heldal, M. (2008). Aquatic phage ecology. In S. T. Abedon (Ed.), *Bacteriophage Ecology* (pp. 251-280). Cambridge, UK: Cambridge University Press.

Thingstad, T. F., and Lignell, R. (1997). Theoretical models for the control of bacterial growth rate, abundance, diversity and carbon demand. *Aquatic Microbial Ecology*, *13*, 19-27.

Townsend, C., Begon, M., and Harper, J. (2003). *Essientials of Ecology*. Oxford, UK: Blackwell.

van Valen, L. (1973). A new evolutionary law. *Evolutionary Theory*, *1*, 1-30.

Vos, M., Birkett, P. J., Birch, E., Griffiths, R. I., and Buckling, A. (2009). Local adaptation of bacteriophages to their bacterial hosts in soil. *Science*, *325*, 833.

Wagner, P. L., and Waldor, M. K. (2002). Bacteriophage control of bacterial virulence. *Infection and Immunity*, *70*, 3985-3993.

Wallace, B. P., and Sabe, V. S. (2009). Environmental and anthropogenic impacts on intra-specific variation in leatherback turtles: opportunities for targeted research and conservation. *Endangered Species Research*, *7*, 11-21.

Wang, I.-N. (2006). Lysis timing and bacteriophage fitness. *Genetics*, *172*, 17-26.

Wang, I.-N., Dykhuizen, D. E., and Slobodkin, L. B. (1996). The evolution of phage lysis timing. *Evolutionary Ecology*, *10*, 545-558.

Weinbauer, M. G. (2004). Ecology of prokaryotic viruses. *FEMS Microbiology Reviews*, *28*, 127-181.

Weinbauer, M. G., and Rassoulzadegan, F. (2004). Are viruses driving microbial diversification and diversity? *Environmental Microbiology*, *6*, 1-11.

Weitz, J. S., Hartman, H., and Levin, S. A. (2005). Coevolutionary arms races between bacteria and bacteriophage. *Proceedings of the National Academy of Sciences of the United States of America*, *102*, 9535-9540.

Whitehead, H. R., and Cox, G. A. (1935). The occurence of bacteriophage in cultures of lactic streptococci, a preliminary note. *New Zealand Journal of Science and Technology*, *16*, 319-320.

Wichman, H. A., Badgett, M. R., Scott, L. A., Boulianne, C. M., and Bull, J. J. (1999). Different trajectories of parallel evolution during viral adaptation. *Science*, *285*, 422-424.

Wichman, H. A., Wichman, J., and Bull, J. J. (2005). Adaptive molecular evolution for 13,000 phage generations: A possible arms race. *Genetics*, *170*, 19-31.

Williams, G. C. (1957). Pleiotropy, natural selection, and the evolution of senescence. *Evolution*, *11*, 411.

Wommack, K. E., and Colwell, R. R. (2000). Virioplankton: viruses in aquatic ecosystems. *Microbiology and Molecular Biology Reviews*, *64*, 69-114.

Wurster-Hill, D. H., Ward, O. G., Davis, B. H., Park, J. P., Moyzis, R. K., and Meyne, J. (1988). Fragile sites, telomeric DNA sequences, B chromosomes, and DNA content in raccoon dogs, Nyctereutes procyonoides, with comparative notes on foxes, coyote, wolf, and raccoon. *Cytogenetics and Cell Genetics*, *49*, 278-281.

Yin, J., and McCaskill, J. S. (1992). Replication of viruses in a growing plaque: a reaction-diffusion model. *Biophysical Journal*, *61*, 1540-1549.

You, L., and Yin, J. (1999). Amplification and spread of viruses in a growing plaque. *Journal of Theoretical Biology*, *200*, 365-373.

Young, R., and Wang, I.-N. (2006). Phage lysis. In R. Calendar and S. T. Abedon (Eds.), *The Bacteriophages* (pp. 104-125). Oxford: Oxford University Press.

In: Bacteriophages in Dairy Processing
Editor: Andrea del Lujan Quiberoni et al.
ISBN 978-1-61324-517-0

Chapter 3

BACTERIOPHAGES IN DAIRY PLANTS

Viviana Beatriz Suárez*, Jorge Alberto Reinheimer and Andrea del Luján Quiberoni

Instituto de Lactología Industrial (UNL-CONICET),
Facultad de Ingeniería Química,
Universidad Nacional del Litoral, Santiago del Estero,Santa Fe, Argentina.

ABSTRACT

Dairy starters are actively growing cultures of lactic acid bacteria (LAB) used in the production of a variety of dairy products including cheese, fermented milks, and cream butter. LAB starters are primarily used because of their capability to produce lactic acid from lactose, although the improvement of sensorial properties, inhibition of undesirable organisms, effects on the texture of dairy products, and contribution to health benefits are also recognized. Several factors can affect the activity of dairy starters, such as the composition of milk as growth medium, chemical inhibitors and bacteriophage attacks. Despite considerable efforts, phage infections of LAB starter cultures remain today as the main cause of delayed lactic acid production during milk fermentation processes in the dairy industry. The goal of this chapter is to make the reader aware of the relevance and implication of bacteriophage attacks in dairy fermentations, showing the LAB species most affected and the dairy fermented products more susceptible to this problem.

INTRODUCTION

Acidification is the most important technological role of lactic acid bacteria (LAB) in the fermentative dairy industry (Tamime 1993). In a fermentative process, a good development of LAB strains is needed to assure a good quality final product. Due to the lactose fermentation, these microorganisms produce lactic acid, which leads to a rapid decrease in pH. In cheese and fermented milk production, the manufacture and characteristics of the curd depend,

* E-mail address: vivisuar@fiq.unl.edu.ar. Tel.: +54 342 4530302; fax: +54 342 4571162

largely, on this factor. In addition, the successful development of starters ensures the control of pathogenic and spoilage microorganisms. Through the production of lactic acid and the generation of more specific metabolites (bacteriocins, antibiotics, hydrogen peroxide), lactic acid bacteria inhibit the development of undesirable microorganisms.

Bacteriophages or "phages" are viruses that infect bacteria (Eubacteria and Archaea). These prokaryotic viruses are present in ecosystems where bacteria have been found, including man-made ecological niches such as food fermentation vats. They are believed to represent the most abundant biological entities in the world with estimates ranging from 10^{30} to 10^{32} total phage particles on earth. It is generally assumed that they outnumber bacteria about 10-fold (Emond and Moineau 2007).

Nowadays, phage attacks are still considered the main cause of delayed lactic acid production during milk fermentation (Moineau 1999, Moineau and Lévesque 2005, Emond and Moineau 2007). Considerable efforts have been aimed at minimizing this problem. Several control strategies, such as sanitizers, thermal and chemical treatments, strain rotation regimes, spontaneous phage resistant mutant strains and phage inhibitory media (PIM) are used in dairy industries (Suárez et al. 2007). Besides, genes that encode natural resistance mechanisms can be introduced into starter strains (Madera et al. 2004). However, phages do represent a real and persistent threat and both researchers and industrial technologists are aware of daily, although unpublished, cases where phage infections actually limit the fermentation process and/or cause product downgrading.

1. Starter Cultures

1.1. General Aspects

A starter culture is defined as any active microbial preparation, intentionally added during a product manufacture, with the aim of initiating desirable changes. Depending on the manufactured product, these microbial preparations can consist of lactic acid bacteria, propionibacteria, surface ripening bacteria and yeast and molds (Frank and Hassan 2001).

In particular, bacterial cultures used in dairy industry may be classified according to:

1. Optimum growth temperature:
    - Mesophilic starters, with growth temperature between 10 - 40°C (optimum around 30°C).
    - Thermophilic starters
      with growth temperature between 40 - 50°C (optimum around 45°C).
2. Starter strain number:
    - Single-strain starters: contain one strain of a certain species.
    - Multiple-strain starters (defined strain starter): contain different known strains of one species.
    - Multiple-mixed-strain starters: contain different defined strains of different species.

3. Starter composition:

- Defined starters: contain of different known strains of one or different species.
- Undefined starters (natural whey and natural milk starter): have an undefined composition because they are obtained from whey or raw milk, under non-aseptic dairy conditions. These starters are commonly used in cheesemaking and, due to their complex composition, better organoleptic characteristics are obtained in the final product.

1.2. Dairy Starter Function

Dairy starters are actively growing cultures of LAB which are added to milk to perform the fermentation process. They are used in the production of a variety of dairy products including cheese, fermented milks, and cream butter. LAB starters are primarily used because of their capability to produce lactic acid from lactose. Moreover, other important functions like the improvement of the sensorial properties, inhibition of undesirable organisms, effects on the texture of dairy products, and contribution to health benefits are recognized.

In the production of fermented milks (including yoghurt) the heat treatment of milk involves high temperatures that strongly reduce the indigenous microflora of raw milk. Thus, the strains added with the starter dominate, being the only responsible for the fermentation process. In the production of cheeses, although the starter maintains the role to conduct the fermentation process, other residual strains from milk and the environment grow together with the starter strains.

In a fermented dairy product manufacture, the essential function of starter cultures is the production of lactic acid as a result of lactose fermentation. The subsequent pH decrease affects a number of aspects of the manufacturing process including the quality, texture, and composition of the products. The lactic acid gives fermented milks the distinctive and fresh acidic flavour. The fermentation process, on the other hand, has to be controlled (e.g. by proper cooling) to avoid excessive acid concentration which could mask more delicate flavours like diacetyl. In cheesemaking, the lactic acid is essential during the coagulation and the texture development of the curd. The pH decrease during this process indirectly affects flavour by controlling the proteolytic activity of both coagulant and natural milk proteinases, and by influencing the biochemical reactions involved in formation of other flavour compounds.

The lactic starter bacteria are involved in the production of aroma compounds, which enhance the organoleptic properties and determine the specific identity of the cultured dairy products. The development of flavour compounds can derive from fermentation of lactose and citrate, from degradation of milk proteins and fat, and from metabolism of amino acid and free fatty acids. The contribution to flavour in the final products depends on a large number of factors (e.g. specific metabolic characteristics of starter strains, milk fermentation conditions, product type and composition, and conditions of storage or ripening).

The acidic condition and the reduced pH of the fermented dairy products, as well as the antimicrobial activity of undissociated lactic acid molecules, prevent the growth or the survival of many spoilage and pathogenic bacteria. The capability of a few strains to produce secondary metabolites having an inhibitory activity (i.e. bacteriocins such as nisin and other inhibitory peptides, hydrogen peroxide, diacetyl) can boost up the preservative effects.

In addition to the influence of pH decrease, which enhances the whey expulsion from the coagulated curd and indirectly affects the cheese texture, most LAB are able to produce exopolysaccharides (EPS). These heteropolysaccharides (HePS) are composed of repeating units of galactose, glucose and/or rhamnose (Rodríguez et al. 2008). When produced in situ, these EPS decrease the syneresis, increase the viscosity and improve the texture in certain fermented dairy products (Duboc and Mollet 2001). As regards their physiological role, EPS from LAB have been claimed to protect the cells from detrimental environmental conditions, such as dehydration, macrophages, antibiotics or bacteriophages, to sequester essential cations and to be involved in adhesion and biofilm formation (Looijesteijn et al. 2001).

1.3. Dairy Starter Microorganisms

The term "lactic acid bacteria" is applied to a group of microorganisms composed of cocci and bacilli, Gram positive, catalase negative, non sporulated, able to ferment carbohydrates, lactic acid producer, acid tolerant, usually non-mobile, unable to reduce nitrates and holder of extremely complex nutritional requirements (Axelsson 2004).

According to these features, several bacterial genera are included in this group. However, the nature of carbon source available in milk (lactose) limits the genera which are usually present in the starter. The genera of lactic bacteria that can produce lactose fermentation are *Lactococcus, Streptococcus, Leuconostoc* and *Lactobacillus.*

Lactococcus

Lactococci are the main mesophilic microorganisms used for acid production in dairy fermentations, of which only *Lactococcus lactis* species have significance in dairy industry. *Lactococcus lactis* cells are cocci that usually occur in chains, although single and paired cells are also found. *Lactococcus lactis* metabolism related to glucose fermentation is homofermentative; when it develops in milk, more than 95% of its final product is lactic acid. As regards the optimum growth temperature, it is a mesophilic microorganism. Two subspecies are usually found in dairy starters, *Lactococcus lactis* subsp. *lactis* and *Lactococcus lactis* subsp. *cremoris. Lactococcus lactis* subsp. *lactis* biovar. *diacetylactis* is used in butter fermentation to produce diacetyl (butter flavour) from citrate fermentation.

Streptococcus

The only species associated to dairy fermented products (cheeses and yoghurt) is *Streptococcus thermophilus.* This microorganism is differentiated from other streptococci (and lactococci) for its heat resistance, ability to grow at 52°C, and ability to ferment only a limited number of carbohydrates (Frank and Hassan, 2001). This species is homofermentative, thermophilic and its cells are grouped in chains. *Strep. thermophilus* is a thermophilic lactic acid bacterium of great technological importance in worldwide cheese production.

In Argentina, it is dominant in autochthonal milk cultures and in commercial starters used for a wide variety of cheeses (Cremoso, Cuartirolo, Port Salut, Holanda, Fontina, Colonia, Edam and Pategrás) (Reinheimer et al. 1997). On the other hand, it is considered the main species in the natural microflora of Italian (Provolone, Pecorino, Montasio, Asiago, Caciocavallo, Crescenza, Caciotta, Mozzarella, etc.) and French-Swiss (Gruyére and Emmental) cheeses made with natural milk starters (Bottazzi 1967, Valles and Mocquot 1972, Giraffa 1993). *Strep. thermophilus*, in association with *Lactobacillus delbrueckii* subsp. *bulgaricus*, is also used in the manufacture of yoghurt, one of the most ancient and best known food fermentation processes (Bottazzi 1988).

Leuconostoc

Leuconostoc is a genus of Gram-positive LAB of great economic importance, remarkable by its ability to produce gas (carbon dioxide) from sugar, flavour compounds (diacetyl, acetate and ethanol) in many dairy products, and dextrans in saccharose containing (dairy) products (Vedamuthu 1994). Some species of *Leuconostoc* (*Leuconostoc mesenteroides*, *Leuconostoc lactis*) are very relevant in the fermentative dairy industry, as they are traditionally used in butter and cream manufacture. *Leuconostoc* uses lactose and citrate and produces lactic acid, acetate, carbon dioxide, ethanol, acetaldehyde, diacetyl, acetoin and 2,3-butanediol, which contribute to organoleptic (flavour and texture) characteristics of butter, cream, and allow the opening in some soft and semi-hard cheeses (Edam and Gouda cheeses) (Hemme and Foucaud-Sheunemann 2004). The production of carbon dioxide by *Leuconostoc* is responsible for eye formation in many artisanal cheeses, usually made with raw milk, or in blue-veined cheeses, such as Roquefort.

In the latter, the openness produced by *Leuconostoc* allows *Penicillium roqueforti* colonization. However, this phenomenon is not well accepted in other types of cheese, where it leads to diminished quality of the product (Vedamuthu 1994, Quiberoni et al. 2008). In starter cultures, *Leuconostoc* spp. is essentially associated to lactobacilli or lactococci cells due to their inability to decrease pH to 5.0-5.5.

Lactobacilli

Lactobacilli are the most acid tolerant LAB able to decrease pH value to 4.0. This genus can be divided in three groups based on fermentation end products. The species in each of these groups can be found in dairy starter cultures, as listed in Table 1. Homofermentative lactobacilli ferment exclusively hexose sugars to mostly lactic acid. They do not ferment pentose sugars or gluconate. *Lb. delbrueckii* subsp. *bulgaricus, Lb. delbrueckii* subsp. *lactis* and *Lb. helveticus*, all thermophilic species (they grow at 45°C) belong to this group. Another member of this group, *Lb. acidophilus*, is not usually a starter culture but is added to dairy foods due to its probiotic benefits. However, fermented milks involving this probiotic species are acidophilus milks. These are fermented by a thermophilic culture of *Strep. thermophilus* combined with *Lb. acidophilus,* requiring prolonged periods of fermentation (>6 h) at 38-45°C to guarantee the formation of a stable gel from coagulated milk (pH < 4.8) (Heller 2001).

Table 1. Characteristics of *Lactobacillus* sp. associated with dairy products

Species	Products	Growth at		Lactic acid isomer	%G+C	Fermentation of [a]					
		15°C	45°C			Glu	Gal	Lac	Mal	Suc	Rib
Homofermentative											
Lb. delbrueckii subsp. *bulgaricus*	Yoghurt, kumiss, kefir, Italian and Swiss cheese	-	+	D	49-51	+	–	+	–	–	–
Lb. delbrueckii subsp. *lactis*	Hard cheese	-	+	D	49-51	+	d	+	+	+	–
Lb. acidophilus	Fermented milk	-	+	DL	34-37	+	+	+	+	+	–
Lb. helveticus	Swiss and Italian hard cheese	-	+	DL	38-40	+	+	+	d	–	–
Facultative heterofermentative											
Lb. casei / paracasei	Hard cheese, Fermented milk	+	-	L	45-47	+	+	d	+	+	+
Obligate heterofermentative											
Lb. kefir	Kefir	+	-	DL	41-42	+	–	+	+	–	+

[a]: Glu, Glucose; Gal, Galactose; Lac, Lactose; Mal, Maltose; Suc, Sucrose; Rib, Ribose; +, over 90% of strains are positive; –, over 90% of strains are negative; d, 1 to 89% of strains are positive.

Amiri et al. (2008) have recently studied a symbiotic acidophilus milk with satisfactory functional food properties. In this case, three different probiotic cultures were employed in combination (*Lb. acidophilus*, *Lb. casei* and *Bifidobacterium bifidum*) and the incorporation of inulin, oat fiber and honey, either singly or combined, resulted in an acceptable probiotic acidophilus milk. *Lb. acidophilus* strains with the optimum combination of functional and technological properties would represent the best candidate to include in the production of a probiotic acidophilus milk.

Facultative heterofermentative lactobacilli ferment hexose sugars, generating only lactic acid or lactic acid, acetic acid, ethanol and formic acid when under glucose limitation. Pentose sugars are fermented to lactic acid and acetic acid. This group includes *Lb. casei / paracasei*, which is not usually found in starter cultures but is associated with health benefits. However, there are commercial products that use *Lb. casei / paracasei* strains with a double function, starter and probiotic microorganism. Yakult and Actimel fermented milks are clear examples of this situation. Yakult was the first commercial probiotic fermented milk developed by Minoru Shirota in Japan in 1935 (Siezen and Wilson 2010). This product results from milk fermentation using *Lactobacillus casei* Shirota strain, which can survive the passage through the stomach and colonize the intestine. On the other hand, Actimel DanActive (Danone) is obtained from fermentation with *Lactobacillus casei* DN 114-001, marketed as *Lactobacillus casei* Defensis or Immunitas(s) and with the traditional yoghurt cultures *Lb. delbrueckii subsp. bulgaricus* and *Strep. thermophilus*. After the introduction of Yakult in Europe in 1993, several other companies released their own *Lb. casei* based product (including Actimel), starting an advertising war about each bacteria advantages. Since 1994, *Lb. casei* based yoghurts have become a common part of West European diets. *Lb. plantarum,* in turn, is a versatile lactic acid bacterium that can be found in a wide range of environmental niches. In consequence, it can be used as starter in dairy, meat and a large variety of vegetable fermentations. It is commonly found in the human gastrointestinal-tract and, therefore, has a demonstrated safety for human consumption and a proven ability to exert a probiotic effect on the consumer (de Vries et al. 2006). This species appears as starter component (combined with other microorganisms) in a high number of traditional dairy products like cheeses and fermented milks (Ercolini et al. 2003, Patrignani et al. 2006, De Angelis et al. 2008, Duan et al. 2008, Mangia et al. 2008, Mathara et al. 2008, Rantsiou et al. 2008). Functional aptitude on health of several *Lb. plantarum* strains is well reported (de Vries et al. 2006, Klarin et al. 2008, Szymański et al. 2008, Troost et al. 2008). For these reasons, *Lb. plantarum* emerges as a species of great interest for the development of new functional foods, based on its probiotic potential as well as on its adaptability and versatility in different substrates that allow assuming the starter role (Carminati et al. 2010).

Obligate heterofermentative lactobacilli ferment hexose sugars to lactic acid, acetic acid (or ethanol) and carbon dioxide; *Lactobacillus kefir*, used in Kefir manufacture, belongs to this group.

1.4. Factors Influencing Starter Activity

Several factors, such as the composition of milk as growth medium, chemical inhibitors and bacteriophage attacks can affect the activity of dairy starters (Mäyrä-Mäkinen and Bigret 2004).

Milk as Growth Medium

Even if lactic acid bacteria are able to grow in milk, milk is not an optimum growth medium since vitamins and non protein nitrogen (NPN) are essential in LAB nutrition and the NPN represents 5 to 7% of the total nitrogen in milk. The constitutive molecules of this fraction play an important role in the nutrition of LAB because of their direct uptake by the cell. However, the concentration of these components is usually excessively low to provide the required nutrients. Moreover, free amino acids are present in milk at insufficient level to allow satisfactory growth of the bacterial cells.

Inhibitory Compounds from Milk

- Lactoperoxidase-hydrogen peroxide system: lactoperoxidase enzyme combines with hydrogen peroxide produced by the aerobic metabolism of microorganisms and this new compound is able to oxidize the tyocianate originated in animal liver. This intermediary compound of oxidation inhibits the enzyme activity of LAB, eokinase in particular. The activity of this system is mainly relevant when raw milk is used in cheese manufacture, since the inhibitor compound is inactivated by heat treatments applied to sanitize milk.
- Agglutinins: they are antibodies from a natural immunization of the animal that inhibit LAB with a high degree of specificity. Homogenization and pasteurization of milk destroy their action, thus producing a delay in the acidification process in cheeses manufactured with raw milk.
- Another inhibitors: other immunoglobulins different to agglutinins can inhibit LAB growth. Leucocytes, also present in milk, show antimicrobial properties, with an action mechanism similar to that of lactoperoxidase-hydrogen peroxide system. Lactoferrin fixes iron and eventually copper and inhibits LAB in the absence of these minerals. Lysozyme, another inhibitory enzyme, can also act, being particularly resistant to thermal treatments.

Antibiotics

The antibiotics used for treatment of mastitis in sick animals are a common cause of delayed development of starter cultures. In general, it is well known that genera *Lactococcus* and *Streptococcus* are more sensitive than *Lactobacillus* and *Leuconostoc.*

Biological Factors

- Microbial interactions: LAB can be inhibited by other LAB in their competition for nutrients present in milk or by production of bacteriocins. Psychrotrophic bacteria like *Pseudomonas*, in turn, release free fatty acids, which inhibit the LAB growth. In contrast, other microorganisms, such as micrococci, can stimulate the growth of *Strep. thermophilus* strains when they generate peptides that are used by this species.
- Bacteriophages: phages are of enormous economic importance for industries using any form of bacterial fermentation. A phage attack has always been a major problem for the dairy industry where lactic acid bacteria are used to ferment milk into a variety of products (e.g. cheese and yoghurt) (Brüssow et al. 1998). Contamination by virulent phages may result in the lysis of bacterial starter strains in the vat, causing slow fermentation or even complete starter failure with a subsequent loss of

the product. Since a big dairy plant processes about half a million litres of milk a day, the infection risks are high and phage attacks can be very costly. These economic implications have led to intensive research on bacteriophages attacking lactic acid bacteria used in the dairy industry. Since the cheese industry, which ferments the largest volumes of milk, uses mainly mesophilic lactococci as bacterial starters, most research has been directed towards lactococcal phages (Emond and Moineau 2007). However, during the past two decades, a significant increase in the consumption of these milk products has greatly augmented the worldwide use of *Strep. thermophilus*. Products based on this bacterium presently have a worldwide market value of approximately 40 billion dollars per year. Therefore, numerous studies from Europe, Canada, the United States, and Argentina have reported the isolation of many virulent and temperate bacteriophages specific of *Strep. thermophilus* (Quiberoni et al. 2010). On the other hand, and for unknown reasons, *Lactobacillus* phage infections remain relatively less frequent as compared to those affecting lactococci and *Streptococcus thermophilus* (Brüssow and Suárez 2006). Nevertheless, the isolation of *Lb. delbrueckii* subsp. *bulgaricus* phages from yoghurt has been repeatedly documented, as well as phages infecting *Lb. helveticus* from various dairy factories (Villion and Moineau 2009). On the other hand, the use of LAB strains with probiotic features (*Lb. plantarum, Lb. acidophilus* and *Lb. casei / paracasei*) as starters becomes associated with a worrying increment in the frequency of bacteriophage infections in dairy plants (Capra et al. 2009).

Dairy fermentations are susceptible to phage infection because raw and pasteurized milk naturally contain lytic phages. Moreover, the fluid nature of milk allows a good dispersion of phages and the perpetual utilization of the same bacterial cultures provides permanent hosts for phage proliferation (Boucher and Moineau 2001). The industry has been dealing with this biological phenomenon for many years now and has relied on a variety of practical approaches to control phages, which include adapted factory design, improved sanitation, adequate ventilation, process changes, improved starter medium, and culture rotation (Moineau 1999). Despite extensive efforts, however, bacteriophage infection of starter LAB cultures remains today the most common cause of slow or incomplete fermentation in the dairy industry.

2. Sources of Phage Contamination

The two key components of any dairy fermentation, namely the milk and the starter, represent the primary source of phages in the dairy plant (Emond and Moineau 2007). Regarding the substrate, pasteurized, thermized (a less stringent heat treatment when compared to pasteurization) or raw milk are generally used to manufacture different types of cheeses and yoghurt. It is well documented that the dairy bacteriophages in general, are resistant to those industrial heat treatments (Quiberoni et al. 1999, Binetti and Reinheimer 2000, Suárez and Reinheimer 2002, Quiberoni et al. 2003, Capra et al. 2004, Madera et al. 2004, Atamer et al. 2009, Briggiler Marcó et al. 2009). Thus, cheese production using raw milk or milk with a soft thermal treatment implies a very high risk of phage contamination.

Since heat treatments routinely applied are insufficient to erradicate phage contamination, it follows that milk will be a continuous source of phages coming into a fermentation facility irrespective of other measures taken to limit phage contamination and spread.

In addition, it is now widely accepted that starter bacteria commonly harbour integrated prophages in their genomes. Under certain environmental conditions, these lysogenic bacteria may be triggered to lytically propagate phages with subsequent bactericidal effects. Numerous lactic acid bacteria carry one or several inducible prophages and / or pseudophages integrated into their genomes (Canchaya et al. 2003). Pseudophages are mutated prophages that have lost part or all of their functionality and are not capable of completing infectious cycles upon induction. Temperate phages, in turn, can disturb the normal process of fermentation when, by mutation, they become virulent phages capable of overcoming superinfection immunity. Another problem derived from the use of lysogenic strains is their potentially spontaneous induction, through which they release phage particles that could infect sensitive strains of starter cultures (Davidson et al. 1990). The risk of prophage induction and its consequences on the final product must be carefully evaluated when developing industrial fermentation processes and selecting the appropriate bacterial strains for fermentation (Emond and Moineau 2007).

It is essential to remark that phage infection problems become important when the starter used in the dairy manufacture contains a low number of strains. Bacterial strains growing in a spontaneous or natural fermentation are in equilibrium with their phages and usually survive the fermentation. The need to standardize the quality of final products from fermentations has led to the replacement of natural starters with commercial ones (defined starter cultures). These contain a low number of strains, a fact that favours the visualization of phage attacks (Suárez et al. 2002). Most defined starters consist of a precise blend of two or three well-characterized strains. These strains are ideally phage-unrelated (not attacked by the same phage isolates) to prevent possible lysis of the complete starter culture and subsequent fermentation failures. When a strain used in a defined starter is found to be sensitive to a particular phage, the priority action is to immediately replace it with a phage-unrelated strain or a phage-resistant derivative that will play a similar technological role in the fermentation process (e.g. acidification, texture or flavour development) (Boucher and Moineau 2001).

It is well reported that lysogeny status is widespread in *Lactococcus lactis* species. More than 300 strains of *L. lactis* have been found to be lysogens, according to the review made by Davidson et al. (1990). On the other hand, Chopin et al. (2001) reported the genetic organization of six prophages present in the genome of *L. lactis* IL1403. More recently, Ventura et al. (2007) showed the presence of six (for *L. lactis* MG1363 strain) and five (for *L. lactis* SK112 strain) prophage-like elements.

It is also well documented that lysogeny is widespread in dairy lactobacilli, several studies having demonstrated high frequencies of lysogenic strains (Séchaud et al. 1988, Davidson et al. 1990, Séchaud et al. 1992, Carminati et al. 1997, Josephsen and Neve 1998, Suárez et al. 2008b). A high lysogeny frequency for probiotic lactobacilli is also reported in bibliography, being particularly described in another Chapter of this volume. In contrast, lysogeny appears to be less common in *Strep. thermophilus* species (Sun et al. 2009). Carminati and Giraffa (1992) found only one inducible strain by MC out of 45 strains tested. However, three temperate phages of *Strep. thermophilus* were investigated at molecular level and their genome sequence are fully (O1205 and Sfi21 phages) or partially (TP-J34) available (Sun et al. 2009).

3. Bacteriophages of LAB

3.1. Lactococcal Phages

All known *L. lactis* phages have a double-stranded genome and a non contractile tail. According to the International Committee on Taxonomy of Viruses, *L. lactis* phages are members of the *Caudovirales* order, an extremely large, morphologically and genetically diverse group that encompasses over 95% of all known phages (Deveau et al. 2006). This order contains three families, namely, *Myoviridae* (with long, contractile tails), *Siphoviridae* (with long, noncontractile tails), and *Podoviridae* (with short tails) (Ackermann 2005) (for a detailed description of phage classification see Chapter 1).

Hundreds of lactococcal phages have already been isolated and characterized worldwide. Originally, 12 phage groups were described (Jarvis et al. 1991), but among them, one appeared extinct (P107), while three others (1483, T187, and BK5-T) were merged with the P335 group, reducing the number of lactococcal phage groups from 12 to 8. However, Deveau et al. (2006) modified this classification to include phages 1706 and Q54, which were unrelated to the other phages, as members of two novel groups. Thus, they are currently classified into 10 species based mainly on stringent DNA-DNA hybridizations, electron microscopy observations and comparative genome analyses. Of the variety of phage species that infect *Lactococcus lactis*, only three -c2, 936, and P335- are commonly found in dairy plants, and are responsible for most milk fermentation failures. All three belong to the family *Siphoviridae*, although 936 and P335 have isometric capsids, whereas c2 has a prolate head (Boucher and Moineau 2001, Madera et al. 2004). Only virulent representatives of the c2 and 936 groups are known, while P335 includes both temperate and lytic viruses. Phages belonging to the 936 quasi-species are most frequently isolated from dairy environments, followed by phages belonging to the c2 group. Madera et al. (2004) reported that the predominant phage group in raw milk was c2 group while the situation changed in whey, when the predominant group was 936-like phages. They hypothesized that this change in "phage composition" could be related with the higher thermal sensitivity (pasteurization) of c2 group when compared with 936 group.

Small isometric-headed phages (936 species) were found to be predominant among lactococcal phages in France, Germany, and New Zealand (Moineau and Lévesque 2005). On the other hand, a survey in Russia showed a predominance of the prolate-headed phages of the c2-species. Either the starter strains or rotation system used, or both, are probably the origin of these differences (Boucher and Moineau 2001). Miklič and Rogelj (2003) characterized 30 phage isolates from Slovenian cheese factories. In this study, the isolated phages also belong to these two species, c2 and 936, no phages from the P335 species being found. On the contrary, in a study reported by Josephsen et al. (1999) most of the phages isolated from a Danish factory were members of the P335 species. The research published by Moineau et al. (1996) revealed that most of the phages isolated from cultured buttermilk plants in the United States belonged to the 936 lactococcal phage species, although still about 10% of them were classified as P335 species. Szczepańska et al. (2007) reported the classification of 33 bacteriophages isolated from whey samples of milk plants located in various regions of Poland, and the 936-type phages appeared to be more common than c2-type phages. Notably, a different situation was found in countries near Poland. In Germany,

Lembke et al. (1980) reported that 60% of the analyzed phages were classified as c2-type. In Argentina, even if *Lactococcus lactis* is not predominant in cheese manufacture, the characterization of four bacteriophages isolated from dairy plants were characterized (Suárez et al. 2008a). Only one phage was classified as a P335-like, while the other three phages analyzed belong to the 936 species, thus confirming the predominance of small isometric-headed phages (936 species), as previously found in many other countries (Moineau and Lévesque 2005).

3.2. Leuconostoc Phages

Since scarce reports in the literature have dealt with problems in dairy manufacture associated with *Leuconostoc* phages, it is not surprising that studies on such phages are lacking. As already known, *Leuconostoc* strains are present in dairy mesophilic mixed starters (i.e. starters cultivated at about 30°C) and although they grow slowly and act later than other lactic acid bacteria, their role on the final texture and aroma of the products is extremely important. The growing behaviour of leuconostocs in milk might explain the low frequency of phage infections to this genus and the difficulty in evidencing their presence in those processes where acidification process is normal. For this reason, unlike phages of *Lactococcus lactis* which have been extensively studied, phages of *Leuconostoc* have received little or no attention.

Only a few authors have reported the isolation and characterization of bacteriophages infecting strains of *Leuconostoc*. Sozzi et al. (1978) have been the first to report the isolation of *Leuconostoc* phages from cheese or other dairy products. Two phages, in particular, which infect the same strain of *Leuconostoc mesenteroides* (strain Pro2 isolated from a commercial starter), were isolated in a cheese factory in Colombia and a cheese and butter factory in Spain.

"Viili," a fermented milk product typical of Finland and some parts of Sweden, is produced with traditional mixed-strain starters containing mesophilic lactic cocci, a few strains of *Leuconostoc cremoris*, and the mold *Geotrichum candidum*, which grows on the surface of the product (Saxelin et al. 1983). Thirteen morphologically different types of phages were found in 90 Viili samples studied by electron microscopy, and among them four of which attacked *Leuc. cremoris* strains.

Despite the significant role of *Leuconostoc* in blue cheese production, only one case of phage infections is documented in the literature: the isolation of a bacteriophage infecting a *Leuc. cremoris* strain used in blue cheese manufacture was specifically reported by Shin (1983). Phage Lc-4 had a latent period of approximately 30 min and a burst size of 130 in yeast glucose citrate broth at 27°C. Survival curves of *Leuc. cremoris* phage Lc-4 in reconstituted skim milk by heat treatment at 63 or 73°C were biphasic. Some years later, Davey et al. (1995) characterized, morphologically and by using of molecular techniques, four bacteriophages infecting the same *Leuconostoc* strain used in several lactic acid casein manufacturing plants. These *Leuconostoc* phages contained a distinctive base plate and collar. The authors also performed a comparison with *Lactococcus* phages in morphology, DNA homology, genome size, restriction enzyme analysis and protein profile. The major protein band for the *Leuconostoc* phages of 23-24 kDa was much smaller than that of the lactococcal small isometric phages (30-34 kDa) and also noticeably smaller than that of the smallest major protein band seen for prolate lactococcal phages (27-30 kDa).

Apart from the dairy environment, twenty bacteriophages specific for *Leuconostoc oenos* were isolated from South African red wines and sugarcane (Nel et al., 1987). Boizet et al. (1992), in turn, reported the characterization of nineteen *Leuconostoc mesenteroides* phages isolated during natural coffee fermentation and dairy products, and eight *Leuconostoc oenos* isolated in France in 1981 from starters used in grape fermentation (Chasselas, Gamay or Sylvaner wine). On the basis of morphology, protein composition, genome size and DNA-DNA hybridization, the nineteen *Leuconostoc mesenteroides* phages analysed were differentiated into six different genetic groups where Group 1, containing twelve phages, appeared to be the most widespread. The authors proposed bcc59a and bcc59b phages as type phages for future classification of newly isolated phages of *Leuconostoc mesenteroides.* Also, six bacteriophages active against *Leuconostoc fallax* strains were isolated from industrial sauerkraut fermentation brines (Barrangou et al. 2002), revealing, for the first time, the existence of phages that are active against *Leuc. fallax*. Morphologically, three of the phages were assigned to the family *Siphoviridae*, and the other three to the family *Myoviridae*. Interestingly, bacteriophages are likely to contribute to the microbial ecology of sauerkraut fermentation and could be responsible for the variability observed in this type of fermentation. A new bacteriophage (phage ggg) and its host, *Leuconostoc gelidum* LRC-BD, were isolated from vacuum-packaged pork loins (Geer et al., 2007). Electron microscopy observation revealed that phage ggg was a member of the *Siphoviridae* and its host range was limited to *Leuc. gelidum* isolates from meats. Authors state that naturally occurring phages may affect the number of *Leuc. gelidum* and other lactic acid bacteria occurring in meats and thereby alter the storage quality or the preservative potential of competitive strains. More recently, Lu et al. (2010) reported the sequence analysis of Φ1-A4, isolated from an industrial vegetable fermentation and specific of *Leuconostoc mesenteroides*. Bioinformatic analysis revealed that Φ1-A4 is a unique lytic phage compared to phages infecting related species of LAB. Approximately 30% of the Φ1-A4 genome is inverted compared to the genome organization of other sequenced phages. The endolysin gene was flanked by two holin genes. The tail morphogenesis module was interspersed with cell lysis genes. The overall amino acid sequences of the phage proteins had little similarity to other sequenced phages. Functional analyses showed that Φ1-A4 clusters with several *Lactococcus* phages. This study represents the first complete genome sequence and genetic characterization of a *Leuc. mesenteroides* phage.

3.3. Streptococcus thermophilus Phages

The first studies reporting the isolation of *Strep. thermophilus* phages were carried out more than 50 years ago in Switzerland (Deane et al. 1953). Since then, at least 345 virulent and temperate *Strep. thermophilus* phages have been described in many countries, mostly in North America and Europe. Over 110 *Strep. thermophilus* phages were also isolated and characterized in Argentina, a country geographically very distant from those regions (Quiberoni et al. 2010). The manufacture of artisanal cheeses in several countries depends on the use of natural undefined starter cultures (natural milk and whey). The need to standardize the technological parameters and the final product quality led to the replacement of those starters for defined ones composed of a low number of strains (Powel et al. 2003). This fact has been crucial in the emergence of the phage problems in cheese industry, being also reinforced for *Streptococcus thermophilus* in view of the fact that selecting an industrial strain for these defined cultures is a long and costly process because as many as 15 phenotypes must be characterized for each strain (Klaenhammer et al. 2002). Consequently, a restricted number of *Strep. thermophilus* strains compose commercial starters worldwide, resulting in an array of related virulent phages (Quiberoni et al. 2010).

Among fermentative dairy processes, cheesemaking is the most sensitive to phage infections because it involves the use of open fermentation vats, soft thermal milk treatments and handling of the product by equipment, utensils and people during manufacture (Forde and Fitzgerald 1999, Suárez et al. 2002). Furthermore, phages may remain in the air for long periods, meaning that aerosolization is one of the main dispersal mechanisms for phage particles (Mullan 2003, Verreault et al. 2008). In contrast, yoghurt production is a process in which the fermented product has minimal exposure to the factory environment and milk undergoes a heat treatment at 90 °C, which kills most phages (Atamer et al. 2009, Capra et al. 2009).

All phages of *Strep. thermophilus* are members of the *Siphoviridae* family and the *Caudovirales* order, characterized by isometric capsids and long, non-contractile tails. To date, neither *Myoviridae* nor *Podoviridae* phages have been reported in *Strep. thermophilus* (for a detailed description of phage classification see Chapter 1).

Strep. thermophilus phages are now assembled into only two distinct groups according to their DNA packaging mechanism (*cos* or *pac*) and the number of major structural proteins. Specifically, Le Marrec et al. (1997) found a strict correlation between the presence of a particular set of major structural phage proteins and the mechanism of DNA packaging, demonstrating that *cos*-containing phages possessed two major structural proteins (32 and 26 kDa), in contrast to *pac*-containing phages which possessed three major structural proteins (41, 25, and 13 kDa).

A great difference in the variability exists (as evidenced by DNA restriction patterns and host range) among *Strep. thermophilus* phages isolated from cheeses, compared with those isolated from yoghurt. Whereas a great diversity of phages isolated from cheese processes was reported (Brüssow et al. 1998, Quiberoni et al. 2003, 2006), the variability of yoghurt-derived phages seems to be quite limited since different authors have apparently described similar phages. It should be noted, however, that there is a strong geographical bias in these collections towards France, Italy, Germany and Switzerland (Brüssow et al. 1998). This fact can be justified taking into account that the cheese industry uses complex starter cultures, integrated in rotation programs. The limited variability of yoghurt-derived phages reflects also the limited number of starter strains used in yoghurt production. Indeed, only a limited number of commercial *Strep. thermophilus* strains possess the industrial phenotypes required for yoghurt manufacture, the capacity of exopolysaccharide production among them.

3.4. *Lactobacillus delbrueckii* (subsp. *bulgaricus* and *lactis*) and *Lactobacillus helveticus* Phages

Lactobacillus phage infections are comparatively less frequent than those affecting lactococci and *Streptococcus thermophilus* (Brüssow and Suárez 2006, Villion and Moineau

2009). Nevertheless, the isolation of *Lb. delbrueckii* subsp. *bulgaricus* phages from yoghurt has been repeatedly documented. Virulent and temperate *Lb. delbrueckii* bacteriophages have been characterized and classified into four groups (*a* to *d*) on the basis of morphology, immunoblotting tests and DNA-DNA hybridizations (Mata et al. 1986, Séchaud et al. 1988) (for a detailed description of phage classification see Chapter 1). Group *a* is made up of *Lb. delbrueckii* (*bulgaricus* and *lactis*) phages, temperate and virulent ones, and comprise two morphological types. Group *b* appeared as very closely related phages, specific for *Lb. delbrueckii* subsp. *bulgaricus* strains. Group *c* includes a small number of prolate-headed bacteriophages which belong only to the species *Lb. delbrueckii* subsp. *lactis* and a few of them were characterized. Finally, group *d* has the temperate phage 0252 as a model, with an isometric head and a fairly long non contractile tail. A virulent bacteriophage of *Lb. delbrueckii* subsp. *bulgaricus* isolated from whey of Swiss cheese was characterized by Chow et al. (1988) on the basis of morphology, genome size, structural proteins and growth kinetics. Forsman et al. (1991) characterized a prolate-heahed bacteriophage of *Lb. delbrueckii* subsp. *lactis*, named JCL1032, based on its homology with isometric-headed phages. On the other hand, Auad et al. (1997) characterized a temperate phage isolated by induction with Mitomycin C (MC) from the collection strain *Lb. delbrueckii* subsp. *bulgaricus* CRL 539. This characterization showed that, morphologically and by the use of dot-blot experiments, phage lb 539 belongs to Group *a*, because its genome hybridized with the genomes of reference phages mv4 (temperate) and LL-H (virulent). More recently, Suárez et al. (2008b) isolated and partially characterized two temperate prolate-headed phages, both of them isolated from the same *Lb. delbrueckii* subsp. *lactis* Cb1 commercial strain used in yoghurt manufacture. In this work, the authors demonstrated that the genome size of phages (named Cb1 / 204 and Cb1 / 342) (53.8 and 58.2 kb, respectively) was higher than that previously reported for temperate and virulent phages specific of *Lb. delbrueckii* (Mata et al. 1986, Forsman and Alatossava 1991, Auad et al. 1997, Quiberoni et al. 2004), and other thermophilic lactobacilli (Capra et al. 2006). The isolated bacteriophages showed three major structural proteins exhibiting molecular weights of 23, 31 and 60 kDa. These protein sizes were more similar to those reported for *Lb. delbrueckii* subsp. *bulgaricus* phages included in Group *b*, even when this phage is morphologically classified as a member of Group *c*. Lately, a new virulent phage (phiLdb) of *Lb. delbrueckii* subsp. *bulgaricus* was isolated from a Chinese yoghurt sample showing slow acidification (Wang et al. 2010). The authors affirm that the differences in the phenotypic and genotypic features of the phage phiLdb compared with others previously studied (Chow et al. 1988, Jarvis 1989, Auad et al. 1997, Alatossava et al. 1998, Quiberoni et al. 2004) demonstrated that it represents a novel *Lb. delbrueckii* subsp. *bulgaricus* phage.

Unlike *Lb. delbrueckii* bacteriophages, *Lb. helveticus* ones remain relatively neglected. To date, no phage specific to this species has been completely sequenced, a few studies having reported its isolation and characterization. Lahbib-Mansais et al. (1988) compared five virulent *Lb. helveticus* phages according to their restriction pattern and their DNA homology. The authors demonstrated that those phages were related to one another, but were unrelated to phages of other lactic acid bacteria species. Séchaud et al. (1989) reported the relationship between the temperate phage 0241 and the virulent phage 832-B1, isolated from cheese whey. This work proposed that the virulent phages found in French dairy factories could originate from prophages harboured by lysogenic starter strains. A later study (Séchaud et al. 1992) compared, by morphology and host range, 23 virulent phages isolated from cheese whey and

12 temperate phages induced by MC from lysogenic starter strains. All phages belonged to *Myoviridae* family. The contractile tails distinguish them from known phages of other termophilic dairy lactobacilli, *Lb. delbrueckii* subsp. *bulgaricus* and *Lb. delbrueckii* subsp. *lactis*, which belong to the *Siphoviridae* family.

3.5. Probiotic Lactobacilli Phages

***Lactobacillus plantarum* Phages**

Several bacteriophages that infect *Lb. plantarum* strains have been isolated from various sources: meat (Trevors et al. 1983, Chibanni-Cennoufi et al. 2004), silage (Caso et al. 1995, Doi et al. 2003), homemade cheese whey (Caso et al. 1995), fermented vegetables (Yoon et al. 2001, Lu et al. 2003), fermented maize and coffee (Chibanni-Cennoufi et al. 2004) and Kefir grains (De Antoni et al. 2010). The same morphology was obtained for the majority of the reported *Lb. plantarum* phages such as JL-1 (Lu et al. 2003), B2 (Nes et al. 1988), SC 921 (Yoon et al. 2001), LP1 and LP2 (Caso et al. 1995), gle (Kakikawa et al. 1996) and FAGK1 and FAGK2 (De Antoni et al. 2010). According to the morphology, all these phages belong to the *Siphoviridae* family. Also, Chibani-Chennoufi et al. (2004) reported the morphology of twenty phages, sixteen of them belonging to the *Siphoviridae* family and the others to the *Myoviridae* family. Another well characterized *Myoviridae* phage is fri (Trevors et al. 1983). All the *Siphoviridae* family members show similar dimensions, with the exception of phage ATCC-B2, with almost twice the values determined for other phages (Nes et al. 1988).

Regarding the genome size, the reported values (34.8 ± 3.2 Kbp) are similar for all *Lb. plantarum* phages belonging to the *Siphoviridae* family. However, the genome sizes of *Myoviridae* phages fri and LP65 are considerably higher (133 and 131 Kbp, respectively; Trevors et al. 1983, Chibani-Chennoufi et al. 2004). *Myoviridae* bacteriophages with large genomes have previously been reported in Gram positive bacteria (Jarvis et al. 1993). The more complex structure of this family of phages could be the reason for this great difference in genome sizes (for a detailed description of phage classification see Chapter 1).

***Lactobacillus casei / paracasei and Lactobacillus acidophilus* Phages**

As stated previously, the available knowledge about lactobacilli phages is limited and only a few have been studied in detail (Séchaud et al. 1988, Moineau and Lévesque 2005). Some of them are virulent *Lb. casei* and *Lb. paracase*i phages isolated from Yakult -a lactic acid beverage fermented with *Lb. casei* (phages PL-1 and J-1; Watanabe et al. 1970, Yokokura 1971)-, cheeses (phage LC-Nu; Tuohimaa et al. 2006), probiotic dairy products (phage MLC-A, Capra et al. 2006) and fresh sewage (phage Lb338-1, Alemayehu et al. 2009). Other authors reported the isolation of temperate phages, FSW (Shimizu- Kadota et al. 2000), A2 (Herrero et al. 1994) and AT3 (Lo et al. 2005). In particular, Capra et al. (2009) have described the isolation and characterization, both morphological and genetic, of phages from the environment of a dairy plant that manufactures a probiotic dairy product. Phages MLC-A, MLC-A2 and MLC-A19 were isolated from intermediate-product samples, MLC-A8, MLC-A17 from end-product samples and MLC-A5 from the starter culture. All these phages were isolated between December 2003 and August 2006.

Morphologically, all known phages of *Lb. casei / paracasei*, with the exception of phage Lb338-1, belong to the *Siphoviridae* family, their dimensions being variable and ranging between 53 and 59 nm for the head width, and a tail with 156 - 262 nm in length and 7 - 10 nm in width. Moreover, phage Lb338-1 shows isometric heads of 85 nm in diameter and, contractile non flexible tails (200 nm non-contractil tail length, 154 nm contractil tail length and 18 nm tail diameter) belonging to *Myoviridae* family. This phage is the only myophage isolated from sewage samples and which attacks a probiotic cheese strain (*Lb. paracasei* 338) (for a detailed description of phage classification see Chapter 1).

If phages infecting *Lactobacillus acidophilus* are considered, the available information is even more limited. This species, considered as probiotic culture, is unable to efficiently grow in milk and therefore it is often used as additive in yoghurts and acidophilus milks. The lack of cell grow during the dairy product manufacture might be a reason for the absence of reports relating virulent phages infecting this species. Besides, a very few temperate phages have been isolated from lysogenic *Lb. acidophilus* strains. Namely, *Lb. acidophilius* ADH, currently reclassified as *Lactobacillus gasseri*, is a lysogenic, bacteriocin-producing, bile-resistant isolate of human origin that exhibits adherence to human fetal intestinal cells (Kleeman and Klaenhammer 1991, Luchansky et al. 1991). A phage, named Φadh, was detected in cell lysates induced by treatment with mitomycin C or UV light. Electron microscopy of virions revealed phage particles with a hexagonal head (62 nm) and a long, non contractile, flexible tail (398 nm) ending in at last five short fibers, being classified within the *Siphoviridae* family (Raya et al. 1989). Bareffot et al. (1990) examined thirty-five strains of *Lactobacillus acidophilus* for the presence of inducible bacteriophages. Treatment of *Lb. acidophilus* CFM11, CFM12, and CFM13 with MC resulted in lysates containing structurally complete phage particles with isometric heads and long concentrically banded tails. One phage, designated Φy8, was spontaneously released from *Lb. acidophilus* Y8, a strain used in yoghurts and acidophilus milks sold in the United States market in the 90's (Kilic et al. 1996). It had a burst size of 100, an elongated prolate head of 39 by 130 nm, a long, flexible but non contractile tail of 300 nm, and a 54.3-kb linear double-stranded DNA. It is noteworthy that even though lysogeny is quite widespread among *Lb. acidophilus*, several authors have failed to find indicator strains suitable for propagating the temperate phages (de Klerk and Hugo 1970, Yokokura et al. 1974, Raya et al. 1989).

CONCLUSION

Phages of lactic acid bacteria were firstly isolated almost 80 years ago and, nowadays, are still one of the more significant problems in environments where lactic fermentations are being conducted. Their complete elimination from industrial plants seems to be impossible, consequently, several strategies must be carried out to minimize their presence and economical impact on the processes.

Since *Lactococcus* is the most used lactic acid bacteria around the world, a lot of information is available on *Lactococcus* phages. Other LAB phages, such as *Leuconostoc*, probiotic lactobacilli and thermophilic lactobacilli ones were isolated to a lesser degree and further information is needed about them to attain a more complete characterization. *Strep. thermophilus* phages are particularly interesting and relevant in several countries, because this species is widely distributed and used as starter in a great variety of cheese and fermented milk manufactures.

Phage infections constitute a subject that must be assumed as a permanent risk in dairy plants and, therefore, the permanent generation of knowledge is crucial to protect industrial lactic fermentations.

REFERENCES

Ackermann, H.-W. (2001). Frequency of morphological phage description in the year 2000. *Archives of Virology, 146,* 843-857.

Ackermann, H.-W. (2005). Bacteriophage classification. In E. Kutter and A. Sulakvelidze (Eds.), *Bacteriophages: biology and applications,* (pp. 67-89). Boca Raton, United States: CRC Press.

Alatossava, T., Forsman, P., Mikkonen, M., Räisänen, L. and Vasala, A. (1998). Molecular genetics and evolution of *Lactobacillus* phage LL-H and its related phages. *Recent Researches and Developments in Agricultural and Biological Chemical, 2,* 345-360.

Alemayehu, D., Ross, R. P., O'Sullivan, O., Coffey, A., Stanton, C., Fitzgerald, G. F. and McAuliffe, O. (2009). Genome of a virulent bacteriophage Lb338-1 that lyses the probiotic *Lactobacillus paracasei* cheese strain. *Gene, 448(1),* 29-39.

Amiri, Z., Khandelwal, P., Aruna, B. and Sahebjamnia, N. (2008). Optimization of process parameters for preparation of symbiotic acidophilus milk via selected probiotics and prebiotics using artificial neural network. *Journal of Biotechnology, 136,* S460.

Arendt, E. K., Lonvaud, A. and Hammes, W.P. (1991). Lysogeny in *Leuconostoc oenos*. *Journal of General Microbiology, 137(9),* 2135-2139.

Atamer, Z., Dietrich, J., Müller-Merbach, M., Neve, H., Heller, K. J. and Hinrichs, J. (2009). Screening for and characterization of *Lactococcus lactis* bacteriophages with high thermal resistance. *International Dairy Journal, 19,* 228-235.

Auad, L., Forsman, P., Alatossava, T., Ruiz Holgado, A. P. and Raya, R. R. (1997). Isolation and characterization of a new *Lactobacillus delbrueckii* subsp. *bulgaricus* temperate bacteriophage. *Journal of Dairy Science, 80,* 2706-2712.

Axelsson, L. (2004). Lactic acid bacteria: classification and physiology. In S. Salminen, A. von Wright and A. Ouwehand (Eds), *Lactic Acid Bacteria, Microbiological and Functional Aspects* (pp. 1-66). New York, United States: Marcel Dekker Inc.

Barefoot, S. F., McArthur, J. L., Kidd, J. K. and Grinstead, D. A. (1990). Molecular evidence for lysogeny in *Lactobacillus acidophilus* and characterization of a temperate bacteriophage. *Journal of Dairy Science, 73,* 2269-2277.

Barrangou, R., Yoon, S.-K., Breidt Jr. F., Fleming, H. P. and Klaenhammer, T. R. (2002). Characterization of six *Leuconostoc fallax* bacteriophages isolated from an industrial sauerkraut fermentation. *Applied and Environmental Microbiology, 68,* 5452-5458.

Binetti, A. G. and Reinheimer, J. A. (2000). Thermal and chemical inactivation of indigenous *Streptococcus thermophilus* bacteriophages isolated from Argentinian dairy plants. *Journal of Food Protection, 63(4),* 509-515.

Boizet, B., Mata, M., Mignot, O., Ritzenthaler, P. and Sozzi, T. (1992). Taxonomic characterization of *Leuconostoc mesenteroides* and *Leuconostoc oenos* bacteriophage. *FEMS Microbiology Letters, 90(3),* 211-216.

Bottazzi, V. (1967). Microbiologia delle culture naturali in latte. *Scienza e Tecnica Lattiero-Casearia, 18(1),* 14-23.

Bottazzi, V. (1988). An introduction to rod-shaped lactic acid bacteria. *Biochimie, 70,* 303-315.

Boucher I., and Moineau, S. (2001). Phages of *Lactococcus lactis*: an ecological and economical equilibrium. *Recent Research and Development in Virology, 3,* 243-256.

Braun, V., Hertwig, S., Neve, H., Geis, A. and Teuber, M. (1989). Taxonomic differentiation of bacteriophages of *Lactococcus lactis* by electron microscopy, DNA-DNA hybridization, and protein profiles. *Journal of General Microbiology, 135,* 2551-2560.

Briggiler Marcó, M., De Antoni, G. L., Reinheimer, J. A. and Quiberoni, A. (2009). Thermal, chemical and photocatalytic inactivation of *Lactobacillus plantarum* bacteriophages. *Journal of Food Protection, 72(5),* 1012-1019.

Brüssow, H. and Suarez J. E. (2006). *Lactobacillus* phages. In R. Calendar. and S. T. Abedon (Eds.), *The bacteriophages - Second Edition* (pp. 653-663). Oxford, United Kingdom: Oxford University Press.

Brüssow, H., Bruttin, A., Desiere, F., Luccini, S. and Foley, S. (1998). Molecular ecology and evolution of *Streptococcus thermophilus* bacteriophages. A Review. *Virus Genes, 16(1),* 95-109.

Canchaya, C., Proux, C., Fournous, G., Bruttin, A., Brüssow, H. (2003). Prophage genomics. *Microbiology and Molecular Biology Reviews,* 67, 238-276.

Capra, M. L., Binetti, A., Mercanti, D. J., Quiberoni, A. and Reinheimer, J. (2009). Diversity among *Lactobacillus paracasei* phages isolated from a probiotic dairy product plant. *Journal of Applied Microbiology, 107,* 1350-1357.

Capra, M. L., Quiberoni, A., Ackermann, H.-W., Moineau, S. and Reinheimer, J. (2006). Characterization of a new virulent phage (MLC-A) of *Lactobacillus paracasei. Journal of Dairy Science, 89,* 2414-2423.

Capra, M. L., Quiberoni, A. and Reinheimer, J. A. (2004). Thermal and chemical resistance of *Lactobacillus casei* and *Lactobacillus paracasei* bacteriophages. *Letters in Applied Microbiology, 38(6),* 499-504.

Carminati, D. and Giraffa, G. (1992). Evidence and characterization of temperate bacteriophage in *Streptococcus salivarius* subsp. *thermophilus* St18. *Journal of Dairy Research, 59,* 71-79.

Carminati, D., Giraffa, G., Quiberoni, A., Binetti, A., Suárez, V. and Reinheimer, J. (2010) Advances and trends in starter cultures for dairy fermentations. In R. Raya, F. Mozzi and G. Vignolo (Eds.) *Biotechnology of Lactic Acid Bacteria* (pp. 177-192). Ames, Iowa, United States: Wiley-Blackwell, John Wiley and Sons.

Carminati, D., Mazzucotelli, L., Giraffa, G. and Neviani, E. (1997). Incidence of Inducible Bacteriophage in *Lactobacillus helveticus* Strains Isolated from Natural Whey Starter Cultures. *Journal of Dairy Science, 80,* 1505-1511.

Caso, J. L., de los Reyes-Gavilán, C. G., Herrero, M., Montilla, A., Rodríguez, A. and Suárez, J. E. (1995). Isolation and characterization of temperate and virulent bacteriophage of *Lactobacillus plantarum*. *Journal of Dairy Science, 78,* 741-750.

Chibanni-Chennoufi, S., Dillmann, M.-L., Marvin-Guy, L., Rami-Shojaei, S. and Brüssow, H. (2004). *Lactobacillus plantarum* bacteriophage LP65: a new member of the SPO1-like genus of the family *Myoviridae. Journal of Bacteriology, 186,* 7069-7083.

Chopin, A., Bolotin, A., Sorokin, A., Ehrlich, S. D. and Chopin. M. (2001). Analysis of six prophages in *Lactococcus lactis* IL1403: different genetic structure of temperate and virulent phage populations. *Nucleic Acids Research, 29,* 644-651.

Chow, J. J., Batt, C. A. and Sinskey, A. J. (1988). Characterization of *Lactobacillus bulgaricus* Bacteriophage ch2. *Applied and Environmental Microbiology, 54(5),* 1138-1142.

Davey, G. P., Ward, L. J. H. and Brown, J. C. S. (1995). Characterisation of four *Leuconostoc* bacteriophages isolated from dairy fermentations. *FEMS Microbiology Letters, 128(1),* 21-25.

Davidson, B. E., Powell, I. B. and Hillier, A. J. (1990). Temperate bacteriophages and lysogeny in lactic acid bacteria. *FEMS Microbiology Letters, 87,* 79-90.

De Angelis, M., de Candia, S., Calasso, M., Faccia, M., Guinee, T., Simonetti, M. and Gobbetti, M. (2008). Selection and use of autochthonous multiple strain cultures for the manufacture of high-moisture traditional Mozzarella cheese. *International Journal of Food Microbiology, 125,* 123-132.

De Antoni, G., Zago, M., Vazek, O., Giraffa, G., Briggiler Marcó, M., Reinheimer, J. and Suárez, V. (2010). *Lactobacillus plantarum* bacteriophages isolated from Kefir grains: phenotypic and molecular characterization. *Journal of Dairy Research, 77(1),* 7-12.

de Klerk, H. C. and Hugo, N. (1970). Phage-like structures from *Lactobacillus acidophilus. Journal of General Virology, 8,* 231-234.

de Vries, M., Vaughan, E., Kleerebezem, M. and de Vos, W. (2006). *Lactobacillus plantarum*-survival, functional and potential probiotic propierties in the human intestinal tract. *International Dairy Journal, 16,* 1018-1028.

Deane, D. D., Nelson, F. E., Ryser, F. C. and Carr, P. H. (1953). *Streptococcus thermophilus* bacteriophage from Swiss cheese whey. *Journal of Dairy Science, 36,* 185-191.

Deveau, H., Labrie, S. J., Chopin, M. C. and Moineau, S. (2006). Biodiversity and classification of lactococcal phages. *Applied and Environmental Microbiology, 72,* 4338-4346.

Doi, K., Zhang, Y., Nishizaki, Y., Umeda, A., Ohmomo, S. and Ogata, S. (2003). A comparative study and phage typing of silage-making *Lactobacillus* bacteriophages. *Journal of Biosciencie and Bioengineering, 95,* 518-525.

Duan, Y., Tan, Z., Wang, Y., Li, Z., Li, Z., Qin, G., Huo, Y. and Cai, Y. (2008). Identification and characterization of lactic acid bacteria isolated from Tibetan Qula cheese. *Journal of General and Applied Microbiology, 54,* 51-60.

Duboc, P. and Mollet, B. (2001) Applications of exopolysaccharides in the dairy industry. *International Dairy Journal, 11,* 759-768.

Emond, E. and Moineau, S. (2007). Bacteriophages and food fermentations. In S. McGrath and D. van Sinderen (Eds.), *Bacteriophage: Genetics and Molecular Biology* (pp. 93-124). Norfolk, United Kingdom: Horizon Scientific Press/Caister Academic Press.

Ercolini, D., Hill, P. and Dodd, C. (2003). Bacterial community structure and location in Stilton cheese. *Applied and Environmental Microbiology, 69(6),* 3540-3548.

Forde, A. and Fitzgerald, G. F. (1999). Bacteriophage defense system in lactic acid bacteria. In W. N. Konings, O. P. Kuipers, J. H. J. Huis in't Veld. (Eds.), *Proceedings of the Sixth Symposium on Lactic Acid Bacteria: Genetics, Metabolism and Applications* (pp. 89-113). Veldhoven, The Netherlands: Kluwer Academic Publishers.

Forsman, P. and Alatossava, T. (1991) Genetic variation of *Lactobacillus delbrueckii* subsp. *lactis* bacteriophages isolated from cheese processing plants on Finland. *Applied and Environmental Microbiology, 57,* 1805-1812.

Frank, J. F. and Hassan, A. N. (2001). Starter Cultures and Their Use. In E. H. Marth and J. L. Steele (Eds.), *Applied Dairy Microbiology* (pp.151-206). New York, United States: Marcel Dekker, Inc.

Giraffa, G. (1993). Fermenti lattici. Starter per formaggi freschi e molli: i batteri lattici termofili. *Latte, 18(4),* 436-444.

Greer, G. G., Dilts, B. D. and Ackermann, H.-W. (2007). Characterization of a *Leuconostoc gelidum* bacteriophage from pork. *International Journal of Food Microbiology, 114,* 370-375.

Heller, K. (2001). Probiotic bacteria in fermented foods: product characteristics and starter organisms. *American Journal of Clinical Nutrition, 73,* 374S-379S.

Hemme, D. and Foucaud-Sheunemann, C. (2004). *Leuconostoc*, characteristics, use in dairy technology and prospects in functional foods. Review. *International Dairy Journal, 14,* 467-494.

Herrero, M., de los Reyes-Gavilán, C. G., Caso, J. L. and Suárez, J. E. (1994). Characterization of ϕ393-A2 a bacteriophage that infects *Lactobacillus casei. Microbiology, 140,* 2585-2590.

Jarvis, A. W. (1989). Bacteriophages of lactic acid bacteria. *Journal of Dairy Science, 72,* 3406-3428.

Jarvis, A. W., Collins, L. J. and Ackermann, H.-W. (1993). A study of five bacteriophages of the *Myoviridae* family which replicate on different grampositive bacteria. *Archives of Virology, 133,* 75-84

Jarvis, A. W., Fitzgerald, G. F., Mata, M., Mercenier, A., Neve, H., Powell, I. B., Ronda, C., Saxelin, M. and Teuber, M. (1991). Species and type phages of lactococcal bacteriophages. *Intervirology, 32,* 2-9.

Josephsen, J. and Neve, H. (1998). Bacteriophages and Lactic Acid Bacteria. In S. Salmine and A. von Wright (Eds.), *Lactic Acid Bacteria. Microbiology and functional aspects* (pp. 385-436). New York, United States: Marcel Dekker, Inc.

Josephsen, J., Petersen, A., Neve, H. and Nielsen, E. W. (1999). Development of lytic *Lactococcus lactis* bacteriophages in a Cheddar plant. *International Journal of Food Microbiology, 50,* 163-171.

Kilic, A. O., Pavlova, S. I., Ma, W. and Tao, L. (1996). Analysis of *Lactobacillus* phages and bacteriocins in American dairy products and characterization of a phage isolated from yogurt. *Applied and Environmental Microbiology, 62,* 2111-2116.

Kakikawa, M., Oki, M., Tadokoro, H., Nakamura, S., Taketo, A. and Kodaira, K. (1996). Cloning and nucleotide sequence of the major capsid proteins of *Lactobacillus* bacteriophage phi gle. *Gene, 175(1-2),* 157-165.

Klaenhammer, T., Altermann, E., Arigoni, F., Bolotin, A., Breidt, F., Broadbent, J., Cano, R., Chaillou, S., Deutscher, J., Gasson, M., van de Guchte, M., Guzzo, J., Hartke, A., Hawkins, T., Hols, P., Hutkins, R., Kleerebezem, M., Kok, J., Kuipers, O., Lubbers, M., Maguin, E., McKay, L., Mills, D., Nauta, A., Overbeek, R., Pel, H., Pridmore, D., Saier, M., van Sinderen, D., Sorokin, A., Steele, J., O'Sullivan, D., de Vos, W., Weimer, B., Zagorec, M. and Siezen, R. (2002). Discovering lactic acid bacteria by genomics. *Antonie Van Leeuwenhoek, 82,* 29-58.

Klarin, B., Wullt, M., Palmquist, I., Molin, G., Larsson, A. and Jeppsson, B. (2008). *Lactobacillus plantarum* 299v reduces colonisation of *Clostridium difficile* in critically ill patients treated with antibiotics. *Acta Anaesthesiologica Scandinavica, 52,* 1096-1102.

Kleeman, E. G., and Klaenhammer, T. R. (1982). Adherence of *Lactobacillus* species to human fetal intestinal cells. *Journal of Dairy Science, 65,* 2063-2069.

Lahbib-Mansais, Y., Mata, M. and Ritzenthaler, P. (1988). Molecular taxonomy of *Lactobacillus phages. Biochimi,e 70(3),* 429-435.

Le Marrec, C., van Sinderen, D., Walsh, L., Stanley, E., Vlegels, E. and Moineau, S. (1997). *Streptococcus thermophilus* bacteriophages can be divided into two distinct groups based on mode of packaging and structural protein composition. *Applied and Environmental Microbiology, 63,* 3246-3253.

Lembke, J., Krusch, U., Lompe, A. and Teuber, M. (1980). Isolation and ultrastructure of bacteriophages of group N (lactic) streptococci. *Zentralblatt für Bakteriologie, Mikrobiologie und Hygiene 1. Abt. Originale B, Hygiene, 1,* 79-91.

Lo, T. C., Shih, T. C., Lin, C. F., Chen, H. W. and Lin, T. H. (2005). Complete genomic sequence of the temperate bacteriophage ϕAT3 isolated from *Lactobacillus casei* ATCC 393. *Virology, 339,* 42-55.

Looijesteijn, P. J., Trapet, L., de Vries, E., Abee, T. and Hugenholtz, J. (2001). Physiological function of exopolysaccharides produced by *Lactococcus lactis*. *International Journal of Food Microbiology, 64,* 71-80.

Lu, Z., Altermann, E., Breidt, F. and Kozyavkin, S. (2010). Sequence Analysis of *Leuconostoc mesenteroides* Bacteriophage Φ1-A4 Isolated from an Industrial Vegetable Fermentation. *Applied and Environmental Microbiology, 76(6),* 1955-1966.

Lu, Z., Breidt, F., Fleming, H. P., Altermann, E. and Klaenhammer, T. R. (2003). Isolation and characterization of a *Lactobacillus plantarum* bacteriophage, ϕL-1, from a cucumber fermentation. *International Journal of Food Microbiology, 84,* 225-235.

Luchansky, J. B., Kleeman, E. G., Raya, R. R. and Klaenhammer T. R. (1989). Genetic transfer systems for delivery of plasmid DNA to *Lactobacillus acidopilus* ADH: conjugation, electroporation, and transduction. *Journal of Dairy Science, 72,* 1408-1417.

Madera, C., Monjardín, C. and Suárez, J. E. (2004). Milk Contamination and Resistance to Processing Conditions Determine the Fate of *Lactococcus lactis* Bacteriophages in Dairies. *Applied and Environmental Microbiology,* 70(12), 7365-7371.

Mangia, N., Murgia, M., Garau, G., Sanna, M. and Deiana, P. (2008). Influence of selected lab cultures on the evolution of free amino acids, free fatty acids and Fiore Sardo cheese microflora during the ripening. *Food Microbiology, 25(2),* 366-377.

Mata, M., Trautwetter, A., Luthaud, G. and Ritzenthaler, P. (1986). Thirteen virulent and temperate bacteriophages of *Lactobacillus bulgaricus* and *Lactobacillus lactis* belong to a single DNA homology group. *Applied and Environmental Microbiology, 52,* 812-818.

Mathara, J., Schillinger, U., Kutima, P., Mbugua, S., Guigas, C., Franz, C. and Holzapfel, W. (2008). Functional properties of *Lactobacillus plantarum* strains isolated from Maasai traditional fermented milk products in Kenya. *Current Microbiology, 56,* 315-321.

Mäyrä-Mäkinen, A. and Bigret, M. (2004). Industrial use and production of lactic acid bacteria. In S. Salminen, A. von Wright, A. Ouwehand (Eds.), *Lactic acid bacteria: Microbiological and functional aspects* (pp. 175-198). New York, United States: Marcel Dekker, Inc.

McGrath, S., Fitzgerald, G. F. and van Sinderen, D. (2004). Starter cultures: Bacteriophage. In P. F. Fox, P. L. H. McSweeney, T. M. Cogan, T. P. Guinee (Eds.), *Cheese: Chemistry, Physics and Microbiology, Volume 1: General Aspects* (Third Edition, pp. 163-190). London, United Kingdom: Elsevier Science Ltd.

Miklič, A. and Rogelj, I. (2003). Characterization of lactococcal bacteriophages isolated from Slovenian dairies. *International Journal of Food Science and Technology, 38(3),* 305-311.

Moineau, S. and Lévesque, C. (2005). Control of bacteriophages in industrial fermentations. In E. Kutler and A. Sulakvelidze (Eds.), *Bacteriophages: Biology and Applications* (pp. 285-296). Boca Ratón, United States: CRC Press.

Moineau, S. (1999). Applications of phage resistance in lactic acid bacteria. *Antoine van Leeuwenhoek, 76,* 377-382.

Moineau, S., Borkaev, M. and Holler, B. J. (1996). Isolation and characterization of lactococcal bacteriophages from cultured buttermilk plants in the United States. *Journal of Dairy Science, 79,* 2104-2111.

Mullan, W. M. A. (2003). Bacteriophage control in cheese manufacture. Available from: http://www.dairyscience.info/phage_control.htm Accessed 18.11.10.

Nel, L., Wingfield, B. D., Van der Meer, L. J. and Van Vuuren, H. J. J. (1987). Isolation and characterization of *Leuconostoc oenos* bacteriophages from wine and sugarcane. *FEMS Microbiology Letters, 44,* 63-67 63.

Nes, I. F., Brendehaug, J. and von Husby, K. O. (1988). Characterization of the bacteriophage B2 of *Lactobacillus plantarum* ATCC 8014. *Biochimie, 70,* 423-427.

Patrignani, F., Lanciotti, R., Mathara, J., Guerzoni, M. and Holzapfel, W. (2006). Potential of functional strains, isolated from traditional Maasai milk, as starters for the production of fermented milks. *International Journal of Food Microbiology, 107,* 1-11.

Poblet-Icart, M., Bordons, A. and Lonvaud-Funel, A. (1998). Lysogeny of *Oenococcus oeni* (syn. *Leuconostoc oenos*) and study of their induced bacteriophages. *Current Microbiology, 36(6),* 365-369.

Powel, I. B., Broome, M. C. and Limsowtin, G. K. Y. (2003). Starter cultures: general aspects. In H. Roginski, J. Fuquay, P. Fox (Eds.), *Encyclopedia of Dairy Science* (pp. 261-268). London, United Kingdom: Academic Press.

Quiberoni, A., Auad, L., Binetti, A., Suárez, V., Reinheimer, J. and Raya, R. R. (2003). Comparative analysis of *Streptococcus thermophilus* bacteriophages isolated from a yogurt industrial plant. *Food Microbiology, 20,* 461-469.

Quiberoni, A., Guglielmotti, D., Binetti, A. and Reinheimer, J. (2004) Characterization of three *Lactobacillus delbrueckii* subsp. *bulgaricus* phages and the physicochemical analysis of phage adsorption. *Journal of Applied Microbiology, 96,* 340-351.

Quiberoni, A., Guglielmotti, D., Reinheimer, J. (2003). Inactivation of *Lactobacillus delbrueckii* bacteriophages by heat and biocides. *International Journal of Food Microbiology, 84,* 51-62.

Quiberoni, A., Guglielmotti, D. and Reinheimer, J. (2008). New and classical spoilage bacteria causing widespread blowing in Argentinean soft and semihard cheeses. *International Journal of Dairy Technology, 61(4)*, 358-363.

Quiberoni, A., Moineau, S., Rousseau G. M., Reinheimer, J. and Ackermann, H.-W. (2010). *Streptococcus thermophilus* bacteriophages. Review. *International Dairy Journal, 20(10),* 657-664.

Quiberoni, A., Suárez, V. B. and Reinheimer, J. A. (1999). Inactivation of *Lactobacillus helveticus* bacteriophages by thermal and chemical treatments. *Journal of Food Protection, 62(8),* 894-898.

Quiberoni, A., Tremblay, D., Ackermann, H.-W., Moineau, S. and Reinheimer, J. A. (2006). Diversity of *Streptococcus thermophilus* phages in a large production cheese factory in Argentina. *Journal of Dairy Science, 89,* 3791-3799.

Rantsiou, K., Urso, R., Dolci, P., Comi, G. and Cocolin, L. (2008). Microflora of Feta cheese from four Greek manufacturers. *International Journal of Food Microbiology, 126,* 36-42.

Raya, R. R., Kleeman, E. G., Luchansky J. B. and Klaenhammer T. R. (1989). Characterization of the temperate bacteriophage phi adh and plasmid transduction in *Lactobacillus acidophilus* ADH. *Applied and Environmental Microbiology, 55,* 2206-2213.

Reinheimer, J. A., Binetti, A. G., Quiberoni, A., Bailo, N. B., Rubiolo, A. C. and Giraffa, G. (1997). Natural milk cultures for the production of Argentinian Cheeses. *Journal of Food Protection, 60,* 59-63.

Rodríguez, C., Van der Meulen, R., Vaningelgem, F., Font de Valdez, G, Raya, R., De Vuyst, L. and Mozzi F. (2008). Sensitivity of capsular-producing *Streptococcus thermophilus* strains to bacteriophage adsorption. *Letters in Applied Microbiology, 46,* 462-468.

Saxelin, M.-L., Nurmiaho-Lassila, E.-L., Merilainen, V. T. and Forsén, R, I. (1986). Ultrastructure and host specificity of bacteriophages of *Streptococcus cremoris, Streptococcus lactis* subsp. *diacetylactis,* and *Leuconostoc cremoris* from Finnish fermented milk "Viili". *Applied and Environmental Microbiology, 52(4),* 771-777.

Séchaud, L., Callegari, M. L., Rousseau, M., Muller, M. C. and Accolas, J. P. (1989). Relationship between temperate bacteriophage 0241 and virulent bacteriophage 832-B1 of *Lactobacillus helveticus*. *Netherlands Milk Dairy Journal, 43,* 261-277.

Séchaud, L., Cluzel, P. J., Rousseau, M., Baumgartner, A. and Accolas, J. P. (1988). Bacteriophages of lactobacilli. *Biochimie, 70,* 401-410.

Séchaud, L., Rousseau, M., Fayard, B., Callegari, M. L., Quenée, P. and Accolas, J.-P. (1992). Comparative Study of 35 Bacteriophages of *Lactobacillus helveticus*: Morphology and Host Range. *Applied and Environmental Microbiology, 58(3),* 1011-1018.

Shimizu-Kadota, M., Kiwaki, M., Sawaki, S., Shirasawa, Y., Shibahara-Sone, H. and Sako, T. (2000). Insertion of bacteriophage phiFSW into the chromosome of *Lactobacillus casei* strain Shirota (S-1): characterization of the attachment sites and the integrase gene. *Gene, 249,* 127-134.

Shin, C. (1983). Some characteristics of *Leuconostoc cremoris* bacteriophage isolated from blue cheese. *Japanese Journal of Zootechnical Science, 54,* 481-486.

Siezen, R. J. and Wilson, G. (2010). Probiotics genomics. *Microbial Biotechnology, 3(1),* 1-9.

Sozzi, T., Poulin, J. M., Maretand, R. and Pousaz, R. (1978). Isolation of a Bacteriophage of *Leuconostoc mesenteroides* from Dairy Products. *Journal of Applied Bacteriology, 44,* 159-161.

Suarez, V. B., Capra, M. L., Rivera, M. and Reinheimer, J. A. (2007) .Inactivation of calcium-dependent lactic acid bacteria phages by phosphates. *Journal of Food Protection, 70(6),* 1518-1522.

Suárez, V., Moineau, S., Reinheimer, J. and Quiberoni, A. (2008a). Argentinian *Lactococcus lactis* bacteriophages: genetic characterization and adsorption studies. *Journal of Applied Microbiology, 104,* 371-379.

Suárez, V., Zago, M., Quiberoni, A., Carminati, D., Giraffa, G. and Reinheimer, J. (2008b). Lysogeny in *Lactobacillus delbrueckii* strains and characterization of two new temperate prolate-headed bacteriophages. *Journal of Applied Microbiology, 105,* 1402-1411.

Suárez, V. B., Quiberoni, A., Binetti, A. G. and Reinheimer, J. A. (2002). Thermophilic lactic acid bacteria phages isolated from argentinian dairy industries. *Journal of Food Protection, 65(10),* 1597-1604.

Sun, X., Van Sinderen, D., Moineau, S. and Heller, K. J. (2009). Impact Of Lysogeny On Bacteria With A Focus On Lactic Acid Bacteria. In H. T. Adams (Ed.), *Contemporary Trends in Bacteriophage Research* (pp. 309-336). New York, United States: Nova Science Publishers, Inc.

Szczepańska, A. K., Hejnowicz, M. S., Kołakowski, P. and Bardowski, J. (2007). Biodiversity of *Lactococcus lactis* bacteriophages in Polish dairy environment. *Acta Biochimica Polonica, 54,* 151-158.

Szymański, H., Armańska, M., Kowalska-Duplaga, K. and Szajewska, H. (2008) *Bifidobacterium longum* PL03, *Lactobacillus rhamnosus* KL53A, and *Lactobacillus plantarum* PL02 in the prevention of antibiotic-associated diarrhea in children: a randomized controlled pilot trial. *Digestion, 78,* 13-17.

Tamime, A. Y. (1993). Modern cheese making: Hard cheeses. In R. K. Robinson (Ed.), *Modern Dairy Technology. Vol. II Advances in Milk Products* (pp. 49-221). London, United Kingdom: Elsevier Science Publisher Ltd., Barking Essex.

Trevors, K. E., Holley, R. A. and Kempton, A. G. (1983). Isolation and characterization of a *Lactobacillus plantarum* bacteriophage isolated from a meat starter culture. *Journal of Applied Bacteriology, 54,* 281-288.

Troost, F., van Baarlen, P., Lindsey, P., Kodde, A., de Vos, W., Kleerebezem, M. and Brummer, R. J. (2008). Identification of the transcriptional response of human intestinal mucosa to *Lactobacillus plantarum* WCFS1 in vivo. *BMC Genomics, 9,* 1-14.

Tuohimaa, A., Riipinen, K. A., Brandt, K. and Alatossava, T. (2006). The genome of the virulent phage Lc-Nu of probiotic *Lactobacillus rhamnosus*, and comparative genomics with *Lactobacillus casei* phages. *Archives of Virology, 151(5),* 947-965.

Valles, E., and Mocquot, E. (1972). Etude sur la technique de préparation de la présure utilisée dans les fabrications traditionelles des fromages de Gruyére de Comté et d'Emmental. *Lait, 52,* 259-282.

Vedamuthu, E. R. (1994). The dairy *Leuconostoc*: Use in dairy products. *Journal of Dairy Science, 77,* 2725-2737.

Ventura, M., Zomer, A., Canchaya, C., O'Connell-Motherway, M., Kuipers, O., Turroni F., Ribbera, A., Foroni, E., Buist, G., Wegmann, U., Shearman, C., Gasson, M. J., Fitzgerald, G. F., Kok, J. and van Sinderen, D. (2007). Comparative analyses of prophage-like elements present in two *Lactococcus lactis* strains. *Applied Environmental Microbiology, 73(23),* 7771-7780.

Verreault, D., Moineau, S. and Duchaine, C. (2008). Methods for sampling airborne viruses. *Microbiology and Molecular Biology Reviews, 72,* 413-444.

Villion, M. and Moineau, S. (2009). Bacteriophages of lactobacillus. *Frontiers in Bioscience, 1(14),* 1661-1683.

Wang, S., Kong, J., Gao, C., Guo, T. and Liu, X. (2010). Isolation and characterization of a novel virulent phage (phiLdb) of *Lactobacillus delbrueckii*. *International Journal of Food Microbiology, 137,* 22-27.

Watanabe, K., Takesue, S., Jin-Nai, K. and Yoshikawa, T. (1970). Bacteriophage active against the lactic acid beverage-producing bacterium *Lactobacillus casei. Applied Microbiology, 20,* 409-415.

Yokokura, T. (1971). Phage receptor material in *Lactobacillus casei* cell wall. I. Effect of L-rhamnose on phage adsorption to the cell wall. *Japanese Journal of Microbiology, 15,* 457-463.

Yokokura, T., Kodaira, S., Ishiwa, H. and Sakurai, T. (1974). Lysogeny in lactobacilli. *Journal of General Microbiology, 84,* 277-284.

Yoon, S.-S., Kim, J.-W., Breidt, F. and Fleming, H. P. (2001). Characterization of a lytic *Lactobacillus plantarum* bacteriophage and molecular cloning of a lysin gene in *Escherichia coli*. *International Journal of Food Microbiology, 65,* 63-74.

In: Bacteriophages in Dairy Processing
Editor: Andrea del Lujan Quiberoni et al.

ISBN 978-1-61324-517-0

Chapter 4

ECOLOGICAL ASPECTS OF PHAGE CONTAMINATION IN NATURAL WHEY AND MILK STARTERS

Domenico Carminati*[*], *Miriam Zago and Giorgio Giraffa
Agriculture Research Council, Fodder and Dairy
Productions Research Centre (CRA-FLC)
Via Lombardo, Lodi, Italy

ABSTRACT

Natural dairy starter cultures are complex ecosystems consisting of undefined mixtures of strains belonging to different microbial species, firstly selected by technological parameters which are different according to the cheesemaking processes or to the conditions for starter culture preparation. Since fermentation systems are not aseptic, these starter cultures are commonly susceptible to the infection by phages naturally present in raw milk, in the environments, or carried as prophages by lysogenic strains. The richness of the microbial composition represents the first natural barrier against the problems caused by phage infection. The coexistence of phage-sensitive and phage-insensitive strains avoids strong variations of starter activity and, in the absence of external factors which could modify the relationships between phages and host strains, serious failures during cheesemaking are restricted. Several ecological interactions between phages and host cells may occur within the natural dairy starter cultures. The selective pressure exerted by phages allows the resistant strains to emerge, then the environment selects the more competitive and more adapted cells to drive the fermentation process. The ability of phages to counteract host resistance represents a mechanism of co-evolution that contributes to the stability of this microbial community. The presence of lysogenic strains has an important ecological role on these natural starter cultures. Lysogeny is widespread in many lactic acid bacteria and the integration of fully functional prophage or defective fragments into the host genome confers superinfection

[*] E-mail address: domenico.carminati@entecra.it. Tel.: + 39 0371 45011; fax: + 39 0371 35579.

immunity, and favors the exchange of genetic material. The transduction events, mainly mediated by temperate phages, allow horizontal gene transfer and have an influence on bacterial diversification. All these phenomena emphasize the ecological role that bacteriophages can play as biodiversity regulators within natural dairy starter cultures, influencing the heterogeneity of bacterial populations and the functioning of these microbial ecosystems.

INTRODUCTION

Natural starter cultures may be classified as starters of undefined composition which contain strains of many microbial species. The production of such cultures is derived from the traditional back-slopping practice (the use of a previous fermented product to inoculate a new one) and/or by application of selective pressures (heat treatment, incubation temperature, low pH). No special precautions are used to prevent contamination from raw milk or from cheesemaking environment, and the control of media and culture conditions during starter reproduction is very limited. These procedures are likely to change the ratio and number of the different strains and/or species in a culture, and natural starters are continuously evolving as undefined mixtures of lactic acid bacteria (LAB). Two subtypes of natural starter cultures are recognized, whey starters and milk starters, depending on the substrate and on the technique used for their production.

Natural whey starter cultures (NWC) are obtained by incubating the cheese whey collected at the end of the cheesemaking process under more or less selective conditions. In the manufacture of Parmigiano Reggiano and Grana Padano cheeses, whey is removed from the cheese vat at the end of cheese making at 53-56°C and it is incubated overnight at a controlled temperature (45°C), or in large containers in which the temperature decreases to 37-40°C, to a final pH of about 3.3. The resulting whey culture is dominated by thermophilic LAB species, mainly *Lactobacillus helveticus* and *Lactobacillus delbrueckii* subsp. *lactis*, with a lower presence of *Streptococcus thermophilus*. Similar whey cultures are used in the production of stretched ("pasta-filata") cheese varieties in Italy, hard cheese varieties in Argentina, and Comte cheese in France. Other types of NWC include deproteinized whey starters used for the manufacture of Pecorino cheese, and deproteinized whey starters with rennet that are used for the manufacture of Swiss-type cheeses (Emmental, Sbrinz, Gruyere) in small cheese factories in the Alps. Invariably, thermophilic lactobacilli dominate cultures produced under selective conditions (high temperature, over 40-42°C) while thermophilic streptococci often dominate cultures incubated at a relatively low temperature (35-40°C), which usually show higher microbial diversity (Reinheimer et al. 1995, Parente and Cogan 2004).

Natural milk starter cultures (NMC) are still used in small cheese-making plants in both Southern and Northern of Italy for the production of traditional cheeses, and less used in Argentina. The selective pressure used for the development of the desired microflora includes thermal treatment of raw milk (62-65°C for 10-15 min) followed by incubation at 37-45°C until the desired acidity is reached. The methodology used to prepare NMC and the microbiological composition of the raw milk used determine their microflora. This is usually dominated by *Strep. thermophilus* but other thermophilic streptococci, lactobacilli, and enterococci may be present (Reinheimer et al. 1997, Parente and Cogan 2004, De Angelis et al. 2008, Carminati et al. 2010). Undefined mesophilic milk cultures, first produced by allowing raw milk to naturally sour and then replicated several time, were used in Northern Europe for making cheese and fermented milks. They contain an unknown number of strains of *Lactococcus lactis* subsp. *lactis* and subsp. *cremoris*, and different species of *Leuconostoc* (Cogan et al. 1991).

The fluctuations in composition of natural starters determine a variable performance in technological properties and this may not be acceptable in modern cheesemaking. For many cheeses, artisanal starters have been replaced by commercial mixed-strain starters (MSS), derived from the "best" natural starters and reproduced under controlled conditions by specialized companies and distributed to cheese plants, which use them to build up bulk starter or for direct vat inoculation. Their composition remains undefined, but the reproduction under more controlled conditions reduces the intrinsic variability associated with the use of artisanal starters (Limsowtin et al. 1996).

Phage Contamination of Natural Starters

Phage contamination of undefined starter cultures is considered a common trait. These cultures are commonly propagated under non-aseptic conditions, without any protection against phage, and are known to be tolerant to phage infection. The presence of a mixture of bacterial strains and/or species makes unlikely that all strains and species will be infected by the phage and so completely dead vats are also unlikely. According to this, addition of cheese whey containing bacteriophages to starter cultures, as a means of increasing phage insensitivity, was employed (Thomas and Lowrie 1975). However, the killing of one or more of the dominant strains may affect the culture composition and could alter the metabolic balance of the culture for the time required to establish a new equilibrium. During this period, the phage infection may cause a shift of dominant strains in mixed-strain cultures and the fermentation efficiency could be altered.

The first observation of phages in thermophilic starter fermentations dates back to the early 1950s when Kiuru and Tybeck (1955) showed that phage outbreaks affected the strains of *Lb. helveticus*, *Lb. delbrueckii* subsp. *lactis*, and *Strep. thermophilus* of the starter used to produce Finnish Emmental cheese. Séchaud (1990), in a comparative study on 35 *Lb. helveticus* bacteriophages, described 23 phages isolated from cheese whey in French factories where artisanal starters were used. From the production of the Italian PDO Montasio cheese, where natural milk starter cultures are used, phages able to grow on *Strep. thermophilus* starter strains were isolated (Tamai et al. 1995). Natural whey starters for production of cooked long-ripened cheeses were found infected by phages active against *Lb. helveticus* when slow acidifying activity was observed (Filosofo et al. 1995). On the contrary, other investigations demonstrated the coexistence of phages and sensitive strains of thermophilic lactobacilli in NWS cultures without evidence of acidifying failures (Zago et al. 2005, 2006b, 2008a). The phages of *Lb. helveticus* and *Lb. delbrueckii* subsp. *lactis* have been characterized according to the capability to form lysis plaques on agar plates, the efficiency of replication on the host strains, the host range, and the *Eco*RI phage-DNA restriction profile. On few phages, other characteristics such as the sensitivity to acidic pH, latent period and burst size, and the behavior of a host under the pressure of different levels of the related phage by conductance measurement technique have been studied. Taken together, these data allowed to demonstrate that different phages were present into the same NWS (Zago et al. 2006b). The relatively low incidence of phage problems when thermophilic starters are used could be explained by several hypotheses: the natural selection of phage-resistant strains during replication of mixed strain cultures; the marked host specificity of thermophilic LAB phages, with *Lactobacillus* not affected by phages that attack *Strep. thermophilus*; the different succession of species during cheese production, where the *Lactobacillus* grow significantly after the cooking stage with limited opportunity for *Lactobacillus* phages to propagate (Heap and Lawrence 1988). For the natural whey starter used in Grana cheesemaking, the microbial complexity was explained by the adaptation of these cultures to different technological and stress factors. Within the latter, it has been attributed to phages an ecological role in selecting, among the different populations, phage-resistant strains able to counteract the loss of the sensitive ones, thus preserving the overall technological performances of the culture. This was partly confirmed by the concomitant finding in the same culture of different sub-populations of both phage-sensitive and phage-resistant strains, these latter being dominant over the whole population (Zago et al. 2008a).

The development of PCR-based methods and the possibility of a direct detection of phages, avoiding the traditional microbiological assays that are time-consuming and relying on the availability of indicator strains, resulted useful for a fast and specific detection of phages in cheese production where undefined starter cultures were used. In 1994, Brüssow et al. developed a PCR method for amplification of a very highly conserved DNA element in *Strep. thermophilus* phage genome that allowed the detection of phages in cheese whey from a factory that produced mozzarella cheese with complex undefined starter mixes. Bruttin et al. (1997) studying the ecology of *Strep. thermophilus* bacteriophages in a mozzarella cheese plant in Italy using an undefined NMC over a 2-year period, found phages in all the whey samples with titers ranging from 10^4 to 10^7 PFU ml^{-1}. Fifty-three NWS cultures from different Grana cheese plants resulted positive for the presence of *Lb. helveticus* phages by PCR amplification of the lysin-encoding *lys* gene, and proliferating phages able to lyse dominant strains isolated from the same NWS were found (Zago et al. 2008b). This finding confirmed previous studies on *Lb. delbrueckii* subsp. *lactis* phages. An optimized PCR assay, for amplification of the gene *g17* encoding for the major tail protein of *Lb. delbrueckii* phage LL-H, allowed the detection of phage DNA in 3 out of 11 NWS cultures tested (Zago et al. 2006a).

In mesophilic cultures of undefined composition, phage contamination with up to 10^8 PFU ml^{-1} were found. In mid-1960 it was noticed that mixed strain mesophilic cultures propagated in factories without any protection against phage (P-starter) were much more phage resistant than similar cultures propagated in laboratory (L-starter) (Heap and Lawrence 1988, Cogan et al. 1991). Stadhouders and Leenders (1984), testing the acidifying activity of P-starters inoculated with phage-containing filtrates obtained from the same starters, did not observe any difference in acidity after 24 h at 20°C. However, testing the acidifying activity after 6 h at 30°C, perturbations were detected. Two mechanisms involved in the phage resistance of P-starters were studied: i) a rapid shift from main sensitive strains to main resistant ones; ii) a presence of a main flora of fast bacteria on which slowly growing phages develop, and a minor flora of bacteria resistant to these phages that protect the sensitive ones by limiting the growth of phages. In a study on a natural starter for Cheddar cheese, used continuously for 5 years, no decrease of starter activity due to bacteriophage attack was observed. When the starter was initially introduced, only one of the strains isolated was inhibited by bacteriophages present in the cheese whey, then new types of bacteriophages gradually developed, and after 5 years bacteriophages attacking 46 of the 62 isolates were detected (Josephsen and Nielsen 1988). From the clotted milk obtained by cultivation of a commercial undefined mixed-strain starter culture (Flora Danica) in sterile milk, indigenous phages able to attack starter bacteria were found. About the half of the isolated strains, belonging to *L. lactis* subsp. *cremoris* and *L. lactis* subsp. *lactis* var. *diacetylactis*, showed various degree of sensitivity to the seven phage types carried by the starter (Lodics and Steenson 1990).

Origin of Phages in Natural Whey and Milk Starters

Cheese-making is a non-sterile process, and phages are virtually always present in the plant. There are three main sources through which phages may affect the natural starters, i) the raw milk, ii) the environment, iii) the starter strains.

i. Raw milk can be an external source of new phages. So, for many traditional cheese productions using natural starter cultures where the cheese milk is not heat treated, phages may enter into the dairy plant. The increase and the accumulation of these new phages depend on the product that is being manufactured and on the type of starter culture used. Raw milk was demonstrated as a source of phage problems in cheese plant using lactococcal starters (Heap et al. 1978). Usually the phage concentration in raw milk is low and only if they find a sensitive host their number may increase. The control of the number of bacteria present in the raw milk was reported as a way to prevent multiplication of naturally occurring free phages (Heap and Lawrence 1988). Where undefined starter cultures are used, the phage population may increase without likely to cause failure in acid production (Svensson and Christiansson 1991). From an ecological survey of cheese factories, phages not active on starter strains used in the plant were detected in the raw milk (McIntyre et al. 1991). Bruttin et al. (1997) demonstrated that raw milk was the most likely source of new *Strep. thermophilus* phages in a mozzarella cheese factory using NMS cultures.

   Thermal treatment of raw milk can allow a control of indigenous phages. However, heating and/or pasteurization treatments normally used in the dairy industry are not able to assure a complete inactivation of viable bacteriophages present in raw milk. The survival is influenced by several factors, including the type and number of phages as well as the type of substrate (Kennedy and Bitton 1987). Dean et al. (1953) found a moderate heat resistance of *Strep. thermophilus* bacteriophage in cheese milk, with 2 to 3 log reductions after 30 min at 61.7°C. Phages of thermophilic streptococci were confirmed more resistant to heat treatments than phages of *L. lactis* and *Lb. helveticus* (Quiberoni et al. 1999, Binetti and Reinheimer 2000, Suárez and Reinheimer 2002). According to these data, the thermal process (62-65°C for 10-15 min) applied to raw milk for the production of thermophilic NMS cultures, with the aim to select the desired lactic microflora, could exert only a partial control on the phages naturally present in the raw milk. Similarly, also the cooking temperatures reached in the manufacture of hard cheese varieties (48-55°C) using natural starter cultures are not higher enough to eliminate the phages carried by raw milk or added with the natural cultures themselves.

ii. Phages are dispersed into the plant environment through spreading and the aerosolisation in the form of little droplets generated by the splashing of whey. The cheese whey is the largest source of phage in a dairy plant, since the host bacteria continue to grow in whey after it has been separated from the curd, consequently phages continue to increase. As the final cheese whey is used for the production of NWS cultures, by incubating overnight the whey drained from the cheese vat, a short-circuit of the phage infection for these cheese productions was determined. When cheese whey is released into the environment (e.g. via irrigation) this practice contributed to the recycling of phages through the plant. McIntyre et al. (1991) investigated the survival of starter bacteria and their bacteriophages following spray irrigation of whey in the cheese plant surroundings, and found that phages were detected infrequently in water, but persisted for up to six weeks in the soil. However, the cheese whey did not appear to be the initial source for new phages. To limit the external source of phage infection several protective measures in the dairy environment have to be adopted. The plant layout and the sanitation practices play key roles in managing bacteriophage levels. Milk reception, production and whey processing areas must be separated from each other because are potentially at high risk. Whey drainage must be designed to prevent unnecessary splashing and recontamination of the floor and equipment. Airflow and ventilation should be under control to prevent back-up circulation of air and airborne phages within the production area. Cleaning and sanitation of all plant equipments and cheesemaker's instruments used in the cheese production are basic conditions to reduce level of phages in the environment (Heap and Harnett 2003). Low levels of phage can often be tolerated without disturb of cheese production, mainly when natural starter cultures are used. However, the application of good cleaning procedures to avoid that whey from previous vats could contaminate a newly filled cheese vat is an effective precaution. Phages are usually killed by sanitizer used to disinfect the environment, but if no correct concentrations are used, some phages can survive. It has been also observed that chemical inactivation of LAB phages can be strain dependent. Among sanitizers, peracetic acid and sodium hypochlorite act quickly and are effective against bacteriophages (Everson 1991, Quiberoni et al. 1999, Binetti and Reinheimer 2000, Suárez and Reinheimer 2002).

iii. Phages can also enter the factory in lysogenic starter strains (i.e. bacteria carrying prophages which occasionally may be released from the bacterial DNA and lead to a phage) or in phage-contaminated starter cultures, known as phage carrying cultures or pseudo-lysogenic cultures. Lysogeny enables the phage to coexist with the host cell. Lysogenic strains in the undefined natural starter cultures can play various roles: act as a source of phages; establish a natural phage barrier as they are immune to similar phages; guarantee a balance of strains that are susceptible or resistant to phage, thus avoiding strong variations of starter activity and serious failures during cheese making. Lysogenic lactococcal strains have been suggested as possible source of phages, but the phage transition from temperate to lytic has proved to be difficult to obtain (Heap et al. 1978, Klaenhammer and McKay 1976). Taxonomic studies reported strong DNA homology between a group of lytic phages and a group of temperate phages (Relano et al. 1987, Josephsen et al. 1990). However, lack of indicator strains for temperate phages and no

DNA hybridization between phages and bacterial DNAs, has led several authors to consider unlikely that temperate phages are an immediate source of lactococcal phages in cheese plant (Heap and Lawrence 1988). On the contrary, other investigations suggested that temperate phages can play a significant role in the development of lytic phages which attack lactobacilli. Lysogeny results widespread in lactobacilli, and the DNA homology between some temperate and virulent phages, the transition from the temperate state to virulence, and the possible spontaneous induction were considered as evidences that lysogenic strains of lactobacilli are a potential source of virulent phages in dairy factories (Shimizu-Kadota et al. 1983, Mata et al. 1986, Séchaud et al. 1988, 1989, 1992). From a study on 74 strains of *Lb. helveticus* isolated from NWS cultures used for the production of Grana and Provolone cheeses, 47% of them resulted inducible with mitomycin C (Carminati et al. 1997). The cell-free culture supernatants produced inhibition zones on double-layer agar plates for 61% of the 74 strains used as sensitive indicators. The presence of complete phage particles able to propagate on a sensitive host was only demonstrated for some of the cell-free culture supernatants. The presence of defective phage particles able to kill sensitive cells without forming plaques and without producing newly synthesized particles was demonstrated. The lysogenic strains resulted more resistant to the phage infection, whereas a higher prevalence of indicator strains was found among the non-inducible strains (Carminati et al. 1997).

Brüssow et al. (1994), studying the lysogeny of *Strep. thermophilus*, identified only 2 lysogenic strains among 100 strains induced by mitomycin, thus concluding that lysogeny in this species is rare and that it is unlikely that lysogenic strains of *Strep. thermophilus* could be a potential source of phage in cheese factories. After a longitudinal survey to study the phage ecology in a mozzarella cheese factory using a complex undefined milk starter culture, Bruttin et al. (1997) did not isolate lysogenic strains of *Strep. thermophilus*.

The phage-carrier state (or pseudo-lysogenic state) of a starter culture is defined as the continuous presence of bacteriophages throughout the growth of a bacterial culture, that neither completely lyses nor acquires full resistance against the phage. The phage carrier state is not a consequence of the presence of lysogenic strains, but is a persistent and incomplete infection of a bacterial population by a lytic phage, which can be present in high numbers (10^7-10^8 PFU ml^{-1}) usually not reached for spontaneous induction of lysogenic strain. This relationship differs from lysogeny because in these cultures the presence of bacteriophages can be eliminated by incubation with phage antisera. (Accolas and Veaux 1983). The phage-carrier state in lactococci was described since the 1940's and was considered widespread for mesophilic mixed-strain cultures. Several strains with a different phage-sensitivity pattern are under constant selective pressure of phages present in the environment. The presence of a phage-sensitive population in conjunction with a genotypically or phenotypically phage-resistant population enable bacteria and phages to coexist in a dynamic equilibrium (Lodics and Steenson 1993). Stadhouders and Leenders (1984) suggested that phages reduce the sensitive strains to a minor flora, while insensitive/resistant strains develop, and the continuous exposure to new phages from the environment allows a rapid shift of the flora. The shift is possible because a resistant flora is present and resistant strains with similar growth rate of the sensitive ones can favor the recovery of the starter activity. Phage-carrying lactococcal mixed-strain starters have several different phage sensitivity patterns. The presence of about 10 to 20% of strains insensitive to any bacteriophages isolated from the same mixed-strain starter has been considered the part of the bacterial population which contributes to the overall resistance of these cultures. These starter cultures lose their phage resistance if they are propagated under aseptic conditions like a selected starter culture (Stadhouders and Leenders 1984). De Vos (1989) demonstrated that a strain which harbors a plasmid-encoded phage resistance mechanism could establish the phage-carrier state with a lytic bacteriophage. Upon subsequent daily transfer of the culture, the phage is always present at levels similar to that initially inoculated without perturbation of starter activity.

Ecological Aspects of Bacteriophage-Host Interactions in Natural Starter Cultures

Phage ecology is the study of the interactions between phages and their environments. These interactions are consequential, particularly to the extent that they affect bacteria. According to Abedon (2006, 2008) phage ecology can be divided into various areas including: organismal ecology, population ecology, community ecology, and ecosystem ecology. i) Phage organismal ecology studies the processes of the phage life cycle and the adaptations that phages employ to overcome physical, chemical, or biological barriers to transmission between hosts. The bacterial cell is a target that impacts on phage phenotype. ii) Phage population ecology studies phage population growth and intraspecific competition, either within bacterial cultures or within individual infected bacteria. The factors that influence the phage population are: the rapidity of progeny maturation and the rate of phage multiplicity, the lysogeny which hinders an active phage replication and causes a delay between infection and progeny maturation, the structure and the quality of the environment which exert-selective pressures and influence host density as well as host physiological state. iii) Phage community ecology studies the interactions among phages and between phages and bacteria, where bacteria are partners in coevolution. In general, each factor that interferes with growth and survival of phages and/or of sensitive bacterial cells could influence the density and the stability of the phage-bacterial community. The presence of phage-resistant bacteria, the ability of phages to counteract host resistance mechanisms, and factors that limit the bacterial growth or the phage productivity contribute to the stability of the communities by avoiding bacterial extinction. Phages can modify bacterial communities by increasing bacterial diversity. The mechanisms can be associated with phage-mediated selection of bacteria, with genetic transduction, and with lysogenic conversion (Kerr et al. 2008). The transfer of genetic material from bacterium to bacterium mediated by phage (transduction) can modify the bacterial genotypes, and therefore increases bacterial genetic diversity. iv) Phage ecosystem ecology concerns the phage impact on nutrient and energy flow within environments. The impact of phages on ecosystems has been discerned from the studies of aquatic phage biology, while is less easily understood in other ecosystems.

Most knowledge on the ecological aspects of phage-host interactions to understand the role of phages on the evolution of microbial communities is available from studies on environmental microbiology (Weinbauer 2004). Laboratory communities of phages and bacteria have been used as model systems to acquire basic information on phage-host interactions, in order to explore how these interactions can affect their evolution and can influence the population dynamics, the community structure, and the evolutionary change in complex natural environments (Bohannan and Lenski 2000, Middelboe 2000). Studies on simple communities of *Escherichia coli* and lytic T-series bacteriophages allowed to demonstrate how phage-resistant bacteria evolve and the role of the environment in determining the rate at which this phenomenon occurs. Most mutations that confer phage-resistant phenotypes often have a cost for the bacterial metabolism. The probability that a cell harboring a resistance mutation will go extinct or persist is a function of the physiological costs of phage-resistance. For example, due to the modification of a phage receptor, which also serves as nutrient uptake receptor, bacterial strains may become resistant against infection but at the same time less competitive for acquiring a specific nutrient. A resistant strain can survive only because the more competitive but phage-sensitive strains are controlled by lytic infection. Once the phage numbers decrease due to the lack of susceptible hosts, the surviving competitive strains may become abundant again (Bohannan and Lenski 2000). Studies under controlled conditions in the laboratory on the behavior of specific marine phage-host system strongly suggested that the effects of phages on bacterial mortality were closely related to the growth conditions of the host cells with the production of phages that is proportional to the growth rate of the host. One-step growth experiments allowed to verify that the average phage burst size decreases and latent period increases when host cells in rich medium were put to grow in poor medium (Middelboe 2000).

Similar laboratory studies on LAB and their phages are not available, and their ecological relationships have been gathered by studying the characteristics of phages and LAB strains isolated from particular dairy ecosystems. Lodics and Steenson (1993) reviewed the knowledge on phage-host interactions in lactococcal mixed-strain starters. These starters were used until the 1970s in small cheese factories without protection from contaminating bacteriophages, and resulted excellent sources of phage-resistant strains. A continuous contact of these starters with phages present in the milk environment allowed to maintain their phage resistance. The presence of a wide heterogeneity among strains of lactococci, with phage-sensitive population of bacterial cells in conjunction with genotypically or phenotypically phage-resistant populations, was considered the key element in establishing and maintaining a phage-carrier state. The bacteriophages select the phage-resistant cells among a population characterized by various phage-resistance mechanisms, the environment allows the more competitive and adapted cells to emerge and to lead successful milk fermentations. The existence of this evolutionary arms race between lactococci and their phages was confirmed by the numerous phage resistance mechanisms that have been found, as reduction in adsorption efficiency, restriction modification systems, abortive infection, and superinfection immunity by prophages. The studies on flora Danica starter strains and coexisting phage populations suggested that phages evolved by evading the resistance mechanisms of the starter strains (Lodics and Steenson 1990).

Interestingly, Bruttin et al. (1994) studying for about two years the *Strep. thermophilus* phage population in a mozzarella cheese factory that used a natural milk starter (*madre naturale*), did not find evidence for an evolution from resident phages but rather a selection of different phages present in the environment. The wide variety of phage types found, in contrast to less variability detected in yoghurt factories which use a small number of well-defined starter cultures, was attributed to the diversity of strains of the undefined natural starters. Phages belonging to different lytic groups showed clearly distinct restriction patterns, and phages isolated over the 2-year survey period infecting the same host showed similar restriction patterns. Also the sequence of a conserved region of phage genome revealed no point mutations in phages of the same type, while nucleotide sequence diversity was observed between the different phage types. The author concluded that phage diversity was unlikely a result of a single phage invasion event and subsequent diversification of the phage during its residence in the factory. The introduction of a defined starter system that was not be able to propagate on the resident phages, and the finding within few days of new phages with host ranges and restriction patterns different from the previous resident phages but identical to those found in the raw milk, led to support the conclusion that the genetic diversity of *Strep. thermophilus* phages found during the survey was present in the environment and was selected by the presence of different hosts. Duplessis and Moineau (2001), studying the phage genetic determinant (anti-receptor) involved in the recognition of *Strep. thermophilus* hosts, demonstrated that phages developed an efficient method to alter their host range by exchanging genetic elements. A variable region within this gene was found to be responsible for host specificity.

Also the relationships established between temperate phages and their homologous hosts are involved in coexistence of phages and bacteria in natural starter cultures. From an ecological perspective, lysogeny can set up as a strategy for survival with an impact on diversification of bacterial community. The integration of phage genome into the host genome confers superinfection immunity against related phages. Some of the genes that are expressed from the prophage in a lysogenic strain are “lysogenic conversion” genes, which alter the properties of the host bacterium. The influx of extra genes during prophage integration into a given bacterial species can play an active and sometimes mutualistic improvement of the competitive fitness of the lysogenic host strains, favoring the selection of strains that are best adapted to a given environment. In many pathogenic bacteria it has been demonstrated that prophage often codes for functions that enhance host virulence and resistance to the immune system, and allows the lysogen to successfully colonize the animal host (Chen et al. 2005). The temperate phages can propagate lytically on sensitive strains, or can re-establish a stable lysogenic relationship with new hosts. However only functional prophages can induce a lytic growth to initiate, but many of the prophage-like entities frequently found in bacterial genomes (defective prophages) are unable to establish the full phage replication cycle. The presence of prophages and/or other phage-related elements, can account for a large fraction of variation among strains within a bacterial species, and confirms that phages are likely to be important vehicles for horizontal gene transfer between bacteria (Casjens 2003).

Lysogeny was found to be a common phenomenon in many LAB species, as confirmed by genome sequence analyses (Desiere et al. 2002, Canchaya et al. 2003, 2004, Ventura et al. 2006). The lysogenic LAB in natural dairy starter can outcompete with the nonlysogenic strains, possibly due to the selective advantage conferred by lysogenic conversion genes of the prophages. If a spontaneous induction happens, a small fraction of the bacterial cells that harbor a prophage can release a temperate phage. Huggins and Sandine (1977) found among 63 strains of lactococci 3 strains able to spontaneously release phages, yielding titers of about 10^3 PFU ml^{-1}. When these strains were grown in mixed culture with indicator strains, phage titers increased to about 10^6 PFU ml^{-1}. They concluded that lysogenic lactic streptococcal strains may serve as a reservoir for phages that attack sensitive strains in mixed- or multiple-strain lactic starter cultures. Many lactococcal strains isolated from mixed strain cultures were induced by mitomycin C treatments and released temperate phages. DNA homology studies demonstrated the presence of regions of homology between some lytic and temperate phages which would allow recombinations, leading the generation of new phages (Lodics and Steenson 1993). The DNA homologies resulted located in the chromosomal region coding for the structural proteins (Relano et al. 1987). Within the same DNA homology group, almost all the phages may possess similar genetic elements (modules) coding for the structural proteins of the head and the tail. A theory of modular evolution for bacteriophages supposed that modules can be exchanged by homologous recombination. The demonstration of a relatedness between virulent and temperate phages has been considered of technological importance because, though did not definitively prove that the temperate phages could be a source of virulent phages in industrial fermentations, it indicated that they may have identical phage ancestors.

Compared to *L. lactis,* lysogeny in *Strep. thermophilus* has not been considered as widespread. Bruttin et al. (1997) did not detect lysogenic strains after a 2-years survey on mozzarella cheese factory using natural milk starters. The diversity of the 12 distinct virulent groups of phages isolated from natural milk starter cultures did not reflect an evolution from a single phage that infected the factory and experienced genetic changes during its residence, but the diversity was naturally present in the environment. However, lysogenic strains from raw milk and artisan mixed strain starters have been detected (Fayard et al. 1993). The corresponding temperate phages were obtained by induction with mitomycin, then multiplied on indicator strains and both characterized. The DNA restriction profiles of seven native temperate phages, i.e. those obtained just after induction of the lysogenic hosts, were noticeably different from those of the corresponding phages once they had been propagated on lysogenic indicators. Also the host range of the latter phages included a large number of sensitive strains. The authors concluded on possible genome rearrangement in infecting temperate phage once in contact with the prophage of the infected indicator cells through modular exchanges between the different genomes as previously attributed by Mercenier (1990). Further studies demonstrated that *Strep. thermophilus* phages, whether temperate or virulent, shared DNA sequence similarity. Virulent phages isolated from a factory after an ecological survey were essentially the result of deletion, gene replacement and rearrangement events in the lysogeny module of temperate phages (Brüssow et al. 1998, Lucchini et al. 1999, Canchaya et al. 2003, Guglielmotti et al. 2009).

Lysogeny is recognized widespread in lactobacilli. Among the *Lactobacillus* species for which the genome sequence is available, prophage sequences were almost always found (Ventura 2006). Prophages can be present in many different forms ranging from inducible prophages to prophages showing apparent deletions, insertions, and rearrangements, to prophage remnants that appear to have lost most of their genome. For *Lb. helveticus*, no complete prophage sequence was found in the recently sequenced DPC 4571 strain (Callanan et al. 2008). However, for this species that is dominant in natural whey starters, lysogeny has been frequently demonstrated by induction studies. Séchaud et al. (1989) comparing the morphological, physiological and biochemical characteristics found a close relationship between the temperate *Lb. helveticus* phage 0241, obtained from a lysogenic strain isolated in the early '60s from an artisanal starter, and the virulent phage 832-B1 isolated 26 years later. The authors suggested that lysogenic strains can be a source of new phages, as previously shown for *Lactobacillus casei* in Japanese dairy plants (Shimizu-Kadota et al. 1983). Induction with mitomycin C of *Lb. helveticus* strains isolated from NWS cultures used for the production of Grana and Provolone cheeses demonstrated a wide presence of lysogenic strains, which were able to release phage and phage-like particles (killer particles that act as a bacteriocin) (Carminati et al. 1997). The lysogenic strains showed a marked phage-resistance confirming the superinfection immunity conferred by the integrated prophage, while the most phage-sensitive strains were found within those that resulted noninducible with mitomycin C. The authors concluded that temperate phages may be involved in the development of lytic phages, and that lysogenic strains can control the evolution of the strain population and have an impact on the diversification of bacterial community. Cluzel et al. (1987) studied the complex and fluctuating strains population arising after lysis of a sensitive *Lb. delbrueckii* subsp. *lactis* strain CNRZ 326 with a temperate *Lb. bulgaricus* phage 0448. Within the secondarily developing culture they found lysogenized cells, which were phage-resistant and inducible with mitomycin C, and non-inducible phage sensitive cells. The lysogenic conversion affected the pattern of phage sensitivity and the lysogenized cells of the strain *Lb. delbrueckii* subsp. *delbrueckii* subsp. *lactis* showed the same host spectra as the original lysogenic *Lb. delbrueckii* subsp. *bulgaricus* strain. This temperate phage showed a significant relationship with virulent phages of both *Lb. delbrueckii* subsp. *bulgaricus* and *Lb. delbrueckii* subsp. *lactis* on the basis of their protein composition, the restriction pattern of their genome, nucleic acid hybridization studies, and serological data. Virulent and temperate phages seemed to share large homologies in their chromosomal region coding for the structural proteins, and these results were confirmed by immunoblotting data as well as by the protein patterns. Therefore, these phages may possess a common module for the structural genes. The non-homologous regions between temperate and virulent phages might be involved in the replication and lysogeny functions. It was suggested that industrial fermentations might provide a dynamic situation in which interactions between lactic bacteria and their bacteriophages could promote phage DNA rearrangements giving rise to new phages derived from one another (Mata et al. 1986).

CONCLUSION

Natural starter cultures are complex dairy ecosystems where bacteriophages may play an active role on microbial composition and population dynamics. The matrices used to prepare these cultures, i.e. milk or whey, are rich in nutrients and are suitable for growth of LAB according to the environmental conditions imposed by the dairy practices. The parameters of the cheesemaking and/or the incubation conditions applied for the preparation and cultivation of these starters, select the more adapted strains which can actively grow. The external factors, like temperature, pH, organic and inorganic compounds, osmotic pressure, affect the growth of host bacteria, as well as the stability of phages and their interactions with host cells, and finally influencing the development of phages (Kennedy and Bitton 1987, Sanders 1987). This picture fits well with the concept of “killing the winner” elaborated from the studies on microbial aquatic ecosystems (Weinbauer and Rassoulzadegan 2004). This concept states that phages keep in check the fastest growing host population, and the increase of phages stops when the host population diminishes and no longer supports phage replication. It means that the more adapted and competitive strains grow as dominants and are under control of phages. This is one of the ecological roles that phages may exert, driving the coevolution, and the generation and maintenance of bacterial diversity (Jessup and Forde 2008).

In addition to the contribution of phage-resistance mechanisms which allow bacteria to escape the selective pressure of phages, favoring the heterogeneity of strain populations, other physiological mechanisms may exert a control on the dynamic of phage-host relationships and explain the possible coexistence of phages and related hosts in the natural dairy cultures, including the appropriate physiological state of the bacteria to allow productive phage infection. When bacterial growth stops, as a consequence of the depletion of nutrients and/or the modification of environmental characteristics (e.g. the drop of the pH), the cells enter in the stationary phase and phages cannot replicate. The virulent phages persist within populations of nongrowing cells, and the cells release phage progeny only after resumption of growth. The preparation of new cultures each day allows the cell growth to start again and consequently also their phages. The variation of cell host concentration is influenced by the competitiveness for the substrate, as well as by the ratio between phages and their target cells. The need of host abundance to sustain phage infection and replication cycles supported the hypothesis of the "numerical refuge", that is the presence of host cells at levels sufficiently high to assure the probability of phage-host contacts (Weinbauer 2004). For example, when new phages enter the natural dairy cultures through raw milk, the numbers are usually low and, only if they find a sensitive and well-growing host strain, the infection cycles can be maintained, leading phages to survive and forcing LAB to develop antiphage strategies. Alternatively, to counteract low host cell concentrations, a broad-host-range (polivalency) can represent a solution for phages to survive when the related host strain is not abundant enough to guarantee phage replication. Polyvalency is an adaptation mechanism to low host cell concentrations suggested for phages isolated from nutrient-poor marine environments (Chibani-Chennoufi et al. 2004). Polivalency has not been described for dairy phages, which are characterized by a narrow host range, whereas it has been reported for *Lactobacillus* phages able to infect two or more ecologically related species in other food and non food ecosystems (Kilic et al. 2001, Lu et al. 2003).

Other mechanisms, whereby phages can increase bacterial diversity thus modifying the bacterial communities, are those involved with transfer of genetic material. Transduction is the horizontal or lateral genes transfer between bacterial cells mediated by phage particles that modify the bacterial genotypes and may alter the metabolic properties of hosts. A generalized transduction is the transfer of bacterial genes between hosts during phage infection. Host DNA fragments are packaged in the phage heads instead of the phage DNA, and delivered into suitable bacteria. Generalized transduction is restricted by the typical low host range, thus allowing the spread of a gene only in a limited population. However, a few data suggested that transduction might be more common and overcomes the host range barriers at the genus level, as demonstrated from several nonrelated marine isolates to *Escherichia coli* and *Bacillus subtilis* (Weinbauer 2004). A specialized transduction occurs when bacterial genes are carried by lysogen-forming temperate phages (lysogenization). Every lysogenization can be considered as a phage conversion, since two properties, lysogeny and superinfection immunity are always conferred. The phage conversion may have effects on other metabolic properties of the host, and these effects may become hereditary once the prophage loses its ability for entering the lytic cycle. It has been estimated that two-third of all sequenced Gram-positive bacteria harbor prophages, and the presence or absence of prophages can account for a large fraction of the variation among individuals within a bacterial species. Phage genomes are mosaics of modules (groups of functionally related genes) that are free to recombine in genetic exchanges between distinct phages infecting the same cell. In the bacterial chromosomes, modular recombination between different phage types were observed at a frequency not encountered in extracellular phages. Prophages serve as anchor points for gene rearrangements, can interrupt genes through insertion, introduce genes that promote host fitness and cause silence of host genes by phage repressors. The various forms of prophage sequences found (inducible prophages, defective prophages, prophage remnants, and isolated phage genes) argue for a dynamic model of phage-bacterial genome coexistence. The presence of phage genomes or their fragments in bacterial genomes confirms that lateral gene transfer is an important factor for bacterial evolution. Specialized transduction contributes to the generation of genetic variety in bacteria populations, which is not only a prerequisite of evolutionary change but also influences the population dynamics of hosts since can also increase the fitness of recipients and thus allow for their survival (Desiere et al. 2002, Canchaya et al. 2003, Casjens 2003, Canchaya et al. 2004).

It can be concluded that natural starter cultures represent complex food microcosms where lactic acid bacteria coexist with phages. Bacteriophages work as biodiversity regulators, since they control the size and the heterogeneity of bacterial populations and have a great influence on the generation of bacterial genetic diversity.

REFERENCES

Abedon, S. T. (2006). Phage ecology. In R. Calendar (Ed.), *The bacteriophages* (pp. 37-46). NewYork, USA: Oxford University Press.

Abedon, S. T. (2008). Phages, ecology, evolution. In S.T. Abedon (Ed.), *Bacteriophage Ecology: population growth, evolution, and impact of bacterial viruses* (pp. 1-27). New York, USA: Cambridge University Press.

Accolas, J.-P. and Veaux, M. (1983). Les bacteriophages des levain lactiques. *La Technique Laitière, 976*, 21-28.

Binetti, A. G. and Reinheimer, J. A. (2000). Thermal and chemical inactivation of indigenous *Streptococcus thermophilus* bacteriophages isolated from Argentinian dairy plants. *Journal of Food Protection, 63*, 509-515.

Bohannan, B. and Lenski, R. (2000). Linking genetic change to community evolution: insights from studies of bacteria and bacteriophage. *Ecology Letters, 3*, 362-377.

Brüssow, H., Bruttin, A., Desiere, F., Lucchini, S. and Foley, S. (1998). Molecular ecology and evolution of *Streptococcus thermophilus* bacteriophages - a review. *Virus Genes, 16*, 95-109.

Brüssow, H., Canchaya, C. and Hardt, W.-D. (2004). Phages and the evolution of bacterial pathogens: from genomic rearrangements to lysogenic conversion. *Microbiology and Molecular Biology Reviews, 68*, 560-602.

Brüssow, H., Freimont, M., Bruttin, A., Sidoti, J., Constable, A. and Fryder, V. (1994). Detection and classification of *Streptococcus thermophilus* bacteriophages isolated from industrial milk fermentation. *Applied and Environmental Microbiology, 60*, 4537-4543.

Bruttin, A., Desiere, F., D'Amico, N., Guérin, J.-P., Sidoti, J., Huni, B., Lucchini, S. and Brüssow, H. (1997). Molecular ecology of *Streptococcus thermophilus* bacteriophages infections in a cheese factory. *Applied and Environmental Microbiology, 63*, 3144-3150.

Callanan, M., Kaleta, P., O'Callaghan, J., O'Sullivan, O., Jordan, K., McAuliffe, O., Sangrador-Vegas, A., Slattery, L., Fitzgerald, G. F., Beresford, T. and Ross, R. P. (2008). Genome sequence of *Lactobacillus helveticus*, an organism distinguished by selective gene loss and insertion sequence element expansion. *Journal of Bacteriology, 190*, 727-735.

Carminati, D., Mazzucotelli, L., Giraffa, G. and Neviani, E. (1997). Incidence of inducible bacteriophage in *Lactobacillus helveticus* strains isolated from natural whey starter cultures. *Journal of Dairy Science, 80*, 1505–1511.

Carminati, D., Giraffa, G., Quiberoni, A., Binetti, A., Suárez, V. and Reinheimer, J. (2010). Advances and trends in starter cultures for dairy fermentations. In F. Mozzi, R. R. Raya and G. M. Vignolo (Eds.), *Biotechnology of Lactic Acid Bacteria: Novel applications* (pp. 177-192). Ames, Iowa, USA: Wiley-Blackwell Publishing.

Canchaya, C., Proux, C., Fournous, G., Bruttin, A. and Brüssow, H. (2003). Prophage genomics. *Microbiology and Molecular Biology Reviews, 67*, 238–276.

Canchaya, C., Fournous, G. and Brüssow, H. (2004). The impact of prophages on bacterial chromosomes. *Molecular Microbiology, 53*, 9–18.

Casjens, S. (2003). Prophages and bacterial genomics: what have we learned so far? *Molecular Microbiology, 49*, 277-300.

Chen, Y., Golding, I., Sawai, S., Guo, L. and Cox, E. C. (2005). Population fitness and the regulation of *Escherichia coli* genes by bacterial viruses. *PLoS Biology, 3* (e229), 1276-1282.

Chibani-Chennoufi, S., Bruttin, A., Dillmann, M.-L. and Brüssow, H. (2004). Phage-host interaction: an ecological perspective. *Journal of Bacteriology, 186*, 3677-3686.

Cluzel, P.-J., Serio, J., Accolas J.-P. (1987). Interactions of *Lactobacillus bulgaricus* temperate bacteriophage 0448 with host strains. *Applied and Environmental Microbiology, 53*, 1850-1854.

Cogan, T.M., Peitersen, N. and Sellars R. L. (1991). Starter systems. *Bulletin International Dairy Federation (FIL-IDF), 263*, 16-23.

Deane, D. D., Nelson, F. E., Ryser, F. C and Carr, P. H. (1953) *Streptococcus thermophilus* bacteriophage from Swiss cheese whey. *Journal of Dairy Science, 36*, 185-191.

De Angelis, M., de Candia, S., Calasso, M., Faccia, M., Guinee, T., Simonetti, M. and Gobbetti, M. (2008). Selection and use of autochthonous multiple strain cultures for the manufacture of high-moisture traditional Mozzarella cheese. *International Journal of Food Microbiology, 125*, 123-132.

Desiere, F., Lucchini, S., Canchaya, C., Ventura, M. and Brüssow, H. (2002). Comparative genomics of phages and prophages in lactic acid bacteria. *Antonie van Leeuwenhoek, 82*, 73-91.

De Vos, W. M. (1989). On the carrier state of bacteriophages in starter lactococci: an elementary explanation involving a bacteriophage resistance plasmid. *Netherland Milk Dairy Journal, 43*, 221-227.

Duplessis, M. and Moineau, S. (2001). Identification of a genetic determinant responsible for host specificity in *Streptococcus thermophilus* bacteriophages. *Molecular Microbiology, 41*, 325-336.

Everson, T. C. (1991). Control of phage in the dairy industry. *Bulletin International Dairy Federation (FIL-IDF), 263*, 24-28.

Fayard, B., Haeflinger, M., Accolas, J.-P. (1993). Interactions of temperate bacteriopohages of *Streptococcus salivarius* subsp. *thermophilus* with lysogenic indicators affect phage DNA restriction patterns and host range. *Journal of Dairy Research, 60*, 385-399.

Filosofo, T., Lombardi, A., Cislaghi, S., Ceccarelli, P., Pirovano, F., Minissi, S., Sozzi, T., Passerini, S., Rusinenti, L. and Rottigni, C. (1995). Casi di sieroinnesti difettosi per infezione fagica. *Il Latte, 20*, 1120-1123.

Guglielmotti, D. M., Deveau, H., Binetti, A. G., Reinheimer, J.A., Moineau, S. and Quiberoni, A. (2009). Genome analysis of two virulent *Streptococcus thermophilus* phages isolated in Argentina. *International Journal of Food Microbiology, 136*, 101-109.

Heap, H. A. and Harnett, J. T. (2003) Bacteriophage in the dairy industry. In H. Roginski, J. W. Fuquay and P. F. Fox (Eds.), *Encyclopedia of Dairy Sciences* (Vol. 1, pp. 136-141). London, UK: Academic Press.

Heap, H. A., Limsowtin, G. K. Y. and Lawrence, R. C. (1978). Contribution of *Streptococcus lactis* strains in raw milk to phage infection in commercial cheese factories. *New Zealand Journal of Dairy Science and Technology, 13*, 16-22.

Heap, H. A. and Lawrence, R. C. (1988). Culture systems for the dairy industry. In: R. K. Robinson (Ed.), *Developments in Food Microbiology* (Vol. 4, pp. 149-185). Amsterdam, NL: Elsevier Applied Science.

Huggins, A. R. and Sandine, W. E. (1977). Incidence and properties of temperate bacteriophages induced from lactic streptococci. *Applied and Environmental Microbiology, 33*, 184-191.

Jessup, C. M. and Forde, S. E. (2008). Ecology and evolution in microbial systems: the generation and maintenance of diversity in phage-host interactions. *Research in Microbiology, 159*, 382-389.

Josephsen, J. and Nielsen, E. W. (1988). Plasmid profiles and bacteriophage sensitivity of bacteria of a Cheddar starter used for five years without rotation. *Milchwissenschaft, 43*, 219-223.

Josephsen, J., Andersen, N., Behrndt, H., Brandsborg, E., Christiansen, G., Hansen, M. B., Hansen, S., Waagner Nielsen, E. and Vogensen, F. K. (1990). An ecological study of lytic bacteriophages of *Lactococcus lactis* subsp. *cremoris* isolated in a cheese plant over a five year period. *International Dairy Journal, 4*, 123-140.

Kennedy, J. E. and Bitton, G. (1987). Bacteriophages in foods. In S. M. Goyal, C. P. Gerba and G. Bitton (Eds.), *Phage ecology* (pp.289-316). New York, USA: John Wiley and Sons, Inc.

Kerr, B., West, J. and Bohannan, B. J. M. (2008). Bacteriophages: models for exploring basic principles of ecology. In S.T. Abedon (Ed.), *Bacteriophage Ecology: population growth, evolution, and impact of bacterial viruses* (pp. 31-63). New York, USA: Cambridge University Press.

Kilic, A. O., Pavlova, S. I., Alpay, S., Kilic, S. S. and Tao, L. (2001). Comparative study of vaginal *Lactobacillus* phages isolated from women in the United States and Turkey: prevalence, morphology, host range, and DNA homology. *Clinical and Diagnostic Laboratory Immunology, 8*, 31-39.

Klaenhammer, T. R. and McKay, L. L. (1976). Isolation and examination of transducing bacteriophage particles from *Streptococcus lactis* C2. *Journal of Dairy Science, 59*, 396-404.

Kiuru, V. J. T. and Tybeck, E. (1955). Characteristics of bacteriophages active against lactic acid bacteria in Swiss cheese. *Suomen Kemistil, 288*, 56-62.

Limsowtin, G. K. Y., Powell, I. B. and Parente, E. (1996). Types of starters. In T.M. Cogan and J.-P. Accolas (Eds.), *Dairy Starter Cultures* (pp. 101-129). New York, USA: VCH Publisher.

Lodics, T. A. and Steenson, L. R. (1990). Characterization of bacteriophages and bacteria indigenous to mixed-strain cheese starter. *Journal of Dairy Science, 73*, 2685-2696.

Lodics, T. A. and Steenson, L. R. (1993). Phage-host interactions in commercial mixed-strain dairy starter cultures: practical significance. A review. *Journal of Dairy Science, 76*, 2380-2391.

Lu, Z., Breidt, F., Plengvidhya, V. and Fleming, H. P. (2003). Bacteriophage ecology in commercial sauerkraut fermentations. *Applied and Environmental Microbiology, 69*, 3192-3202.

Lucchini, S., Desiere, F., Brüssow, H. (1999). Comparative genomics of *Streptococcus thermophilus* phage species supports a modular evolution theory. *Journal of Virology, 73*, 8647-8656.

Mata, M., Trautwetter, A., Luthaud, G. and Ritzenthaler, P. (1986). Thirteen virulent and temperate bacteriophages of *Lactobacillus bulgaricus* and *Lactobacillus lactis* belong to a single DNA homology group. *Applied and Environmental Microbiology, 52*, 812-818.

McIntyre, K., Heap, H. A., Davey, G. P. and Limsowtin, G. K. Y. (1991). The distribution of lactococcal bacteriophage in the environment of a cheese manufacturing plant. *International Dairy Journal, 1*, 151-214.

Mercenier, A. (1990). Molecular genetics of *Streptococcus thermophilus*. *FEMS Microbiology Review, 87*, 61-78.

Middelboe, M. (2000). Bacterial growth rate and marine virus-host dynamics. *Microbial Ecology, 40*, 114-124.

Parente, E. and Cogan, T. M. (2004). Starter cultures: general aspects. In P. F. Fox, P. L. H. McSweeney, T. M. Cogan and T. P. Guinee (Eds.), *Cheese: Chemistry, Physics and Microbiology* (Vol. 1, pp. 123-147). London, UK: Elsevier Academic Press.

Powell, I. B., Broome, M. C. and Limsowtin, G. K. Y. (2003). Starter cultures: general aspects. In H. Roginski, J. W. Fuquay and P. F. Fox (Eds.), *Encyclopedia of Dairy Sciences* (Vol. 1, pp. 261-268). London, UK: Academic Press.

Quiberoni, A., Suárez, V. B. and Reinheimer, J. A. (1999). Inactivation of *Lactobacillus helveticus* bacteriophages by thermal and chemical treatments. *Journal of Food Protection, 62*, 894-898.

Reinheimer, J. A., Suárez, V. B., Bailo, N. and Zalazar, C. (1995). Microbiological and technological characteristics of natural whey cultures for Argentinian hard cheese production. *Journal of Food Protection, 58*, 796–799.

Reinheimer, J. A., Binetti, A., Quiberoni, A., Bailo, N., Rubiolo, A. and Giraffa, G. (1997). Natural milk cultures for the production of Argentinian cheeses. *Journal of Food Protection, 60*, 59–63.

Relano, P., Mata, M., Bonneau, M. and Ritzenthaler, P. (1987). Molecular characterization and comparison of 38 virulent and temperate bacteriophages of *Streptococcus lactis*. *Journal of General Microbiology, 113*, 3053-3063.

Sanders, M. E. (1987). Bacteriophages of industrial importance. In S. M. Goyal, C. P. Gerba and G. Bitton (Eds.), *Phage ecology* (pp.211-244). New York, USA: John Wiley and Sons, Inc.

Séchaud, L. (1990). Caractérisation de 35 bactériophages de *Lactobacillus helveticus*. Thesis Universités Paris VII et XI, ENSIAA de Massy.

Séchaud, L., Callegari, M. L., Rousseau, M., Muller, M. C. and Accolas, J. P. (1989). Relationship between temperate bacteriophage 0241 and virulent bacteriophage 832-B1 of *Lactobacillus helveticus*. *Netherland Milk and Dairy Journal, 43*, 261-277.

Séchaud, L., Cluzel, P. J., Rousseau, M., Baumgartner, A. and Accolas, J. P. (1988). Bacteriophages of lactobacilli. *Biochimie, 70*, 401-440.

Séchaud, L., Rousseau, M., Fayard, B., Callegari, M. L., Quénée, P. and Accolas, J. P. (1992). Comparative study of 35 bacteriophages of *Lactobacillus helveticus*: morphology and host range. *Applied and Environmental Microbiology, 58*, 1011-1018.

Shimizu-Kadota, M., Sakurai, T. and Tsuchida, N. (1983). Prophage origin of a virulent phage appearing on fermentations of *Lactobacillus casei* S-1. *Applied and Environmental Microbiology, 45*, 669-674.

Stadhouders, J. and Leenders, G. J. M. (1984). Spontaneously developed mixed-strain cheese starters. Their behaviour towards phages and their use in the Dutch cheese industry. *Netherland Milk and Dairy Journal, 38*, 157-181.

Suárez, V. B. and Reinheimer, J. A. (2002). Effectiveness of thermal treatments and biocides in the inactivation of Argentinian *Lactococcus lactis* phages. *Journal of Food Protection, 65*, 1756-1759.

Svensson, U. and Christiansson, A. (1991). Methods for phage monitoring. *Bulletin International Dairy Federation (FIL-IDF), 263*, 29-39.

Tamai, M., Basso, A., Zilio, F., Lombardi, A. and Corradini, C. (1995). L'infezione fagica nella produzione del formaggio Montasio. 1. Isolamento di batteriofagi di *Streptococcus salivarius* subsp. *thermophilus* da sieri di fine lavorazione. *Scienza e Tecnica Lattiero-Casearia, 46*, 380-392.

Thomas, T. D. and Lowrie, R. J. (1975). Starters and bacteriophages in lactic acid casein manufacture. II. Development of a controlled starter system. *Journal of Milk and Food Technology, 38*, 275-278.

Ventura, M., Canchaya, C., Bernini, V., Altermann, E., Barrangou, R., McGrath, S., Claesson, M. J., Li, Y., Leahy, S., Walker, C. D., Zink, R., Neviani, E., Steele, J., Broadbent, J., Klaenhammer, T. R., Fitzgerald, G. F., O'Toole, P. W. and van Sinderen, D. (2006). Comparative genomics and transcriptional analysis of prophages identified in the genomes of *Lactobacillus gasseri*, *Lactobacillus salivarius*, and *Lactobacillus casei*. *Applied and Environmental Microbiology, 72*, 3130-3146.

Weinbauer, M. G. (2004). Ecology of prokaryotic viruses. *FEMS Microbiology Reviews, 28*, 127-181.

Weinbauer, M. G. and Rassoulzadegan, F. (2004). Are viruses driving microbial diversification and diversity? *Environmental Microbiology, 6*, 1-11.

Zago, M., Comaschi, L., Neviani, E. and Carminati, D. (2005). Investigation on the presence of bacteriophages in natural whey starters used for the production of Italian long-ripened cheeses. *Milchwissenschaft, 60*, 171-174.

Zago, M., De Lorentiis, A., Carminati, D., Comaschi, L. and Giraffa, G. (2006a). Detection and identification of *Lactobacillus delbrueckii* subsp. *lactis* bacteriophages by PCR. *Journal of Dairy Research, 73*, 146-153.

Zago, M., Rossetti, L., Fornasari, M.E., De Lorentiis, A., Bonioli, A., Perrone, A., Giraffa, G. and Carminati, D. (2006b). Identificazione e caratterizzazione di batteriofagi di lattobacilli termofili isolati da innesti naturali. *Scienza e Tecnica Lattiero-Casearia, 57*, 525-534.

Zago, M., Rossetti, L., Bonvini, B., Remagni, M. C., Perrone, A., Fornasari, M. E., Carminati, D. and Giraffa, G. (2008a). Il sieroinnesto di Grana Padano: una comunità di batteri e batteriofagi. *Scienza e Tecnica Lattiero-Casearia, 59*, 277-286.

Zago, M., Rossetti, L., Reinheimer, J., Carminati, D. and Giraffa, G. (2008b). Detection and identification of *Lactobacillus helveticus* bacteriophages by PCR. *Journal of Dairy Research, 75*, 196-201.

In: Bacteriophages in Dairy Processing
Editor: Andrea del Lujan Quiberoni et al.
ISBN 978-1-61324-517-0

Chapter 5

INFECTIVE CYCLE OF DAIRY BACTERIOPHAGES

Daniela M. Guglielmotti, Diego Javier Mercanti and Mariángeles Briggiler Marcó*

Instituto de Lactología Industrial (UNL-CONICET),
Facultad de Ingeniería Química,
Universidad Nacional del Litoral, Santiago del Estero
Santa Fe, Argentina.

ABSTRACT

Phages follow two strategies to survive: they can either infect and lyse the host cell (virulent/lytic phages) or insert its genome into the host chromosome (prophages). In either case, the first step is the adsorption of virions to specific attachment sites (receptors) on the cell wall surface. Adsorption velocity and efficiency depend upon a series of external factors, such as the presence of divalent ions (Ca^{2+}, Mg^{2+}, Mn^{2+}), pH, temperature, inorganic salts, and the physiological state of bacterial cells. In general, both divalent cations and physiological state of the host show little or no influence on the rate of adsorption, whereas maximum adsorption is in general attained at the optimal cell growth temperature and pH close to neutrality. Adsorption interference is regarded as a significant mechanism of phage resistance in lactic acid bacteria (LAB), and thus research on this topic is very valuable. These studies are directed to characterize the chemical nature of phage receptors and encompass the use of diverse chemical compounds and enzymes. When receptors are carbohydrate in nature, the use of lectins and either simple or complex saccharides in competition and desorption assays could be very useful. Injection of phage DNA into the cytoplasm comes after irreversible attachment step of adsorption. Unlike adsorption, the DNA injection process was found to be highly dependent on Ca^{2+}. Besides, temperature and physiological state of the host had an important influence as well. During the latent phase of a lytic cycle, phage DNA is replicated and phage genes are transcribed in strict order. Proteins of tailed phages are assembled through separate pathways into heads, tails and fibers. Once phage virions are

* E-mail address: mbriggi@fiq.unl.edu.ar. Tel.: +54 342 4530302; fax: +54 342 4571162

fully assembled, they must escape from the host cell so as to persist in nature. To do this, lytic phages must degrade or at least compromise the peptidoglycan layer of the cell wall. Likewise the rest of tailed bacteriophages, those infecting LAB lyse the host using two phage-encoded factors: a small hydrophobic protein that permeabilizes the membrane (holin), allowing the second one (lysin) to gain access to the cell wall. Genes encoding for either holin or lysin were found to be highly conserved among phages infecting LAB.

INTRODUCTION

Life Cycle: Phage Multiplication

As well as other viruses, phages need to infect specific bacterial cells in order to propagate. Phages that infect and lyse the host cell are known as "virulent/lytic phages", while those that insert their genome into the host bacterial chromosome are known as "temperate phages" (Neve 1996). Either lytic or lysogenic cycles begin with the adsorption of a phage particle to the cell wall surface of a sensitive cell, also called "host cell" (Figure 1).

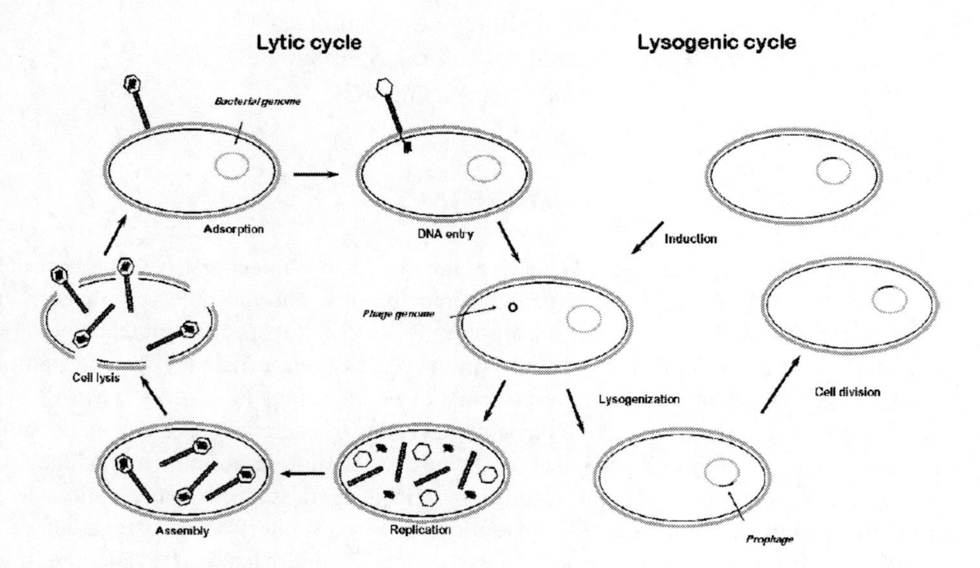

Figure 1. Life cycles of bacteriophages.

Once the phage genome is internalized into the cell, a decision is made; if the phage follows a lytic cycle, it manages the cellular metabolism in order to construct many new phage particles that are released from the host (with or without cell lysis) so as to begin a new infective cycle. Alternatively, the phage could adopt a rather different lifestyle, by inserting its genome within the bacterial chromosome and remaining inactive (as a prophage) over successive cell divisions (lysogenic cycle). However, some circumstances could alter this latent state and a lytic cycle is induced. Equilibrium between lytic and lysogenic lifestyles, as well as importance of lysogeny in probiotic lactobacilli will be comprehensively discussed in Chapter 6. This chapter will deal with the lytic cycle.

1. Phage Adsorption

Adsorption is a highly specific event, and depends on the presence of specific attachment sites known as "receptors" placed on the bacterium's cell wall surface. Phages of lactic acid bacteria (LAB) adsorb to cell wall by their tails, which generally possess additional structures like fibers or spikes that give stability to this union (Neve 1996, Kutter and Goldman 2008). Tailed phages adsorb in a tail-first orientation leaving the empty capsid on the surface of the host bacterium (Ackermann and DuBow 1987).

Adsorption may involve multiple steps, such as *reversible* binding by fibers responsible for initial host recognition and proper baseplate positioning, followed by *irreversible* binding of a different tail protein to a secondary receptor molecule, as seen for coliphage T1 and for the T4-like phages (Kutter et al. 2005). For the lactococcal c2-like phages, the first reversible binding step is to a carbohydrate component of the cell wall, followed by a second irreversible binding step to a 901-amino acids membrane protein named Phage Infection Protein (PIP) (Valyasevi et al. 1991). Some studies suggested an irreversible mechanism for the earlier steps of the infective cycle. With this regard, Quiberoni and co-workers (2004) found that adsorption was an efficient process for all *Lactobacillus delbrueckii* phages on their hosts, as over 89% of phage particles were adsorbed after 30 min of incubation at 42°C. The irreversibility of binding was also demonstrated for Φ kh on *Lactococcus lactis* subsp. *cremoris* KH (Valyasevi et al. 1994) and five *Streptococcus thermophilus* bacteriophages (Binetti et al. 2002). In contrast, the initial adsorption step on the cell wall was a reversible process for seven phages on *L. lactis* subsp. *lactis* C2 (Monteville et al. 1994).

1.1. Influence of Physiological and Environmental Factors

Adsorption velocity and efficiency may vary for a given phage/host system depending on external factors and host physiological state (Guttman et al. 2005). There are some recognized factors affecting the phage adsorption process, such as the presence of divalent ions (Ca^{2+}, Mg^{2+}, Mn^{2+}), the physiological state of bacterial cells, pH and temperature. The inorganic salts probably act by neutralising the net negative charges on the host cell and its phage so that initial contact is facilitated (Mullan 2002). The adsorption of some phages to lactococci has been reported to be reduced by rennet, increased temperature or by growth on a different host. Although the differences were often small, the combination of rennet with elevated temperature (37°C) substantially reduced adsorption for several phages (Mullan 2002). In many cases, cell lysis of LAB can be achieved in absence of calcium ions, but its presence can accelerate the process. This behaviour was reported for phages of *Strep. thermophilus* (Binetti et al. 2002), *Lb. helveticus* (Quiberoni and Reinheimer 1998, Séchaud et al. 1988), *Lb. casei/paracasei* (Capra et al. 2006b) and *Lb. plantarum* (Briggiler Marcó et al. 2010). Similarly, the presence of calcium ions accelerated cell lysis of *Lb. delbrueckii* strains (Quiberoni et al. 2004, Wang et al. 2010). In the same way, calcium was not necessary to complete the infection cycle of phages QF12 and QP4 of *Lactococcus* (Suárez et al. 2008). In contrast, this cation was indispensable for completion of the vegetative cycle of Φ ATCC 15807-B1 of *Lb. helveticus* (Quiberoni and Reinheimer 1998) and for Φ PL-1 of *Lb. casei* (Watanabe et al. 1993). Similar behaviour was reported for Φ YAB on *Lb. delbrueckii* Ab_1

and for Φ Ib_3 on *Lb. delbrueckii* LB-Ib_3 (Quiberoni et al. 2004). Phages QF9 and CHD of *Lactococcus* had an absolute requirement for calcium ions to efficiently complete its lytic cycle in liquid medium (Suárez et al. 2008). Regarding to plaque formation, calcium improved it for phages of *Strep. thermophilus* (Binetti et al. 2002) and *Lb. delbrueckii* (Quiberoni et al. 2004, Wang et al. 2010). Similar results were observed for *Lb. helveticus* Φ ATCC 15807-B1 (Quiberoni and Reinheimer 1998), phages of *Lb. casei/paracasei* (Capra et al. 2006b) and *Lb. plantarum* (Briggiler Marcó et al. 2010). However, the presence of calcium had no effect on plaque size for four phages of *Lactococcus* (Suárez et al. 2008) and for Φ hv on *Lb. helveticus* ATCC 15807 (Quiberoni and Reinheimer 1998).

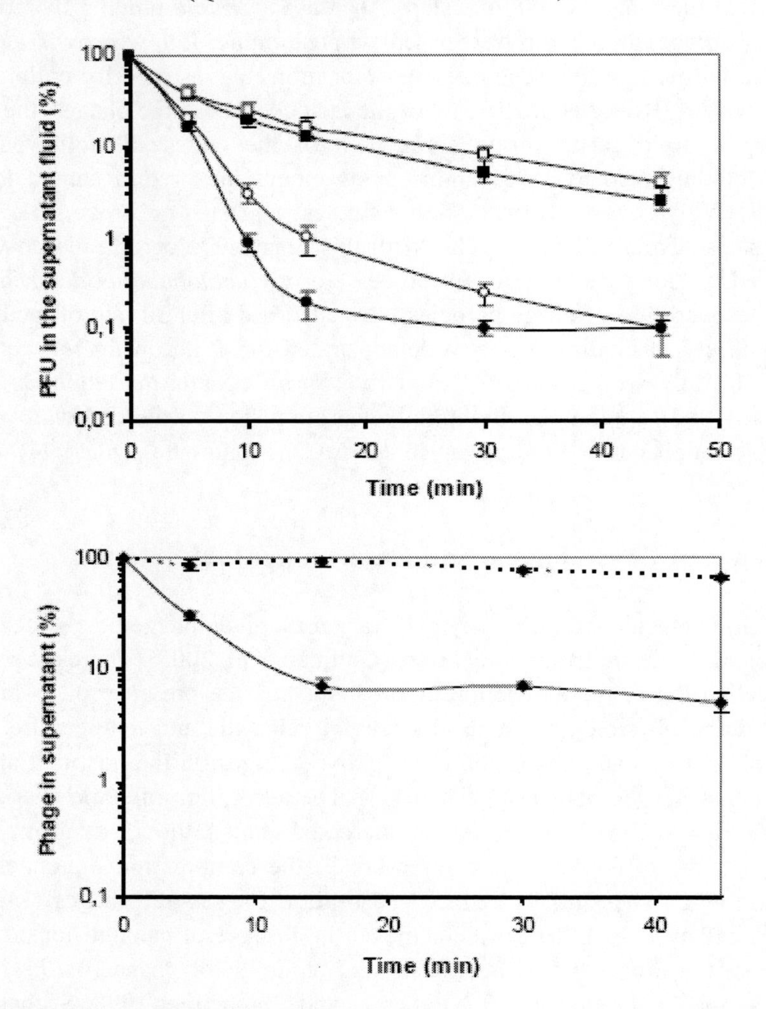

Figure 2. a) Adsorption kinetics of phages hv (•) and ATCC 15807-B1 (■) on *Lactobacillus helveticus* ATCC 15807 in MRS broth with (•, ■) and without (○, □) Ca^{2+} (10 mmol l^{-1}) at 37°C (Data adapted from Quiberoni and Reinheimer 1998). b) Adsorption of phage MLC-A on *Lactobacillus paracasei* A, in MRS broth with (—) and without (- - -) Ca^{2+} (10 mmol l^{-1}) at 37°C (Data adapted from Capra et al. 2006a).

In general, calcium ions were not required to complete phage adsorption on LAB (Figure 2-a). This behaviour was reported for phages CHD, QF9, QF12 and QP4 of *L. lactis* isolated from whey samples in Argentina (Suárez et al. 2008) and for *Lb. helveticus* phages (Quiberoni and Reinheimer 1998, Séchaud et al. 1988). With respect to *Lb. delbrueckii*, Quiberoni et al. (2004) reported that for three phages isolated from yoghurt plants and one collection phage, more than 94% of the initial viral particles were absorbed within 45 min of incubation. Also, calcium ions did not influence the adsorption of Φ phiLdb (*Lb. delbrueckii* subsp. *bulgaricus*) isolated from Chinese yoghurt (Wang et al. 2010). The same behaviour was observed for Argentinian *Strep. thermophilus* phages (Binetti et al. 2002) and for *Lb. plantarum* phages (Briggiler Marcó et al. 2010). Capra et al. (2006b) reported that calcium ions did not influence adsorption of two collection phages of *Lb. casei/paracasei*. However, an influence of calcium ions in the adsorption of Φ MLC-A on *Lb. casei* A was observed (Figure 2-b), since in their absence only 37% of phage particles were adsorbed (Capra et al. 2006a). On the contrary, the adsorption of Φ PL-1 on *Lb. paracasei* A was faster in the absence of calcium (Capra et al. 2006b).

The physiological state of a cell can substantially change the concentration of particular cell surface molecules and thus the efficiency of infection by certain phages (Kutter and Goldman 2008). After cells of *Lb. helveticus* were thermally killed, phages ATCC 15807-B1 and hv successfully adsorbed, demonstrating the thermostable nature of the phage receptors and the independence of this process for the physiological state of the cells (Quiberoni and Reinheimer 1998) (Figure 3-a). Similar behaviour was observed for four phages of *Lb. delbrueckii* (Quiberoni et al. 2004) and for six phages of *Strep. thermophilus* (Binetti et al. 2002). In the same way, a study by electron microscopy study revealed that Φ PL-1 successfully adsorbed on cells of *Lb. casei* which had been killed by keeping cell suspensions in boiling water (Watanabe et al. 1993). For three phages of *Lactococcus* it was observed a slightly higher adsorption rate on native cells than on heat-treated ones (Suárez et al. 2008). In contrast, adsorption values were lower when a thermal treatment was used on *Lb. paracasei/casei* cells. This fact indicates that adsorption depends on the physiological state of the cells and/or that phage receptors might have a thermosensitive nature (Capra et al. 2006a, Capra et al. 2006b) (Figure 3-b). Likewise, adsorption of *Lb. plantarum* phages after 45 min on non-viable cells ranged between 77% and 91%, whereas on viable cells it reached 99% for all phages (Briggiler Marcó et al. 2010).

The diminished phage adsorption observed on thermally treated cells could be linked to a disorganization of phage receptor sites and/or to the cell physiological state. Although phage binding is not expected to be an energy-dependent process, several authors have investigated this issue and obtained dissimilar results. Chloramphenicol (a protein synthesis inhibitor) was used to study this phenomenon in *Lb. plantarum*. Immediately after antibiotic treatment, the non-proliferating cells, which are expected to maintain almost the original ATP content (Hahn et al. 1955, Watanabe et al. 1991), were able to adsorb the phages suggesting that adsorption on this bacterial species is not a energy-dependent process (Briggiler Marcó et al. 2010). When several energy inhibitors were used, it was demonstrated that adsorption of Φ fd to *Escherichia coli* male cells was not energy-dependent (Yamamoto et al. 1980). However, adsorption of Φ PRD1 to *Salmonella typhimurium* strain DS88 was drastically reduced when NaN_3, NaF and arsenate (all inhibitors that considerably decreased the intracellular concentration of ATP) were used (Daugelavicius et al. 1997).

Generally, phage adsorption reached maximum rates at the optimal temperature range for the growth of bacterial cells. With this regard, Quiberoni and co-workers (2004) have observed that even when phages adsorbed on *Lb. delbrueckii* cells at 0°C, adsorption rate reached its maximum at 37°C. Similar behaviour was observed for *Lb. helveticus* (Quiberoni and Reinheimer 1998) and *Strep. thermophilus* phages (Binetti et al. 2002). Sometimes, the effect of the temperature can be highly dependent on the phage/strain system studied.

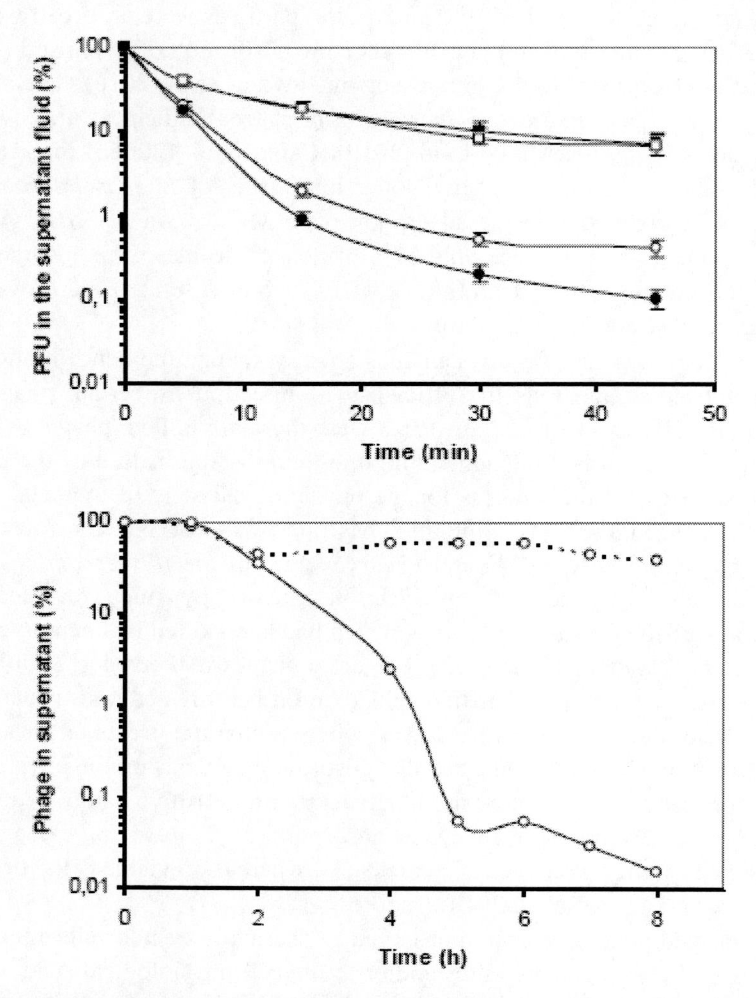

Figure 3. a) Adsorption kinetics of phages hv (•) and ATCC 15807-B1 (■) on viable (•, ■) and non-viable (○, □) cells of *Lactobacillus helveticus* ATCC 15807 at 37°C in MRS-Ca broth (Data adapted from Quiberoni and Reinheimer 1998). b) Influence of cell-thermal treatment on the adsorption kinetics of phage PL-1 on *Lactobacillus paracasei* A in MRS-Ca broth on viable (—) and non-viable (- - -) cells at 37°C (Data adapted from Capra et al. 2006b).

Figure 4 shows the influence of temperature on the adsorption of *Lb. casei/paracasei* phages. Concerning this, Capra and co-workers (2006b) observed that in general, optimal adsorption was achieved at 37°C (Φ PL-1/*Lb. paracasei* A system) or 30°C (Φ J-1 on all its sensitive strains, including *Lb. paracasei* A). In some other cases, phage adsorption can be independent from incubation temperature, for example for Φ PL-1/*Lb. casei* ATCC 393 (Capra et al. 2006b) and Φ PL-1/*Lb. paracasei* ATCC 27092 systems (Watanabe and Takesue 1972), which maintained the adsorption rates along the temperature range studied. Watanabe et al. (1982) showed that, although the initial decrease in the amount of free state phages (reversible adsorption) was not affected by lowering the reaction temperature during the adsorption process, the formation of irreversibly bound phage-cell complexes was in this case inhibited.

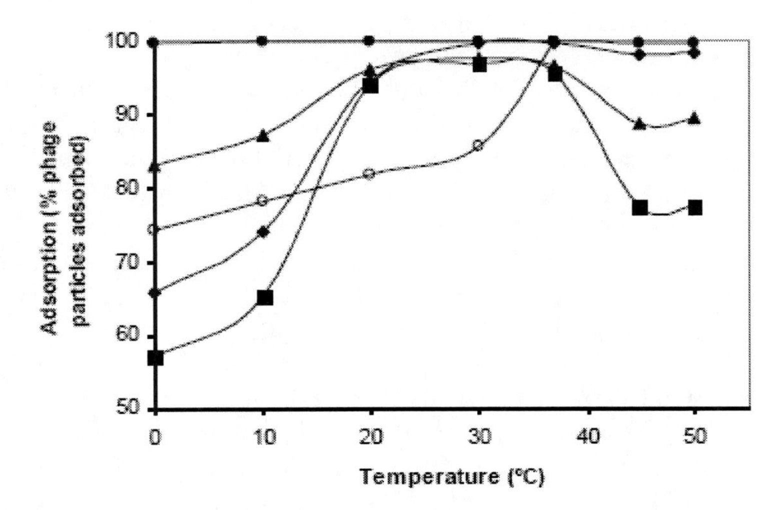

Figure 4. Influence of temperature on the adsorption (after 30 min) of phage J-1 on Lactobacillus casei ATCC 27139 (♦), Lb. casei ATCC 393 (▲) and Lb. paracasei A (■), and phage PL-1 on Lb. casei ATCC 393 (•), and (after 6 h) phage PL-1 on Lb. paracasei A (○) in MRS-Ca broth (Data adapted from Capra et al. 2006b).

The pH of the medium can also affect the adsorption of phages infecting LAB. In general, the highest adsorption rates were found between pH 5 and 7 (Binetti et al. 2002, Capra et al. 2006a, 2006b, Suárez et al. 2008, Briggiler Marcó et al. 2010). It is noteworthy that maximum adsorption rates were achieved at the pH of the milk, near neutrality. Most phages showed to be sensitive to extreme acidic or alkaline conditions. This could be consequence of the loss of phage viability, since it has been shown that either at pH 3 or lower, or at pH 10 or higher, the number of viable phage particles was notably reduced (Auad et al. 1997, Binetti et al. 2002, Quiberoni et al. 2004, Capra et al. 2006b), thus affecting adsorption rates. Nonetheless, some phages showed to be particularly resistant to extreme pH conditions, as they exhibited high infectivity rates even at pH 4 and pH 11 (Briggiler Marcó et al. 2010).

1.2. Phage Receptors

Phage receptors are quite different on Gram-negative bacteria (a variety of lipopolysaccharide –LPS– components and outer membrane proteins), and on Gram-positive bacteria (peptidoglycan elements, embedded teichoic acids and lipoteichoic acids, and associated proteins). Components of any secreted external matrix (polysaccharide K, S-layers) can also be initial recognition targets. The specific phage adhesions are often highly complex and versatile, processes that can involve quite different receptors for different bacteria. The adhesion regions are the most variable parts of phage genome (Kutter et al.

2005). In Gram-negative bacteria many proteins, oligosaccharides and lipopolysaccarides can act as receptors for some phages. The more complex murein of Gram-positive bacteria offers a very different set of potential binding sites. Many phages require clusters of one specific kind of molecule present in high concentration to properly position the phage tail for surface penetration (Guttman et al. 2005). Considering the significance of phage resistance in LAB by mechanisms that interfere with phage adsorption, research in order to characterize and identify the exact nature and accessibility of the receptor sites is very valuable. Depending on the sensitive strain/phage system, several methodologies to investigate the nature of phage receptors are available to be applied. Firstly, to identify phage receptors in LAB, whole bacterial cells or purified cell walls can be used. When purified cell walls of *Strep. thermophilus* (Quiberoni et al. 2000, Binetti et al. 2002), *Lb. delbrueckii* (Quiberoni et al. 2004), *Lb. casei* (Capra et al. 2009) and *Lb. plantarum* (Douglas and Wolin 1971, Briggiler Marcó et al. 2011) were treated with several chemical and enzymatic compounds to see if they had an effect on phage adsorption, phage receptors were revealed as carbohydrate in nature. In the same way, phage receptors located on the cell wall polysaccharides of several lactobacilli were reported (Sijstma et al. 1988, Schäfer et al. 1991, Gopal and Reilly 1995, Valyasevi et al. 1994). Also, certain *L. lactis* phages have been described to first adsorb to cell wall carbohydrates and then to a plasma membrane protein called PIP, which leads to injection of phage DNA (Monteville et al. 1994). After comparing the cell wall composition of a phage sensitive wild type *Lb. plantarum* strain with those of their phage resistant mutant counterparts, it was demonstrated that glycosylated cell wall teichoic acids were required for adsorption (Douglas and Wolin 1971). On the contrary, phage receptors located in the membrane protein fraction were reported for *L. lactis* subsp. *lactis* (Valyasevi et al., 1991, Arendt et al. 1993) and for *Lb. helveticus* (Ventura et al. 1999). When saccharides were used at a final concentration of 500 mM to identify the specific compounds involved in phage receptors, extremely variable results were reported for *Lb. helveticus* (Quiberoni and Reinheimer 1998), *Lb. delbrueckii* (Quiberoni et al. 2004), *Lb. casei* (Capra et al. 2009) and *Lb. plantarum* (Briggiler Marcó et al. 2011). However, it was reported that *Strep. thermophilus* phages were inactivated by glucosamine, N-acetylglucosamine, rhamnose and ribose (Quiberoni et al. 2000, Binetti et al. 2002), whereas D-galactosamine and rhamnose impeded Φ J-1 to infect *Lb. casei* cells (Yokokura 1977). Moreover, lipoteichoic acids (LTAs) produced phage inactivation in *Lb. delbrueckii* (Räisänen et al. 2004, Räisänen et al. 2007). In few cases bacterial genes required for phage adsorption have been identified by insertional mutagenesis or complementation of a defective receptor gene in phage resistant mutants (Geller et al. 1993, Dupont et al. 2004). Assays of competition and desorption with saccharides can be carried out to complement the results obtained from inactivation experiments. Using these methodologies, it was possible to demonstrate that L-rhamnose was a constituent of phage receptors in *Lb. casei* (Watanabe and Takesue 1975, Capra et al. 2009). The same saccharide produced also an inactivation of 99% of Φ kh on *L. lactis* subsp. *cremoris* KH (Valyasevi et al. 1990). Rhamnose inhibited adsorption of seven phages on *L. lactis* subsp. *lactis* C2 (Monteville et al. 1994). Valyasevi et al. (1994) reported that the Φ sk1 directly adsorbed to rhamnose and a glucose moiety on *L. lactis* subsp. *lactis* C2 cells. However, it was not possible to obtain conclusive results for *Lb. plantarum* (Briggiler Marcó et al. 2011). Assays with saccharides added to mid-exponential-phase cultures demonstrated that Φ sk1 stopped cell growth of *L. lactis* subsp. *lactis* C2 within 5 h when rhamnose was used (Valyasevi et al. 1994). Rhamnose also prevented Φ kh infection on *L. lactis* subsp.

cremoris KH (Valyasevi et al. 1990). Also, if phage receptors are carbohydrate in nature, lectins (1 mg/ml) can be used to identify the specific saccharides involved, as these compounds may block phage adsorption if their binding sugar and phage receptor sites are identical or in proximity (Valyasevi et al. 1994). Sijtsma et al. (1988) reported that phage adsorption to *L. lactis* subsp. *cremoris* SK112 was reduced when cells were incubated with concanavalin A (ConA, specific for glucose) prior to the addition of phages. Similarly, two lectins with the same galactose specificity but with different molecular weight showed different inactivation levels on *L. lactis* subsp. *cremoris* KH (Valyasevi et al. 1994). Ishibashi et al. (1980) reported that the most effective inhibitor of adsorption of Φ PL-1 on *Lb. casei* strain ATCC 27092 was a lectin (hemagglutinin SHA) from *Streptomyces* that is specific for L-rhamnose. On the other hand, working with lectins it was observed that SHA and ConA were able to generate an immediate desorption of the previously adsorbed phages when they were added 15 min after adsorption. Also, phenyl galactoside, a competitive inhibitor of SHA, was assayed in order to test if the inhibitory action of lectins on phage adsorption was reversible. When phenyl galactoside was added to the mixture of cells, SHA and phage, a partial reversion of phage adsorption was observed (Ishibashi et al. 1982). On the contrary, no conclusive results using lectins were obtained for *Lb. plantarum* (Briggiler Marcó et al. 2011).

2. Injection/Penetration of Phage DNA

After the irreversible attachment step of adsorption, the phage genome passes through the tail into the host cell, while the empty or "ghost" viral particle remains on the cell surface (Neve 1996). This process was traditionally described as *injection*. However, it is not actually an injection process because involves mechanisms of DNA transfer specific for each phage. In general, the tail tip uses an enzymatic mechanism for penetrating the peptidoglycan layer and then touching or penetrating the inner membrane to eventually release the DNA directly into the cell; the binding of the tail also releases a mechanism that has been blocking the exit of phage DNA from the capsid until properly positioned on a potential host (Kutter and Goldman 2008). The permeability of the cell membrane is changed during viral entry. This could be observed as simultaneous K^+ efflux and Ca^{2+} or Mg^{2+} influx during the early stage of Φ LL-H (*Lb. delbrueckii*) infection (Räisänen 2007).

The influence of calcium ions, ATP, some protein-synthesis inhibitors (chloramphenicol and erythromycin), temperature and cell physiological state on the process of genome injection of Φ PL-1 in *Lb. casei* ATCC 27092 was studied. This mechanism was dependent on calcium ions and a high intracellular concentration of high-energy compounds such as ATP (Watanabe and Takesue 1972, Watanabe et al. 1979). Also, both chloramphenicol and erythromycin inhibited the injection of PL-1 DNA. The inhibitory effect of chloramphenicol on DNA injection was observed only when it was added before DNA injection. The authors suggested that proteins synthesized during the early stages of infection and prior to complete DNA transfer itself are required to transfer DNA from Φ PL-1 to its host strain (Watanabe et al. 1991). When cells of *Lb. casei* ATCC 27092 were infected by Φ PL-1 at 0°C, the relative number of ghost particles among the adsorbed phages did not increase during the incubation period. The same behaviour was observed when cell suspensions were inactivated by incubation in boiling water for 5 min or by exposing them to UV light. These results demonstrated that the process of DNA ejection was temperature-dependent and required a living host cell, suggesting that an enzymatic reaction using energy is required for the process (Watanabe et al. 1987).

Poranen et al. (2002) proposed that the energy required to drawn the DNA into the cell may come from conformational changes of the virion structure due to a receptor binding, or that the energy could be captured by the virion in the form of an internal pressure as a viral genome is packaged into a capsid. The velocity of phage DNA transport is variable. Slow DNA transport could be connected to transcription of phage genes or translation of certain phage proteins. Different mechanisms have been suggested to explain how phage DNA is transported through membranes (Räisänen 2007).

3. Synthesis of Viral Components

Once inside the cell, phage DNA is potentially susceptible to host exonucleases and restriction enzymes. Nevertheless, phages utilize some mechanisms to surpass bacterium defences, as circularizing their DNA rapidly by means of sticky ends or terminal redundancies, having the linear ends protected, inhibiting host nucleases or using an odd nucleotide for protection (Kutter and Goldman 2008).

Phage DNA replication often leads to accumulation of head-to-tail arrays of phage genomes. The host's bacterial DNA is often subject to degradation during the course of the phage infection. The consequence of these two phenomena is that capsids of many phages occasionally package pieces of host DNA instead of phage DNA, making a transducing particle (i.e., a vehicle that transmits a piece of DNA from one bacterium to another). This is because the phage particle containing the piece of bacterial DNA can attach to an uninfected bacterium and have its DNA content taken up by the recipient cell. If the introduced donor DNA recombines with the chromosome of the recipient cell, then genes carried on the donor DNA will be incorporated into that recipient. Originally discovered in 1952 by Zinder and Lederberg this process is called generalized *transduction*, because any portion of the genome can, in principle, be packaged and transferred to a recipient cell. In practice, some phages are much better than others in carrying out transduction, which is a consequence of the characteristics of the specific phage life cycle (Kutter and Goldman 2008).

During the latent phase of a lytic cycle, phage DNA is replicated and the phage genes are transcribed in strict order. Proteins of tailed phages are assembled through separate pathways into heads, tails and fibers (Ackermann and Dubow 1987).

4. Assembling of the New Phage Particles

The DNA is packaged into preassembled icosahedral protein shells called procapsids. In most phages, their assembly involves complex interactions between specific scaffolding proteins and the major head structural proteins, followed by proteolytic cleavage of both the scaffolding and the N-terminus of the main head proteins. Before or during packaging, the head expands and becomes more stable, with increased internal volume for allocating the

DNA. Located at one vertex of the head there is a portal complex that serves as the starting point for head assembly, the docking site for the DNA packaging enzymes, a conduit for the passage of DNA and, for myoviruses and siphoviruses, a binding site for the phage tail, which is assembled separately (Räisänen 2007). The so-called concatemeric DNA is cut by heedful cutting during the packaging process into DNA-free proheads. All cuts occur either precisely at *cos*-sites, or the first cut occurs at the packaging sequence *pac* and then the cutting process continues by imprecise heedful measuring leading to terminal redundancy (Black 1989). Phage heads and tails are joined together to finish phage maturation and phages are released due to cell bursts or lysis as a result of cell wall degradation by a phage lysozyme (Ackermann and DuBow 1987).

5. Cell Lysis

5.1. General Aspects

During the final steps of phage infection cycle, once phage virions are assembled, they must escape from the host cell in order to infect another cell so as to keep the infection and persist in nature. Virulent bacteriophages are known to liberate their progeny from their host by two different ways. The simplest mechanism corresponds to the family *Inoviridae*: filamentous phages that are continuously extruded from the host by means of a sophisticated secretion apparatus, without cell lysis. Among this family, M13, fd and f1 (all infecting *E. coli*) are the best characterized (Russel et al. 1997). Filamentous phages contain single-stranded DNA in the genome and infect only Gram-negative bacteria. As they escape the objective of this chapter, they will not be further discussed.

Alternatively, non-filamentous lytic phages abruptly finish the infection with the lysis of the cell. However, this is not an easy task to achieve, especially because of the presence of a protective rigid wall that impedes cell rupture in low-osmolarity media. This cell wall is a huge sac surrounding the cell formed by chains of alternating sugars –N-acetylglucosamine (NAG) and N-acetyl muramic acid (NAM)– cross-linked by short peptides thus forming a two-dimensional sheet, and is the site of attack of β-lactamic antibiotics. The cell wall of Gram-positive bacteria is much thicker than that of Gram-negative bacteria and contains characteristic components such as teichoic acids, lipoteichoic acids and glycolipids (Vasala et al. 1995). Lytic phages must degrade or at least compromise this peptidoglycan layer to achieve lysis, and could do it following two diverse general mechanisms (Young et al. 2000). The first one is limited to the families *Microviridae*, *Alloleviviridae* and *Leviviridae*, several simple phages of Gram-negative hosts having a single gene responsible of lysis. Three completely different types of single lysis genes are known, for which protein E of the (ss)-DNA φX174 (*Microviridae*), protein L of the (ss)-RNA Φ MS2 (*Leviviridae*) and protein A_2 of the (ss)-RNA Φ Qβ (*Alloleviviridae*) represent the prototypes (Bernhardt et al. 2000, Young et al. 2000). No muralytic enzyme activity was detected in the supernatants after infection with these phages, and according to sequence analysis these lysis genes would not encode for an enzyme with such an activity. Alternatively, gene φX174 *E* was postulated to act as a "protein antibiotic" by inhibiting murein synthesis (Bernhardt et al. 2000).

5.2. Lysis in Lactic Acid Bacteria

Tailed bacteriophages with double stranded (ds) DNA, including all known bacteriophages infecting LAB, use another strategy for lysing involving two factors: an endolysin and a holin. The former is a soluble phage-encoded protein responsible for the degradation of the peptidoglycan layer and thus the rupture of the cell wall. Although endolysin is a generic name applied to any phage enzyme able to degrade the cell wall and hence promoting cell lysis, they can be subsequently classified, depending on the chemical bond they cleave, into (i) lysozyme-like enzymes (muramidases), which hydrolyze N-acetylmuramyl-1,4-b-N-acetylglucosamine bonds, (ii) lytic transglycosylases, (iii) N-Acetylmuramyl-L-alanine amidases, which hydrolyze the bond between N-acetylmuramic acid and L-alanine, and (iv) endopeptidases that hydrolyze the peptidoglycan stem or bridge peptides (Vasala et al. 1995). It was postulated that teichoic acids or lipoteichoic acids contribute to the regulation of the activity of bacterial autolysins (Fischer et al. 1981, Cleveland et al. 1975). Although all lysins degrade the peptidoglycan layer, teichoic acids or lipoteichoic acids substantially vary among the cell walls of different bacterial species. This fact could explain the strict host specificity generally observed for endolysins. Indeed, they could be active on Gram-negative bacteria but completely ineffective on Gram-positive bacteria and, besides, it is difficult to find a lytic enzyme with a broad spectrum among Gram-positive bacterial species. When endolysins are externally applied as purified recombinant proteins, they can access and degrade the cell wall of Gram-positive bacteria from outside, thus acting as exolysins (Borysowski et al. 2006). Consequently, the availability of pure lytic proteins in dairy biotechnology serves as an efficient approach either to reduce the activity of specific contaminants or to induce the release of intracellular enzymes with desired activity during and after dairy fermentations (Vasala et al. 1995). In this sense, they were claimed to be helpful also in the treatment of systemic infections caused by Gram-positive bacteria pathogens, such as *Strep. pneumoniae*, *Strep. pyogenes* and *Staphylococcus aureus*, with several advantages over classical antibiotics: higher specificity, low probability of resistance development, high activity, rapid lysis efficiency and retaining of activity after antibody neutralization (Wang and Lu 2009).

Interestingly, most phage lysins lack a secretory signal sequence and must be assisted by a helper lysis factor: the holin. Holins comprise one of biology's most diverse classes of functional homologs. They are small hydrophobic phage-encoded proteins that inserts into the membrane, producing a permeabilization that allows the endolysin to reach the cell wall and gain access to the murein. Holins can be grouped into two main classes: class I, larger and with three potential transmembrane domains, and class II, usually smaller and having two transmembrane domains (Young et al. 2000). Holins could either accelerate or delay the lysis taking into consideration physiological and environmental conditions. In fact, a lysis carried out too fast will yield a low number of new virions unable to maintain the infection cycle, but a large delay would instead reduce the opportunities for new explosive infection cycles (Abedon 1990, Abedon 2005). A further fine regulation of lysis timing is provided by complementary proteins acting as holin inhibitors. Lastly, two other auxiliary lysis proteins named *Rz* and *Rz1* (and their equivalents) have also been studied, but *Rz1* encodes an outer membrane lipoprotein, and it and its *Rz* companion have never been described in phages of Gram-positive hosts. Although the function of these proteins is not clear, they appear to be involved in the link between the cell wall and the outer membrane (Young et al. 2000, Summer et al. 2007). Given the advantages provided by this multiple-component system, it is alleged to be the result of a more sophisticated evolution process in (ds)DNA phages.

Although phage-mediated cell lysis has not been deeply studied in LAB, there are several reports of predicted lysins for the *Lactobacillus* phages φadh (Henrich et al. 1995), LL-H (Vasala et al. 1995), mv1 (Boizet et al. 1990), mv4 (Dupont et al. 1993), PL-1 (Kashige et al. 2000) and φg1e (Oki et al. 1996), for the *Lactococcus* phages φLC3 (Birkeland 1994), Tuc2009 (Arendt et al. 1994), φvML3 (Shearman et al. 1994), and r1t (van Sinderen et al. 1996), for the *Streptococcus* phages Cp-1 (García et al. 1990), ALQ13.2 and φAbc2 (Guglielmotti et al. 2009), for the *Leuconostoc* φ1-A4 (Lu et al. 2010) and for the *Oenococcus* phages fOg44 (Parreira et al. 1999) and φ10MC (Gindreau and Lonvaud-Funel 1999).

Only two of the four known classes of lysins were found in phages of *Lactobacillus*: phages LL-H, φadh and mv1 encoding a muramidase-like endolysin (Henrich et al. 1995, Vasala et al. 1995, Boizet et al. 1990) and Φ PL-1 encoding an amidase-like endolysin (Kashige et al. 2000). The concomitant expression of holin and lysin of φadh did not cause lysis when expressed on *E. coli*, which was unexpected since either of them used separately could effectively lyse *E. coli* when expressed along with their counterpart of Φ lambda or *Bacillus subtilis* φ29 (Henrich et al. 1995).

Muramidase Mur of Φ LL-H was found to have a relatively ample host spectrum activity, as it was active on the cell wall of *Lb. delbrueckii* and taxonomically related species such as *Lb. delbrueckii* subsp. *lactis*, *Lb. delbrueckii* subsp. *bulgaricus*, *Lb. acidophilus* and *Lb. helveticus*, but also on *Pediococcus damnosus* cell walls. However, it was ineffective against *Lb. casei* strain 1/3 (Vasala et al. 1995). Other broad-spectrum lytic proteins isolated from phages of LAB are Mur-LH and Lys_{gaY}. The former is an N-acetylmuramidase isolated from the temperate φ-0303 of *Lb. helveticus* CNRZ 303 strain and active mainly on thermophilic lactobacilli but also on lactococci, pediococci, *B. subtilis*, *Brevibacterium linens*, and *Enterococcus faecium* (Deutsch et al. 2004). The latter, purified from the *Lb. gasseri* φgaY, was capable of lysing many Gram-positive bacterial species, including lactobacilli, lactococci, enterococci, micrococci, and staphylococci (Sugahara et al. 2007).

Surprisingly, experimental results obtained by São-José et al. (2000) evidenced the presence of a signal peptide in the *Oenococcus oeni* Φ fOg44 lysin (Lys44), which would therefore be an exolysin. This constituted the first found evidence of this kind for a bacteriophage lysin. Furthermore, database searches and protein sequence alignments supported the theoretical presence of secretory lysins in other known *O. oeni* and *L. lactis* phages. Later, Kakikawa et al. (2002) discovered the presence of a signal-peptide-like domain also in the N-terminal 26-amino acids region of *Lb. plantarum* φg1e Lys precursor, which is post-translationally processed as a signal sequence rendering a mature-Lys with noticeably higher lytic activity. The authors also found that maturation and translocation of φg1e Lys across the cytoplasmic membrane relies on the secretion system of the bacterial host rather than on membrane lesions provoked by holin.

From the comparative genomic analyses of diverse phages infecting *Strep. thermophilus*, it was observed that *cos*-type phages DT1 (Tremblay and Moineau 1999), Sfi19 (Lucchini et al. 1999), Sfi21 (Bruttin et al. 1997), and 7201 (Stanley et al. 2000), as well as *pac*-type phages O1205 (Stanley et al. 1997) and Sfi11 (Lucchini et al. 1998), share extensive DNA sequence similarity in the replication module and lysis cassette. With respect to the sequence of *Strep. thermophilus* Φ 2972 (Lévesque et al. 2005), though the genome region containing the genes necessary for host cell lysis was found to be highly conserved with respect to previously sequenced *Strep. thermophilus* phages, one exception was the putative holin-encoding gene (ORF25). The most recent sequence data reported for phages of *Strep. thermophilus* correspond to ALQ13.2 and φAbc2, two lytic phages isolated from dairy plants in Argentina (Guglielmotti et al. 2009). Interestingly, genes encoding lysin were highly conserved with respect to all previously sequenced *Strep. thermophilus* phages, despite their geographically distant origin. Phage φ1-A4, infecting for *Leuconostoc mesenteroides* and recently isolated from an industrial vegetable fermentation (Lu et al. 2010), showed a large-scale (30%) genome inversion affecting the lysis cassette, which is embedded in the tail morphogenesis module, making it unique among lytic phages infecting related species of LAB.

5.3. Growth Curve of Phages

There are certain kinetic parameters of phage infection which are characteristic of each phage-host system and can be determined studying the one-step growth curve. In this experiment, phages added to a mid-exponential culture are allowed to adsorb during 10-30 min approximately. Then, cells are harvested by centrifugation, resuspended and diluted in growth medium. At intervals, bacteriophage counts are done (Capra et al. 2006a). At the beginning of the experiment, the number of lysis plaques remains constant because they are formed only by phage released later by infected unlysed cells. This phase is called the *latent period*. Afterwards, the number of lysis plaques increases provided that infected cells have lysed (rise time). When all infected cells have lysed, phage concentration remains constant. The ratio of phage produced to the initial number of infective centers is defined as the *burst size*, and represents a measure of the efficiency of phage production (Maloy et al. 1994), whereas the time before the increase in the number of lysis plaques occurs is the lysis time (or burst time).

Figure 5 shows the growth-step curve for Φ MLC-A on *Lb. casei* A. For this system, the latent and burst periods were estimated in 30 min and 135 min respectively, whereas the burst size was 69 ± 4 PFU/infective center (Capra et al. 2006a). Suárez et al. (2002) found large burst size values and remarkably short latent periods for lytic phages infecting *Strep. thermophilus* isolated in dairy plants in Argentina. Likewise, recently sequenced *Strep. thermophilus* phages ALQ13.2 and φAbc2 showed high burst size values (Guglielmotti et al. 2009). Excepting phages fri (*Lb. plantarum*) and 328-B1 (*Lb. casei*), *Strep. thermophilus* phages showed burst size values higher than other LAB (Table 1).

Analysis based upon animal foraging theory showed that low host density selects for long latent periods and vice versa, whereas the better the conditions of the hosts are, the shorter the latent period will be, despite the same host densities (Wang et al. 1996).

Table 1. Kinetic parameters of the phage infection in LAB

Specie	Phage	Sensitive strain	Latent period (min)	Rise period (min)	Burst time (min)	Burst size (PFU infective center^{-1})	Source	Reference
Lactococcus	c6A	C6	25	-	-	124 ± 8	-	Powell and Davidson 1985
	1706	-	85	-	-	160	French cheese production	Garneau et al. 2008
	asccΦ28	ASCC99-319	44	-	-	121 ± 18	Australian whey sample	Kotsonis et al. 2008
	Tk1	*L. lactis* subsp. *diacetylactis*	15 ± 1.0	25 ± 3.0	-	80 - 85	Dahi whey	Masud et al. 2009
	RK5		20 ± 3.0	20 ± 2.65	-	50 - 70		
	KS1		10 ± 2.64	20 ± 4.65	-	90 - 100		
	KS2		15 ± 3.60	20 ± 1.73	-	65 -75		
	1358	*L. lactis* SMQ-388	90 ± 1.3	-	-	72 ± 2	Cheese whey	Dupuis and Moineau 2010
		L. lactis SMQ-382	90 ± 3	-	-	77 ± 36		
Streptococcus thermophilus	DT1	SMQ-301	25	-	-	276 ± 36	Mozzarella whey	Tremblay and Moineau 1999
	ALQ13.2	ST13.2	30-40	-	165	370	Cremoso cheese	Guglielmotti et al. 2009
	φAbc2	Abc2	30-40	-	170	350		
Lactobacillus delbrueckii	BYM	*Lb. delbrueckii* subsp. *bulgaricus* YSD V	<40	-	80	23	Yoghurt plant	Quiberoni et al. 2004, Guglielmotti et al. 2006
	Ib3	*Lb. delbrueckii* subsp. *lactis* Ib3	<40	-	80	27		
	YAB	*Lb. delbrueckii* subsp. *lactis* Ab1	<40	-	60	48		
	phiLdb	*Lb. delbrueckii* subsp. *bulgaricus* ATCC 11842	45	-	75	56 ± 2	Chinese yoghurt sample	Wang et al. 2010

Table 1. (Continued)

Specie	Phage	Sensitive strain	Latent period (min)	Rise period (min)	Burst time (min)	Burst size (PFU infective center^{-1})	Source	Reference
Lactobacillus helveticus	328-B1	CNRZ 328	94	86	-	87	French Emmenthal cheese whey	de los Reyes Gavilán et al. 1990
		CNRZ 1095	97	93	-	33		
Lactobacillus plantarum	B2	ATCC 8014	75	-	-	12-14	Anaerobic sewage sludge	Nes et al. 1988
	fri	A	75	75	-	200	Commercial meat starter culture	Trevors et al. 1983
	JL-1	MU45	35	40	-	22	Commercial cucumber fermentation	Lu et al. 2003
Lactobacillus casei/paracasei	393-A2	*Lb. casei* ATCC 393	140	30	-	180-200	Artisanal cheese whey	Herrero et al. 1994
	MLC-A	*Lb. paracasei* A	30	-	135	69 ± 4	Probiotic dairy product	Capra et al. 2006a
	J-1	*Lb. casei* ATCC 27139	45	-	240	160	Yakult	Capra et al. 2006b
	J-1	*Lb. paracasei* A	60	-	225	160		
	J-1	*Lb. casei* ATCC 393	45	-	145	35		
	PL-1	*Lb. casei* ATCC 393	90	-	230	80	Yakult	
	PL-1	*Lb. paracasei* A	5	-	120	32		
	CL1	*Lb. paracasei* A	30	-	120	148	Temperate phages	Capra et al. 2010
	CL2	*Lb. paracasei* A	30	-	120	85		

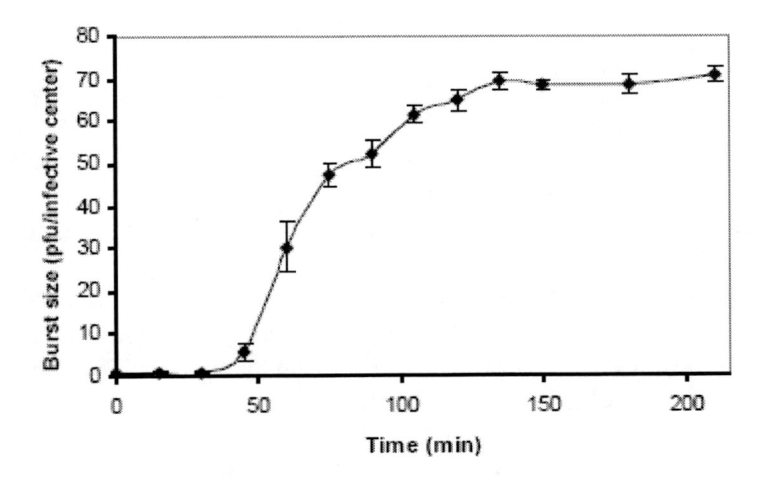

Figure 5. One-step growth curve of phage MLC-A on *Lactobacillus casei* A (Data adapted from Capra et al. 2006a).

CONCLUSION

Once a phage has infected a bacterial cell, it could follow either a lytic or a lysogenic cycle. The former involves common steps to all phage/host systems, which can be basically divided into adsorption, injection of nucleic acids into the cytoplasm, synthesis of viral components, assembling of new infective viral particles and cell lysis. However, extrinsic or intrinsic factors influencing each step can markedly vary from one system to another. Adsorption was found to be decisive on the specificity for a phage to infect a particular bacterial strain. Presence of divalent cations, pH and temperature, as well as physiological state of the host, showed an enormous influence on the adsorption process of LAB phages. The internalization of DNA is an enzymatic process for which different mechanisms have been proposed. It entails sophisticated mechanisms, uses calcium and high-energy compounds such as ATP, and is also specific for each phage. Several mechanisms permit phage DNA to survive in the interior of the host, such as genome circularization and inhibition of host endonucleases.

After synthesis of viral components, new phage particles are assembled and escape from the cell to begin a new cycle. All known LAB phages cause the lysis of the cell by means of a refined mechanism using two classes of proteins: holins and endolysins; the study of these proteins at molecular level has been of great value to determine the degree of relatedness among different LAB phages.

Finally, the velocity at which the entire process takes place, as well as the number of virions released per infected cell, are also typical features of every phage/host system studied. Research focused on the factors influencing each mentioned step is essential to develop anti-phage strategies directly applicable on dairy industry.

REFERENCES

Abedon, S. T. (1990). Selection for lysis inhibition in bacteriophage. *Journal Theoretical Biology*, *146*, 501-511.

Abedon, S. T. (2005). Phage Ecology. In S. T. Abedon, and R. L. Calendar (Eds.), *The Bacteriophages* (second edition, pp. 37-48). New York, NY: Oxford University Press.

Ackermann, H.-W., and DuBow, M. S. (1987). *Viruses of Prokaryotes* 1 and 2. Boca Raton, FL: CRC Press.

Arendt, E. K., Daly, C., Fitzgerald, G. F. and van de Guchte, M. (1994). Molecular characterization of lactococcal bacteriophage Tuc2009 and identification and analysis of genes encoding lysin, a putative holin, and two structural proteins. *Applied and Environmental Microbiology*, *60*, 1875-1883.

Arendt, E. K., van de Guchte, M., Coffey, A. G., Daly, C. and Fitzgerald, G. F. (1993). Molecular genectis of bacteriophages of lactic acid bacteria. *Lait*, *73*, 1901-1198.

Auad, L., Ruiz Holgado, A. A. P., Forsman, P., Alatossava, T. and Raya, R. R. (1997). Isolation and characterization of a new *Lactobacillus delbrueckii* ssp. *bulgaricus* temperate bacteriophage. *Journal of Dairy Science*, *80*, 2706-2712.

Bernhardt, T. G., Roof, W. D. and Young, R. (2000). Genetic evidence that the bacteriophage phi X174 lysis protein inhibits cell wall synthesis. *Proceeding of the National Academy of Science USA 97*, 4293-4302.

Binetti, A., Quiberoni, A. and Reinheimer, J. (2002). Phage adsorption to *Streptococcus thermophilus*: Influence of environmental factors and characterization of cell-receptors. *Food Research International*, *35*, 73-83.

Birkeland, N. K. (1994). Cloning, molecular characterization, and expression of the genes encoding the lytic functions of lactococcal bacteriophage φLC3: a dual lysis system of modular design. *Canadian Journal of Microbiology*, *40*, 658-665.

Black, L. W. (1989). DNA packaging in dsDNA bacteriophages. *Annual Review of Microbiology*, *43*, 267-292.

Boizet, B., Lahbib-Mansais, Y., Dupont, L., Ritzenthaler, P. and Mala, M. (1990). Cloning, expression and sequence analysis of an endolysin encoding gene of *Lactobacillus bulgaricus* bacteriophage mv1. *Gene*, *94*, 61-67.

Borysowski, J., Weber-Dabrowska, B. and Górski, A. (2006). Bacteriophage endolysins as a novel class of antibacterial agents. *Experimental Biology and Medice (Maywood)*, *231*, 366-377.

Briggiler Marcó, M., Reinheimer, J. A. and Quiberoni, A. (2010). Phage adsorption to *Lactobacillus plantarum*: Influence of physiological and environmental factors. *International Journal of Food Microbiology*, *138*, 270-275.

Briggiler Marcó, M., Reinheimer, J., and Quiberoni, A. (2011). Characterization of phage receptors in Lactic Acid Bacteria. In: D. A. Medina, and A. M. Laine (Eds.), *Food Quality: Control, Analysis and Consumer Concerns*. New York, NY: Nova Science Publishers, Inc. (in press).

Bruttin, A., Desiere, F., Lucchini, S., Foley, S. and Brüssow. H. (1997). Characterization of the lysogeny DNA module from the temperate *Streptococcus thermophilus* bacteriophage Sfi21. *Virology*, *233*, 136-148.

Capra, M. L., Mercanti, D., Reinheimer, J. and Quiberoni, A. (2010). Characterisation of three temperate phages released from the same *Lactobacillus paracasei* commercial strain. *International Journal of Dairy Technology*, *63*, 396-405.

Capra, M. L., Quiberoni, A., Ackermann, H.-W., Moineau, S. and Reinheimer, J. A. (2006a). Characterization of a new virulent phage (MLC-A) of *Lactobacillus paracasei. Journal of Dairy Science*, *89*, 2414-2423.

Capra, M. L., Quiberoni, A. and Reinheimer, J. (2006b). Phages of *Lactobacillus casei / paracasei*: response to environmental factors and interaction with collection and commercial strains. *Journal of Applied Microbiology*, *100*, 334-342.

Capra, M. L., Quiberoni, A., and Reinheimer, J. A. (2009). Characterization of receptor sites for bacteriophages PL-1, J-1 and MLC-A using two strains of *Lactobacillus casei.* In H. T. Adams (Ed.), *Contemporary trends in bacteriophage research* (pp. 391-406). New York, NY: Nova Science Publishers, Inc.

Cleveland, R. F., Höltje, J.-V., Wicken, A. J., Tomasz, A., Daneo-Moore, L. and Shockman, G. D. (1975). Inhibition of bacterial wall lysins by lipoteichoic acids and related compounds. *Biochemical and Biophysical Research Communications*, *67*, 1128-1135.

Daugelavicius, R., Bamford, J. and Bamford, D. (1997). Changes in host cell energetics in response to bacteriophage PRD1 DNA entry. *Journal of Bacteriology*, *179*, 5203-5210.

de los Reyes-Gavilan, C., Limsowtin, G. K. Y., Sechaud, L., Veaux, M. and Accolas, J-P. (1990). Evidence for a plasmid-linked Restriction-Modification system in *Lactobacillus helveticus*. *Applied and Environmental Microbiology*, *56*, 3412-3419.

Deutsch, S. M., Guezenec, S., Piot, M., Foster, S. and Lortal, S. (2004). Mur-LH, the broad-spectrum endolysin of *Lactobacillus helveticus* temperate bacteriophage phi-0303. *Applied and Environmental Microbiolology*, *70*, 96-103.

Douglas, L. J. and Wolin, M. J. (1971). Cell wall polymers and phage lysis of *Lactobacillus plantarum*. *Biochemistry*, *10*, 1551-1555.

Dupont, K., Janzen, T., Vogensen, F. K., Josephsen, J. and Stuer-Lauridsen, B. (2004). Identification of *Lactococcus lactis* genes required for bacteriophage adsorption. *Applied and Environmental Microbiology*, *70*, 5825-5832.

Dupont, L., Vasala, A., Mikkonen, M., Boizet-Bonhoure, B., Alatossava, T. and Ritzenthaler, P. (1993). Evolution of the lysis modules of two related *Lactobacillus delbrueckii* bacteriophages. GenBank accession number Z26590.

Dupuis, M-E. and Moineau, S. (2010). Genome organization and characterization of the virulent lactococcal phage 1358 and its similarities to *Listeria* phages. *Applied and environmental microbiology*, *76*, 1623-1632.

Fischer, W., Rösel, P. and Koch, H. U. (1981). Effect of alanine ester substitution and other structural features of lipoteichoic acids on their inhibitory activity against autolysins of *Staphylococcus aureus*. *Journal of Bacteriology*, *146*, 467-475.

García, P., García, J. L., García, E., Sánchez-Puelles, J. M. and López, R. (1990). Modular organization of the lytic enzymes of *Streptococcus pneumoniae* and its bacteriophages. *Gene*, *86*, 81-88.

Garneau, J. E., Tremblay, D. M. and Moineau, S. (2008). Characterization of 1706, a virulent phage from *Lactococcus lactis* with similarities to prophages from other Firmicutes. *Virology*, *373*, 298-309.

Geller, B. L., Ivey, R. G., Trempy, J. E. and Hettinger-Smith, B. (1993). Cloning of a chromosomal gene required for phage infection of *Lactococcus lactis* subsp. *lactis*. *Journal of Bacteriology*, *175*, 5510-5519.

Gindreau, E. and Lonvaud-Funel, A. (1999). Molecular analysis of the region encoding the lytic system from *Oenococcus oeni* temperate bacteriophages φ10MC. *FEMS Microbiology Letters*, *171*, 231-238.

Gopal, P. K. and Reilly, K. I. (1995). Molecular architecture of the Lactococcal cell surface as it relates to important industrial properties. *International Dairy Journal*, *5*, 1095-1111.

Guglielmotti, D. M., Deveau, H., Binetti, A. G., Reinheimer, J. A., Moineau, S. and Quiberoni A. (2009). Genome analysis of two virulent *Streptococcus thermophilus* phages isolated in Argentina. *International Journal of Food Microbiology*, *136*, 101-109.

Guglielmotti, D. M., Reinheimer, J. A., Binetti, A. G., Giraffa, G., Carminati, D. and Quiberoni, A. (2006). Characterization of spontaneous phage-resistant derivatives of *Lactobacillus delbrueckii* commercial strains. *International Journal of Food Microbiology*, 111, 126-133.

Guttman, B., Raya, R., and Kutter, E. (2005). Basic phage biology. In E. Kutter, and A. Sulakvelidze (Eds.), *Bacteriophages: Biology and Applications* (pp. 29-66). Boca Raton, FL: CRC Press.

Hahn, F. E., Weisseman Jr., C. L. and Hopps, H. E. (1955). Mode of action of chloramphenicol. III. Action of chloramphenicol on bacterial energy metabolism. *Journal of Bacteriology*, *69*, 215-223.

Henrich, B., Binishofer, B. and Blasi, U. (1995). Primary structure and functional analysis of the lysis genes of *Lactobacillus gasseri* bacteriophage φadh. *Journal of Bacteriology*, *177*, 723-732.

Herrero, M., de los Reyes Gavilán, C. G., Caso, J. L. and Suárez, J. E. (1994). Characterization of 393-A2, a bacteriophage that infects *Lactobacillus casei*. *Microbiology*, *140*, 2585-2590.

Ishibashi, K., Takesue, S. and Watanabe, K. (1980). Effect of some lectins on the adsorption of PL-1 phage to *Lactobacillus casei*. *Agricultural and biological chemistry*, *44*, 1689-1691.

Ishibashi, K., Takesue, S., Watanabe, K. and Oishi, K. (1982). Use of lectins to characterize the receptor sites for bacteriophage PL-1 of *Lactobacillus casei*. *Journal of General Microbiology*, *128*, 2251-2259.

Kakikawa, M., Yokoi, K. J., Kimoto, H., Nakano, M., Kawasaki, K., Taketo, A. and Kodaira, K. (2002). Molecular analysis of the lysis protein Lys encoded by *Lactobacillus plantarum* phage φg1e. *Gene*, *299*, 227-234.

Kashige, N., Nakashima, Y., Miake, F. and Watanabe, K. (2000). Cloning, sequence analysis, and expression of *Lactobacillus casei* phage PL-1 lysis genes. *Archives of Virology, 145*, 1521-1534.

Kotsonis, S. E., Powell, I. B., Pillidge, C. J., Limsowtin, G. K., Hiller and A. J. Davidson, B. E. (2008). Characterization and genomic analysis of phage asccΦ28, a phage of the family *Podoviridae* infecting *Lactococcus lacti*. *Applied and environmental microbiology*, *74*, 3453-3460.

Kutter, E., and Goldman, E. (2008). Introduction to Bacteriophages. In E. Goldman, and H. L. Green (Eds.), *Practical Handbook of Microbiology* (Second Edition, pp. 685-707). Boca Raton, FL: CRC Press.

Kutter, E., Raya, R., and Carlson, K. (2005). Molecular mechanisms of phage infection. In E. Kutter and A. Sulakvelidze (Eds.), *Bacteriophages: Biology and Applications* (pp. 165-222). Boca Raton, FL: CRC Press.

Lévesque, C., Duplessis, M., Labonté, J., Labrie, S., Fremaux, C., Tremblay, D. and Moineau, S. (2005). Genomic organization and molecular analysis of the virulent bacteriophages 2972 infecting an exopolysaccharide-producing *Streptococcus thermophilus* strain. *Applied and Environmental Microbiology, 71*, 4057-4068.

Lu, Z., Altermann, E., Breidt, F. and Kozyavkin S. (2010). Sequence analysis of *Leuconostoc mesenteroides* bacteriophage φ1-A4 isolated from an industrial vegetable fermentation. *Applied and Environmental Microbiology, 76*, 1955-1966.

Lu, Z. Breidt Jr., F., Fleminga, H. P., Altermannb, E. and Klaenhammer, T. R. (2003). Isolation and characterization of a *Lactobacillus plantarum* bacteriophage, Φ JL-1, from a cucumber fermentation. *International Journal of Food Microbiology, 84*, 225- 235.

Lucchini, S., Desiere, F. and Brüssow, H. (1998). The structural gene module in *Streptococcus thermophilus* bacteriophage Sfi11 shows a hierarchy of relatedness to *Siphoviridae* from a wide range of bacterial hosts. *Virology, 246*, 63-73.

Lucchini, S., Desiere, F. and Brüssow, H. (1999). The genetic relationship between virulent and temperate *Streptococcus thermophilus* bacteriophages: whole genome comparison of cos-site phages Sfi19 and Sfi21. *Virology, 260*, 232-243.

Maloy, S. R., Cronan, J. E., and Freidfelder, D. (1994). Phage biology. In *Microbial Genetics* (Second edition, pp.81-100). Boston, Jones and Bartlett Publishers, Inc.

Masud, T., Latif, A. and Hameed, A. (2009). Characterization of four new *Lactococcus lactis* bacteriophages isolated from Dahi whey. *International Journal of Dairy Technology, 72*, 107-111.

Monteville, M., Ardestani, B. and Geller, B. (1994). Lactococcal bacteriophages require a host wall carbohydrate and a plasma membrane protein for adsorption and ejection of DNA. *Applied and Environmental Microbiology, 60*, 3204-3211.

Mullan, W. M. A. (2002). Bacteriophages for lactic acid bacteria. [On-line]. Available from: http://www.dairyscience.info/bacteriophages-for-lactic-acid-bacteria-with-particular-emphasis-on-lactococci.html. Accessed: 4 January, 2011.

Nes, I. F., Brendehaug, J. and von Husby, K. O. (1988). Characterization of bacteriophage B2 of *Lactobacillus plantarum* ATCC 8014. *Biochimie, 70*, 423-427.

Neve, H. (1996). Bacteriophage. In T. M. Logan, and J.-P. Accolas (Eds.), *Dairy Starter Cultures* (pp. 157-190). New York, NY: VCH Publishers, Inc.

Oki, M., Kakikawa, M., Yamada, K., Taketo, A. and Kodaira, K.-I. (1996). Cloning, sequence analysis, and expression of the gene encoding lytic functions of bacteriophage φg1e. *Gene, 176*, 215-223.

Parreira, R., Sao-Jose, C., Isidro, A., Domingues, S., Vieira, G. and Santos, M. A. (1999). Gene organization in a central DNA fragment of *Oenococcus oeni* bacteriophage fOg44 encoding lytic, integrative and non-essential functions. *Gene, 226*, 83-93.

Poranen, M., Daugelavičius, R. and Bamford, D. H. (2002). Common principles in viral entry. *Annual Review of Microbiology, 56*, 521-538.

Powell, I. B. and Davidson, B. E. (1985). Characterization of streptococcal bacteriophage c6A. *Journal of General Virology, 66*, 2737-2741.

Quiberoni, A., Guglielmotti, D., Binetti, A. and Reinheimer, J. (2004). Characterization of three *Lactobacillus delbrueckii* subsp. *bulgaricus* phages and the physicochemical analysis of phage adsorption. *Journal of Applied Microbiology*, *96*, 340-351.

Quiberoni, A. and Reinheimer, J. A. (1998). Physicochemical characterization of phage adsorption to *Lactobacillus helveticus* ATCC 15807 cells. *Journal of Applied Bacteriology*, *85*, 762-768.

Quiberoni, A., Stiefel, J. I. and Reinheimer, J. A. (2000). Characterization of phage receptors in *Streptococcus thermophilus* using purified cell walls obtained by a simple protocol. *Journal of Applied Microbiology*, *89*, 1059-1065.

Räisänen, L. (2007). Phage-host interactions in *Lactobacillus delbrueckii*: host recognition and transcription of early phage genes. Faculty of Science, Department of Biology, University of Oulu, Finland Acta Univ. Oul. A 484.

Räisänen, L., Draing, C., Pfitzenmaier, M., Schubert, K., Jaakonsaari, T., von Aulock, S., Hartung, T. and Alatossava, T. (2007). Molecular interaction between lipoteichoic acids and *Lactobacillus delbrueckii* phages depends on D-alanyl and α-glucose substitution of poly(glycerophosphate) backbones. *Journal of bacteriology*, *189*, 4135-4140.

Räisänen, L., Schubert, K., Jaakonsaari, T. and Alatossava, T. (2004). Characterization of lipoteichoic acids as *Lactobacillus delbrueckii* phage receptor components. *Journal of bacteriology*, *186*, 5529-5532.

Russel, M., Linderoth, N. A. and Sali, A. (1997). Filamentous phage assembly: variation on a protein export theme. *Gene*, *192*, 23-32.

São-Jose, C., Pariera, R., Vieira, G., Santos, M. A. (2000). The N-terminal region of the *Oenococcus oeni* bacteriophage fOg44 lysin behaves as a bona fide signal peptide in *Escherichia coli* and as a *cis*-inhibitory element, preventing lytic activity on oenococcal cells. *Journal of Bacteriology*, *182*, 5823-5831.

Schäfer, A., Geis, A., Neve, H. and Teuber, M. (1991). Bacteriophage receptors of *Lactococcus lactis* subsp. *diacetylactis* F7/2 and *Lactococcus lactis* subsp. *cremoris* Wg2-1. *FEMS Microbiology Letters*, *78*, 69-74.

Séchaud, L., Callegari, M. L., Rousseau, M., Muller, M. C. and Accolas, J. P. (1988). Relationship between temperatt bacteriophage 0241 and virulent phage 832-B1 of *Lactobacillus helveticus*. *Netherlands Dairy Journal*, *43*, 261-277.

Shearman, C. A., Jury, K. L. and Gasson, M. J. (1994). Controlled expression and structural organization of a *Lactococcus lactis* bacteriophage lysin encoded by two overlapping genes. *Applied and Environmental Microbiology*, *60*, 3063-3073.

Sijtsma, L., Sterkenburg, A. and Wouters, J. (1988). Properties of the cell walls of *Lactococcus lactis* subsp. *cremoris* SK110 and SK112 and their relation to bacteriophage resistance. *Applied and Environmental Microbiology*, *54*, 2808-2811.

Stanley, E., Fitzgerald, G. F., Le Marrec, C., Fayard, B. and van Sinderen, D. (1997). Sequence analysis and characterization of O1205, a temperate bacteriophage infecting *Streptococcus thermophilus* CNRZ1205. *Microbiology*, *143*, 3417-3429.

Stanley, E., Walsh, L., van der Zwet, A., Fitzgerald, G. F. and van Sinderen, D. (2000). Identification of four loci isolated from two *Streptococcus thermophilus* phage genomes responsible for mediating bacteriophages resistance. *FEMS Microbiology Letters*, *182*, 271-277.

Suárez, V. B., Quiberoni, A., Binetti, A. and Reinheimer, J. A. (2002). Thermophilic lactic acid bacteria phages isolated from Argentinian dairy industries. *Journal of Food Protection, 65*, 1597-1604.

Suárez, V., Moineau, S., Reinheimer, J. and Quiberoni, A. (2008). Argentinean *Lactococcus lactis* bacteriophages: genetic characterization and adsorption studies. *Journal of Applied Microbiology, 104*, 371-379.

Sugahara, K., Yokoi, K. J., Nakamura, Y., Nishino, T., Yamakawa, A., Taketo A. and Kodaira, K. I. (2007). Mutational and biochemical analyses of the endolysin Lys_{gaY} encoded by the *Lactobacillus gasseri* JCM 1131^T phage φgaY. *Gene, 404*, 41-52.

Summer, E. J., Berry, J., Tran, T. A., Niu, L., Struck, D. K. and Young, R. (2007). Rz/Rz1 lysis gene equivalents in phages of Gram-negative hosts. *Journal of Molecular Biology, 373*, 1098-1112.

Tremblay, D. and Moineau, S. (1999). Complete genomic sequence of the lytic bacteriophage DT1 of *Streptococcus thermophilus*. *Virology, 255*, 63-76.

Trevors, K., Holley, A and Kempton, A. G. (1983). Isolation and characterization of a *Lactobacillus plantarum* bacteriophage isolated from a meat starter culture. *Journal of Applied Bacteriology, 54*, 281-288.

Valyasevi, R., Sandine, W. E. and Geller, B. L. (1990). The bacteriophage kh receptor of *Lactococcus lactis* subsp. *cremoris* KH is the rhamnose of the extracellular wall polysaccharide. *Applied and Environmental Microbiology, 56*, 1882-1889.

Valyasevi, R., Sandine, W. E. and Geller, B. L. (1991). A membrane protein is required for bacteriophage c2 infection of *Lactococcus lactis* subsp. *lactis* C2. *Journal of Bacteriology, 173*, 6095-6100.

Valyasevi, R., Sandine, W. and Geller, B. (1994). *Lactococcus lactis* ssp. *lactis* C2 bacteriophage sk1 receptor involving rhamnose and glucose moieties in the wall. *Journal of Dairy Science, 77*, 1-6.

van Sinderen, D., Karsens, H., Kok, J., Terpstra, P., Ruiters, M. H. J., Venema, G. and Nauta, A. (1996). Sequence analysis and molecular characterization of the temperate lactococcal bacteriophage r1t. *Molecular Microbiology, 19*, 1343-1355.

Vasala, A., Välkkilä, M., Caldentey, J. and Alatossava, T. (1995). Genetic and biochemical characterization of the *Lactobacillus delbrueckii* subsp. *lactis* bacteriophages LL-H lysin. *Applied and Environmental Microbio*logy, *61*, 4004-4011.

Ventura, M., Callegari, M. L. and Morelli, L. (1999). Surface layer variations affecting phage adsorption on seven *Lactobacillus helveticus* strains. *Annali in Microbiology and Enzimology, 49*, 45-53.

Wang, I.-N., Dykhuizen, D. E. and Slobodkin, L. B. (1996). The evolution of phage lysis timing. *Evolutionary Ecology, 10*, 545-558.

Wang, S., Kong, J., Gao, C., Guo, T. and Liu, X. (2010). Isolation and characterization of a novel virulent phage (phiLdb) of *Lactobacillus delbrueckii*. *International Journal of Food Microbiology, 137*, 22-27.

Wang, Y. and Lu, C. (2009). Bacteriophage lysins: progress and perspective--a review. *Wei Sheng Wu Xue Bao, 49*, 1277-1281.

Watanabe, K. and Takesue, S. (1972). The requirement for calcium in infection with *Lactobacillus* phage. *Journal of General Virology, 17*, 19-30.

Watanabe, K. and Takesue, S. (1975). Use of L-rhamnose to study irreversible adsorption of bacteriophage PL-1 to a strain of *Lactobacillus casei*. *Journal of General Virology*, *28*, 29-35.

Watanabe, K., Fukuzaki, T., Hayashida, M. and Nakashima, Y. (1987). Electron microscopic study of the process of DNA ejection from the head of PL-1, a *Lactobacillus casei* phage. *Journal of General Virology*, *68*, 569-572.

Watanabe, K., Takesue S. and Ishibashi, K. (1979). Adenosine triphosphate content in *Lactobacillus casei* and the blender-resistant phage-ell complex-forming ability of cells on infection with PL-1 phage. *Journal of General Virology*, *42*, 27-36.

Watanabe, K., Takesue, S., Ishibashi, K. and Nakahara, S. (1982). A computer simulation of the adsorption of *Lactobacillus* phage PL-1 to host cells: some factors affecting the process. *Agricultural and Biological Chemistry*, *46*, 697-702.

Watanabe, K., Shirabe, M., Nakashima, Y. and Kakita, Y. (1991). The possible involvement of protein synthesis in the injection of PL-1 phage genome into its host, *Lactobacillus casei*. *Journal of General Microbiology*, *137*, 2601-2603.

Watanabe, K., Shirabe, M., Fukuzaki, T., Kakita, Y., Nakashima, Y. and Miake, F. (1993). Electron microscope studies on the host cell energy requirement for injection of PL-1 phage DNA into *Lactobacillus casei*. *Current Microbiology*, *26*, 293-298.

Yamamoto, M., Kanegasaki, S. and Yoshikawa, M. (1980). Effects of temperature and energy inhibitors on complex formation between *Escherichia coli* male cells and filamentous phage fd. *Journal of General Microbiology*, *119*, 87-93.

Yokokura, T. (1977). Phage receptor material in *Lactobacillus casei*. *Journal of general microbiology*, *100*, 139-145.

Young, I., Wang, I. and Roof, W. D. (2000). Phages will out: strategies of host cell lysis. *Trends in Microbiology*, *8*, 120-128.

Zinder, N. D. and Lederberg, J. (1952). Genetic exchange in *Salmonella*. *Journal of Bacteriology*, *64*, 679-699.

In: Bacteriophages in Dairy Processing
Editor: Andrea del Lujan Quiberoni et al.
ISBN 978-1-61324-517-0

Chapter 6

LYSOGENY IN PROBIOTIC LACTOBACILLI

María Luján Capra and Diego Javier Mercanti*
Instituto de Lactología Industrial (UNL-CONICET),
Facultad de Ingeniería Química,
Universidad Nacional del Litoral, Santiago del Estero,
Santa Fe, Argentina

ABSTRACT

Bacteriophages show several life cycles: lytic (as a virulent phage), lysogenic (as a temperate phage or prophage) and persistent infections (pseudolysogeny, phage-carrier state). The "lysogenic decision" of a temperate phage to either follow a lytic cycle or insert its genome into host chromosome and continue to replicate in a dormant stage (prophage) is made just after infection, and depends upon host density and environmental factors. Bacterial and phage genomes are not completely independent entities; they underwent a co-evolution which favored both. During evolution, bacteria tend to eliminate unessential prophage DNA giving place to the following sequence: inducible prophages - defective prophages (with deletions, insertions and/or rearrangements) - prophage remnants (with massive loss of the original phage genome) - isolated prophage genes. The latter, called "lysogenic conversion genes" (LCG) confer a competitive advantage to the host cell. Hence, lysogeny greatly contributed to the generation of diversity, being responsible for almost all genetic differences found among strains of the same species. Lysogeny is more the rule than the exception and many bacteria harbour more than one prophage (poly-lysogeny). Prophages and LCG were first described on pathogenic bacteria, where they are known to encode for diverse virulence factors, but later they were postulated to help in the adaptation of any bacterial strain to their ecological niche. This obviously applies to gut commensal bacteria, which are constantly under selection pressure and include probiotic strains. Among lactic acid bacteria, most lactococcal strains are lysogenic and poly-lysogenic, but lysogeny was also found to be highly widespread in lactobacilli, which encompasses most probiotic strains currently

* E-mail address: diegojav78@gmail.com. Tel.: +54 342 4530302; fax: +54 342 4571162

used by dairy industries. In particular, phage infections in probiotic lactobacilli are especially important due to singular and unique characteristics ascribed to each strain. In addition, lysogeny could potentially affect strain functionality, either by improving or counteracting its probioticity. Finally, it is known that lactobacilli play an active role in the maintaining of mucosal health. Several studies have correlated the alteration of this equilibrium with the presence of virulent phages or furthermore, with prophages that are induced under certain circumstances, causing lactobacilli depletion.

Introduction

Bacteriophages show several life cycles: lytic (as a virulent phage), lysogenic (as a temperate phage or prophage) and persistent infections (pseudolysogeny, phage-carrier state). In the lytic cycle, phage redirects the host metabolism towards the production of new phage particles, which are released during the lysis of the cell. Lytic phages are perhaps better described as predators than parasites, since they do not make prokaryotes "sick" but lyse them. Alternatively, in the lysogenic cycle phage genomes typically remain in the host in a dormant stage (prophage) and replicate along with host chromosome during cell division, but a lytic cycle can be induced under certain circumstances. Finally, in persistent infections phages multiply in a fraction of the host population. Lysogenic and persistent infections actually represent a parasitic interaction, although mutualism or even symbiosis might better describe some forms of lysogeny (Weinbauer 2004).

1. Lysogeny

Early phage researchers in 1920s and 1930s claimed to have found phages irregularly associated with bacterial stocks, and considered that bacteria were able to spontaneously generate them. Phages were thus long and widely regarded as some sort of "ferment" or enzyme rather than real viruses. In 1950, Lwoff and Gutmann, through microscopic observations of *Bacillus megaterium*, demonstrated that even though individual cells continue to divide in phage-free medium without signs of phage production, some cells could occasionally and spontaneously lyse and liberate phages. Lwoff named the hypothetical intracellular state of the phage genome a *prophage*, and later accomplished a uniform induction of lysogenic cells and lytic growth of phages by challenging cells with ultraviolet light, an agent that perturbs the host (Guttman et al. 2005).

In the lysogenic cycle, the genome of the temperate phage typically remains in the host in a dormant stage (prophage) and replicates along with the host chromosome into daughter cells during cell multiplication. Under circumstances that perturb the host, these phages can escape by excision and the lytic cycle is induced. The "lysogenic decision" of whether or not to establish a prophage state is made by the temperate phage just after infection. In this sense, temperate phages differ from lytic ones in that a choice actually exists in their life cycle (Lodics and Steenson 1993). Lysogeny is thought to be advantageous when the host abundance, i.e., the food resource, is lower than that necessary to sustain lytic infection, or when the destruction rate of free phages is too high to allow for lytic replication. Both explanations assume that the rate of successful encounters between phage and hosts is too low to sustain lytic phage production. Consequently, lysogeny would predominate at low host density whereas lytic infection would prevail in the opposite situation (Weinbauer 2004).

Prophages are not silent units within bacterial chromosomes; in fact, they can be induced by numerous known external factors. Although most of them are found in natural systems (sunlight, hydrogen peroxide, temperature changes, polyaromatic hydrocarbons, trichloroethylene), mitomycin C (MC) and UV-C radiation (254 nm) are the most powerful agents used to induce the lytic cycle in lysogenized bacterial isolates. Data from several studies indicate that MC is often a better inducing agent for natural bacterial communities than UV-C radiation (Weinbauer 2004), and is for that reason considered the first agent of choice for induction assays. Moreover, it is important to point out that MC enters the cytoplasm much more easily in lactic acid bacteria (LAB) than in *Escherichia coli* due to the multiple pores left by the peptidoglycan-based wall of gram-positive bacteria (Soberón et al. 2007), thus resulting highly effective to induce *Lactobacillus* strains. On the other hand, temperate phages can also be spontaneously released (Davidson et al. 1990). Prophages can generate, though at relatively low levels, derived virulent phages (Shimizu-Kadota et al. 1985, 2000). Moreover, a prophage released by a bacterial strain is considered not able to infect the same strain (which is true in most cases), but frequently attacks related ones. However, recent evidence indicates that remarkably heat-resistant siphophages C_L1 and C_L2, isolated from noninfected lysed-cultures of *Lactobacillus paracasei* A (a commercial dairy strain), were capable to lyse this strain (Capra et al. 2010).

The observation that the treatment of a culture with an inductor leads to both cell lysis and appearance of phage particles and plaque forming units is only a first indication for lysogeny. Nevertheless, suitable indicator strains are extremely difficult to find and so, a negative result does not prove the absence of lysogeny. Furthermore, a major reason for this difficulty is likely to be the process of superinfection immunity, in view of the widespread occurrence of lysogeny (Davidson et al. 1990). A formal proof of lysogeny would require that the lysogen can be cured of its prophage, the cured derivative is susceptible to infection by the temperate phage and it can also be relysogenized (Davidson et al. 1990). Whereas the first formal report of lysogeny for a lactococcal strain was made in 1980, the corresponding for *Lactobacillus* strains first appeared in *Lb. casei* S-1 (Shimizu-Kadota and Sakurai 1982) and then in *Lb. acidophilus* ADH (currently reclassified as *Lb. gasseri* ADH, Raya et al. 1989). Alternative approaches to demonstrate phage presence in induced cultures have been to visualize them by electron microscopy (Terzaghi and Sandine 1981, Durmaz et al. 2008, Capra et al. 2010) or to isolate viral DNA from the filtered supernatants of the induced cultures (Ventura et al. 2004, Capra et al. 2010, Mercanti et al. 2011). Afterwards, however, the presence of prophages within the bacterial chromosome should be confirmed by Southern blot (Jarvis 1984, Wiederholt and Steele 1993, Ventura et al. 2006, Capra et al. 2010).

2. Persistent Infections

Persistent infections refer to those particular situations where a phage is capable to subsist and multiply by means of lytic infection but without producing the lysis of the entire bacterial host population. Pseudolysogeny and phage-carrier state cases will be subsequently described.

2.1. Pseudolysogeny

Pseudolysogeny can be defined as an approximately steady state between phages and hosts that is often found in the environment. This relationship may be the result of a mixture of resistant and sensitive host cells or a mixture of lytic and temperate phages (Weinbauer 2004). Other authors (Lodics and Steenson 1993, Josephsen and Neve 1998) consider that pseudolysogeny results from a permanent infection of a starter culture by a virulent phage that persists in low titres in the culture. The partial nature of this infection would be due to a continuous balance of phage-sensitive and phage-resistance variants within the host population. Fortunately, it was observed that these permanent infected cultures are protected from phages able to attack the corresponding phage-free cultures, therefore constituting a powerful anti-phage mechanism of defense (Klaenhammer 1984).

A pseudolysogenic relationship was reported for *Lactococcus lactis* subsp. *cremoris* SK11 (Klaenhammer and Fitzgerald 1994) due to the presence of a plasmid which conferred adsorption resistance. To sustain this state, the coexistence of sensitive and resistant bacteria within the pure culture population of the strain is indispensable. Capra et al. (2010), in turn, postulated a likely pseudolysogenic state between virulent phages C_L1 or C_L2 coexisting in a metastable balance with their host *Lb. paracasei* A.

In summary, different life cycles and strategies seem to be enclosed by the term pseudolysogeny, and its role in the environment remains largely unknown (Weinbauer 2004). Pseudolysogeny can be readily differentiated from true lysogeny by treatment of the culture with antisera reactive against the phage (Klaenhammer 1984) or by repeated single-colony isolation (Lodics and Steenson 1993), thereby eliminating the phage. However, it has not been investigated in detail and is still poorly understood. Probably, lysogeny has evolved from lytic phages by co-adaptation as a mean to stabilize the interactions between phages and hosts, allowing for the survival of both. In this context, pseudolysogeny could be seen as an intermediate evolutionary step (Weinbauer 2004).

2.2. Phage-Carrier State

Although pseudolysogeny and carrier state are widely used as synonyms (Klaenhammer 1984, Josephsen and Neve 1998), the latter is more strictly used for bacteria with a plasmid-like prophage. Mechanisms for establishing the carrier state include a reduced success rate of infection conferred by limited available receptors or by enzymatic loss of receptors or superinfection immunity of a temperate phage, which mutates at a fixed frequency to lytic phages, thus allowing for the simultaneous presence of phages and bacteria in a culture (Weinbauer 2004). Plasmids may be the key in helping cells establish the phage-carrier state since many phage resistance mechanisms are plasmid-encoded. Loss of such plasmids from some cells provides a subpopulation of phage-sensitive hosts in the culture that allow phage propagation. On the contrary, the presence of resistant hosts deters total destruction of the culture. Therefore, both populations promote the establishment of a phage-carrier state (Lodics and Steenson 1993).

In 1952, Graham et al. reported an equilibrated existence of sensitive lactic *Streptococcus* bacteria, resistant bacteria and the associated phage, resulting in the incomplete lysis of the sensitive strain, thereby continually providing substratum for the associated bacteriophage. This "carrier" state differs from lysogenesis in that the bacteria are easily separated from the bacteriophage by a simple plating and re-isolation procedure. Arendt et al. (1991b) reported the occurrence of a special type of a phage-carrier state for a culture of *Leuconostoc oenos*. Although exposing *Leuc. oenos* 58N to MC led to release of phage P58II, a phage-free culture of the strain could be obtained after repeated subculture. Moreover, DNA/DNA hybridization experiments failed to show homology between the DNA of phage P58II and the chromosomal DNA of *Leuc. oenos* 58N.

3. Phage and Bacteria Co-Evolution: Role of Lysogeny in the Generation of Diversity

Bacterial and phage genomes should not be regarded as completely independent entities; in fact they underwent a co-evolution during which neither microbes nor phages could have survived without each other. Microbial DNA was continuously bombarded with parasitic DNA over evolutionary time, being bacteriophages the main responsible agents of horizontal DNA transfer (Canchaya et al. 2003b). In this sense, lysogeny contributes a substantial share of the mobile DNA of their bacterial hosts and seem to influence the short-term evolution of bacteria (Ventura et al. 2006), which, together with other ways of lateral DNA transfer such as conjugation and transformation, changed a tree-like to a web-like representation of the phylogenetic relationship between bacteria (Canchaya et al. 2003a). From the bacteriophage point of view, lysogeny is beneficial given that prophages do not have to find a host to survive; they can exploit the repair system of the host and are protected inside the host cell (Weinbauer 2004). Moreover, co-infection of cells with two or more phages can be beneficial due to the exchange of genetic material, a process that can occasionally contribute to phage diversity and thus enhance the chance to adapt to the environment (Weinbauer 2004). Labrie and Moineau (2007) have demonstrated that a lactococcal phage (ul36) extensively evolved by large-scale homologous and non-homologous recombination events with the inducible prophage present in the host strain. The phage ul36 was used to sequentially infect the wild-type strain and two isogenic derivates (with genes encoding for different Abi phage resistance mechanisms). Four phage mutants were isolated; one of them exchanged as much as 79% of its genome compared to the core genome of phage ul36. Thus, natural phage defense mechanisms and prophage elements found in bacterial chromosomes contribute significantly to the evolution of the lytic phage population.

Alternatively, from the bacterial point of view one could think that the presence of bacteriophages would negatively affect the fitness of the bacteria, due to either the metabolic burden necessary to replicate extra DNA or cell lysis produced by prophage induction (Canchaya et al. 2003b). However, lysogeny is more the rule than the exception and many bacteria harbour more than one prophage (poly-lysogeny), and as many as 18 prophage genome elements (16% of the entire genome) were reported for the food pathogen *E. coli* O157:H7 strain Sakai (Hayashi et al. 2001). In fact, prophage-like elements and prophage remnants have been identified in almost all bacterial genomes sequenced so far (Canchaya et al. 2003b). The only plausible explanation to this fact is that prophages must in some way be useful for the bacteria in order to resist selection pressure. Actually, superinfection exclusion genes of the prophage protect both the cell (against phage infection) and the prophage itself by excluding superinfecting phage DNA from competing with the resident prophage DNA for the same host (Weinbauer 2004). But what is more important, phages can profoundly influence genomes of their bacterial hosts through various mechanisms. Infection of *Campylobacter jejuni* with a specific virulent phage leads to major rearrangement in surviving bacteria that are associated with the acquisition of phage resistance, but also reduce *Campylobacter jejuni* capability to colonize chicken intestine (Nechaev and Severinov 2008). Certain genes carried by bacteriophages change the phenotype by providing a competitive advantage to the host cell; they are called "lysogenic conversion genes" (LCG). The process of lysogenization could be considered as the incorporation into the bacterial genome of molecular time bombs potentially able to kill, after an incidental induction, the host cell (Lawrence et al. 2001), which therefore will try to inactivate the prophage induction process. Natural selection favours strains with mutations in the prophage DNA, and lysogenic cells tend to eliminate unessential prophage DNA over time, giving place to the following switch along evolution: inducible prophages - defective prophages (with deletions, insertions and/or rearrangements) - prophage remnants (with massive loss of the original phage genome) - isolated prophage genes. In terms of evolution, the tendency of lysogeny is to turn lytic phages into temperate ones. *In silico* analysis of the bacterial genomes revealed that most prophages are in fact defective and supported the idea of this dynamic phenomenon of gradual decay (Canchaya et al. 2003b). LCG were first studied in pathogenic bacteria, where they encoded toxins and other virulence factors such proteins that alter antigenicity and proteins required for intracellular survival and host cell attachment (Boyd and Brussow 2002). The lost or acquisition of extra cellular toxin expressions, virulent factors and enzymes in *Staphylococcus aureus* (Lee and Iandolo 1986) and *Streptococcus pneumoniae* (Loeffler and Fischetti 2006) depends on the prophage presence in the host chromosome. Bacteriophage-encoded toxins were reported for *E. coli* (Huang et al. 1987), *Pseudomonas aeruginosa* (Nakayama et al. 1999), *S. aureus* (Betley and Mekalanos 1985) and *Streptococcus pyogenes* (Weeks and Ferretti 1984, Goshorn and Schlievert 1989), among others. The latter is a human pathogen and belongs to the LAB group, comprising from harmless to life-threatening strains, feature that seems to be related with prophage presence within the bacterial genome (Ferretti et al. 2001). However, some elements present on the eight prophages found on the highly pathogenic *Strep. pyogenes* strain M1 resemble prophages harboured by dairy starters *Strep. pyogenes* and *L. lactis* (Desiere et al. 2001). On the other hand, LCG were postulated to be also functional on nonpathogenic bacterial strains, helping in their adaptation to their ecological niche. Likewise pathogenic bacteria, gut commensals are constantly under selection pressure and must overcome mucosal IgA response (Brüssow and Hendrix 2002). *L. lactis* IL1403, a typical dairy strain, harbour six prophage genomes (Chopin et al. 2001), and multiple prophages genomes with LCG that resemble those of prophages of *Strep. pyogenes* were found on sequenced *Lactobacillus* commensals (Brüssow and Hendrix 2002). Further research might finally change the perception of viruses as nasty parasites, to consider them instead as catalysts of information transfer and sustainers of bacteria mediated cycling of energy and matter. As a result, viruses may turn out to be rather benign than malign for ecosystem functioning (Weinbauer 2004).

Analogously, it could be thought that prophage presence in the chromosome of probiotic bacteria is related with the expression level of a particular probiotic property. This possibility will be further discussed.

4. LYSOGENY IN LAB

Ackermann and DuBow (1987) reported that 47% of all isolated bacteria were lysogenic. In LAB, the first report confirming the isolation of a lysogenic strain was that of Reiter in 1949, who worked with three *L. lactis* strains (Davidson et al. 1990). In fact, the majority of lactococci are lysogenic (Jarvis 1989) and poly-lysogenic strains (with more than one prophage within their genomes) also occur (Davidson et al. 1990).

Lysogeny was long thought to be co-responsible for phage infections. Particularly, lactococcal phages have been reclassified recently in 10 genetically distinct groups of dsDNA and tail-containing phages (Labrie and Moineau 2007) and 3 of them (936-, c2- and P335-especies) dominate the industrial phage ecology. Most phage isolates were virulent, though the P335 species is heterogeneous and comprises both virulent and temperate phages (Brüssow 2001).

Oddly, lysogenic *Strep. thermophilus* strains are rare among industrial strains (Davidson et al. 1990), as only 1-2.2% of these were induced by MC (Carminati and Giraffa 1992, Brüssow and Kutter 2005). Southern hybridization with DNA of the two major classes of *Strep. thermophilus* phages confirmed a low lysogeny rate. Only a single survey reported a 10% rate of lysogenic cells by hybridization (Fayard et al. 1993). In addition, less than 1% of *Strep. thermophilus* phages from major strain collections are temperate phages, though the genome maps of the major virulent *Strep. thermophilus* phages betray their origin from temperate parental phages. The prevalence of virulent phages in the collections might therefore represent an adaptation to the abundance of host cells in the dairy environment: serial passage of a temperate *Strep. thermophilus* phage quickly resulted in its replacement by a virulent derivative mutant (Brüssow and Kutter 2005).

Therefore, the role of lysogenic strains of LAB as a source of lytic phages is not always clear. In general, lytic and temperate phages of lactic streptococci are not closely related by DNA homology. However, some DNA homology groups contain both temperate and lytic phages, suggesting that these phages may have a common ancestor. In some instances, lytic phages evolve from one another, but it is possible that DNA from temperate phages contributes to phage evolution. DNA studies on *Lactobacillus* phages have shown a closer relationship between temperate and lytic phages (Jarvis 1989).

Lysogeny is widespread in strains of *Leuc. oenos* used for malolactic fermentations in wine (Arendt et al. 1991a, 1991b, Cavin et al. 1991). *Leuconostoc* spp. commonly constitute a small proportion of the strains in mixed dairy mesophilic starter cultures, where they are thought to contribute to the final texture and aroma of the product. Two unrelated insertion sequence (IS) elements, have been discovered: one in *Leuc. mesenteroides* (IS*1165*) and the other in *Leuc. lactis* (IS*1070*). Plasmids are a common component of the *Leuconostoc* genome as well (Davidson et al. 1996). Recently, the isolation, characterization and sequencing of the first *Leuc. pseudomesenteroides* temperate phage was reported (Jang et al. 2010).

Finally, lysogeny is highly widespread in lactobacilli (Séchaud et al. 1988, Davidson et al. 1990, Villion and Moineau 2009). Given its significance for the purpose of this chapter, lysogeny will be thoroughly discussed in the next section.

5. Lysogeny in Lactobacilli

In 1970, Sakurai et al. found that a strain of *Lb. salivarius* isolated from animal stool liberated phage-like particles following lysis upon treatment with MC or UV irradiation. Additionally, three lysogenic strains of *Lb. salivarius* and of *Lb. casei* out of 110 strains of lactobacilli tested released temperate phages that were detected by electron microscopy after UV induction. Other authors (Tohyama et al. 1972, Tohyama 1973) had also studied and characterized temperate phages released by MC-treatment from *Lb. salivarius* isolated from human feces. A few years later, Yokokura et al. (1974) had already established the common occurrence of lysogeny in diverse species of lactobacilli; they found that 40 out of 148 strains were lysed after MC- treatment, and observed phage-like particles in 30 of them by electron microscopy, whereas indicator strains were found for 10 induced prophages (9 from *Lb. casei* and one from *Lb. acidophilus*). The high frequency of inducible lysogenic strains in *Lb. casei* (15 out of 16 strains studied) is noteworthy, especially since such a high frequency was observed again later (Stetter 1977). More recently, Mercanti et al. (2011) tested the presence of inducible prophages on *Lactobacillus* of the *casei* group of diverse origin, including commercial, collection and dairy-isolated strains. The percentage of MC-inducible strains reached 83% (globally) and 91% taking into account only commercial strains, which evidenced the high potential risk of their use at industrial scale. However, none of the strains could be successfully induced after treatment with different concentrations of hydrogen peroxide. This study also analyzed prophage diversity (DNA restriction profiles) and discussed it in view of strains relatedness (percentage of homology by RAPD-PCR analysis). Interesting remarks were that the same prophage was found in a group of highly related strains, but not in all of them, as well as in a strain with noticeable lesser homology. Notably, not all of the strains with high homology within a RAPD-PCR cluster contained MC-inducible prophages. In this work, two new virulent phages (i*Lp*1308 and i*Lp*84) were isolated from filtered supernatants of inductions, and were capable to lyse the majority of the commercial strains, with a host range very similar to that of previously studied *Lb. casei* phage MLC-A (Capra et al. 2006).

The first known temperate phage of *Lb. helveticus* (phage 0241) was shown to be related to virulent phage 832-B1 isolated from cheese whey, thus reinforcing the hypothesis that virulent phages could originate from prophages harbored by lysogenic starter strains. Furthermore, Séchaud et al. (1992) found that 37% of *Lb. helveticus* strains tested were inducible by MC and, among them, 57% harboured temperate phages that were active on indicator strains. Moreover, they concluded that lysogenic strains could be the major source of phages in French cheese factories. Some years later, the presence and function of lysogenic strains of *Lb. helveticus* in natural whey starter cultures was investigated (Carminati et al. 1997). Although the authors admitted that temperate phages may be involved in the development of lytic phages attacking lactobacilli, they assumed that the high and unknown number of strains present on the natural starter presumably makes possible the replacement of phage sensitive strains, thus avoiding strong variations of starter activity and, consequently, serious failures during cheese making. Another interesting fact to be studied in the case of *Lb. helveticus* strains is the relationship between lysis and lysogeny, considering that strain lysis positively influences cheese proteolysis during cheese ripening. In 2003, Deutsch et al. demonstrated that particularly in the case of *Lb. helveticus* CNRZ 303, able to release Φ 0303

-either spontaneously or after MC-treatment-, the extensive lysis observed in cheese was mainly due to autolysin activity and not to prophage induction.

The first two temperate phages of *Lb. bulgaricus*, 0449 and 0448, arose from strains LT1 and LT4 respectively, and were isolated in the same yoghurt factory. Both strains were lysogenic and inducible by MC and UV irradiation. Five indicator strains (four of *Lb. lactis* and one of *Lb. bulgaricus*) were found to propagate well the two phages in a liquid medium, but plaque formation on bacterial lawns was less readily. The two prophages are closely related and morphologically similar (Cluzel et al. 1987a). Moreover, Mata et al. (1986) have demonstrated that the two temperate phages 0448 and 0449 (in their study named mv4 and mv1, respectively) show almost identical protein, serological and DNA patterns. In addition, both phages showed also a significant relationship with 11 virulent phages of *Lb. lactis* and *Lb. bulgaricus*. Virulent and temperate *Lb. delbrueckii* phages have been characterized and classified in four groups (*a* to *d*) on the basis of their morphology, immunoblotting tests and DNA-DNA hybridizations (Mata et al. 1986). Further work on phage 0448 (mv4) strongly suggested that this phage could trigger a lysogenic conversion of certain hosts, which involved changes in colony morphology and in the phage sensitivity pattern of these bacterial hosts (Cluzel et al. 1987b). In a second study of temperate and lytic phages (Lahbib-Mansais et al. 1988) 2 out of 48 strains of lactobacilli subjected to MC induction produced temperate phages able to attack *Lb. lactis* LKT (the propagating host also for phage mv4). These 2 temperate and 4 lytic phages isolated during abnormal fermentations, were partially homologous by DNA-DNA hybridizations. They could be classified in phage group *a*, the largest group that comprised temperate and virulent phages of both subspecies (*bulgaricus* and *lactis*). Thus, some virulent phages isolated during abnormal fermentations and some temperate phages isolated by induction of starter strains were genetically closely related. Close relationships were demonstrated between *pac*-site virulent LL-H and temperate mv4 phages of *Lb. delbrueckii*. Over the structural genes, both phages shared up to 92% nt sequence identity (Vasala et al. 1993). Interestingly, the virulent phage LL-H still contained remnants of a phage integrase and phage attachment sites, demonstrating that it was recently derived from a temperate phage (Brüssow 2001). *Lb. delbrueckii* phage JCL1032 has a prolate head and can integrate its genome into two different sites in the choromosome of *Lb. delbrueckii* subsp. *lactis* ATCC 15808 (also a host strain for phages LL-H and mv4), even though at low efficiency. Phage JCL1032 is thus now considered a temperate phage (Villion and Moineau 2009). Temperate phage lb539 was released after MC-treatment of *Lb. delbrueckii* subsp. *bulgaricus* strain CRL 539 (Auad et al. 1997). Dotblot experiments showed that the genome of phage lb539 hybridized with the genomes of phages mv4 and LLH, which are type phages of group *a* of *Lb. delbrueckii* phages (according to the classification of Mata et al. 1986). However, restriction enzyme patterns and morphological features showed phage lb539 to be distinct from phages mv4 and LL-H. Among the *Lb. delbrueckii* subsp. *bulgaricus* strains tested as potential indicators for phage lb539, only mv4 prophage-cured derivative CNRZ1004 was found to be sensitive. Therefore, resistance of LT4 cells (the mv4 lysogenic strain) to phage lb539 attack suggests either that both phages may possess a homoimmune system to superinfection or that phenotypic expression of the lysogenic conversion phenomenon occurred as described for LT4 cells (Auad et al. 1997). In addition, physical mapping and partial genetic characterization of phage lb539 was made (Auad et al. 1999). The presence of homologous regions between the genomes of phage lb539 and some others of *Lb. delbrueckii*, supports the modular evolution theory of bacteriophages

(Alatossava et al. 1995, Auad et al. 1999). Suárez et al. (2008) investigated the frequency of lysogeny in 16 *Lb. delbrueckii* strains which presented a DNA amplification fragment coding for the *Lb. delbrueckii* subsp. *lactis* integrase gene. For 37% of the supernatants obtained from the MC-treated strains, it was possible to detect phage particles or killing activity. Host strains were found for 19% of the MC-treated supernatants and only two active bacteriophages were isolated. Even if in lactobacilli a high percentage of strains are inducible, complete phage particles able to propagate on host strains are less frequently present. From the commercial strain *Lb. lactis* Cb1, it was possible to isolate active viral particles, which were able to propagate on two different indicator strains. Phages that were able to propagate on *Lb. delbrueckii lactis* 204 and *Lb. delbrueckii bulgaricus* 342 were named Cb1(204) and Cb1(342), respectively. Both of them were prolated-headed and belonged to group *c*. Different DNA restriction patterns and one-step growth curves were obtained for each phage, while the structural protein profiles and packaging sites were identical. Further studies were carried out to investigate thermal and chemical resistances of both phages and their comparison with those obtained for virulent phages specific for *Lb. delbrueckii* (Ebrecht et al. 2010). Results obtained showed that temperate phages possess lower resistance than virulent ones to thermal treatments. Regarding chemical treatments with some commercial biocides typically used in plants and laboratories, their efficiency was variable and dependent on the phages considered.

Over the last few years, several new *Lactobacillus* bacteriophage genomes have been sequenced and sequence data is currently available for phages of diverse species (*Lb. johnsonii*, *Lb. gasseri*, *Lb. plantarum*, *Lb. casei*, *Lb. delbrueckii* and *Lb. rhamnosus*) (Ventura et al. 2006, Durmaz et al. 2008, Zhang et al. 2010). Still, the sequencing of more phage genomes (either lytic or temperate) infecting *Lactobacillus* species is necessary to better understand their behaviour. The amount of prophage sequences in *Lactobacillus* genomes is variable, ranging from poly-lysogenic strains, such as *Lb. plantarum* WFCS1 (Kleerebezem et al. 2003) and *Lb. johnsonii* NCC 533, to strains containing only prophage remnants, such as *Lb. acidophilus* NCFM. In contrast, some strains that have been subjected to extensive industrial processing such as *Lb. helveticus* DPC4571 do not contain any prophage-like sequence (Ventura et al. 2006). A low DNA homology was generally found among phages infecting distinct species of *Lactobacillus* (Desiere et al. 2000), suggesting the existence of a barrier that hampers the transfer of phage DNA across *Lactobacillus* species (Brüssow 2001). Nevertheless, Zhang et al. (2010) recently provided a comparative analysis of Lcazh1, a 12.5-kb long prophage remnant from *Lb. casei* strain Zhang, in which most genes shared similarities and strong synteny with the prophage Lp3, harboured by *Lb. plantarum* WCSF1, thus being the first data supporting the possibility of lateral DNA transfer between different species of *Lactobacillus*.

Botstein (1980) proposed a modular theory for the evolution of viruses (including bacteriophages), by which related genes are grouped into modules according to their functionality, and these modules freely recombine during genetic exchanges between distinct phages infecting a given host. This theory could explain why hosts with rather different nucleotide sequence are sometimes infected by highly related bacteriophages. It appears that known prophages of LAB share a common gene arrangement throughout their genomes, with their modules organized according to the following well-conserved array: left attachment site (attL)-lysogeny-DNA replication-transcriptional regulation-DNA packaging-head-joining-tail-tail fiber-lysis modules-right attachment site (attR) (Canchaya et al. 2003b). The structural genes arrange in a particularly well-conserved order within a cluster, and hence it was proposed to base the taxonomy of *Siphoviridae* on the structural organization of this cluster (excluding the tail fiber genes) (Proux et al. 2002). This system permits a classification on three different major forms of head gene clusters, represented by the *cos*-site *Strep. thermophilus* phage Sfi21, the *pac*-site *Strep. thermophilus* phage Sfi11, and the *cos*-site *L. lactis* phage r1t.

In *Lb. johnsonii* strain NCC533 three prophage sequences were identified (Lj965, Lj928 and Lj771), but none was spontaneously induced nor could be recovered after MC-induction. Predicted gene map of Lj965, the largest identified prophage sequence, covered the late gene cluster (DNA packaging, morphogenesis, and lysis modules) of a typical *pac*-site temperate *Siphoviridae* and head and tail morphogenesis modules of *pac*-site *Strep. thermophilus* phages (Desiere et al. 2000). Apparently, both the 38- and 40-kb-long Lj928 and Lj965 are complete prophages within the genome of the strain and represent distinct lineages of Sfi11-like *pac*-site *Siphoviridae* unrelated at the DNA sequence level (Ventura et al. 2004).

Caso et al. (1995) investigated lysogeny on *Lb. plantarum* and found that only 12.7% of the supernatants of cultures treated with MC inhibited growth of other strains. Out of that 12.7%, two phages were able to produce plaques on a sensitive indicator strain. However, the authors were unable to further study the released temperate phages due to the instability of phage suspensions, even in the presence of protecting compounds. In addition, Φ LP2 was obtained from the whey of a homemade cheese and further work demonstrated its temperate nature. The temperate *pac-type* siphophage gle lysogenizes *Lb. plantarum* Gle, which was isolated from plant material in Japan. It was the second sequenced phage of *Lactobacillus* (Kodaira et al. 1997). The double-stranded DNA is composed of 42,259 bp, and encodes for sixty-two possible open reading frames (ORF) as well as several potential regulatory sequences. Functions were putatively assigned to several ORFs. An about 1,000-bp region present in Φg1e DNA was inferred to function as a promoter/repressor system for the Φ g1e lysogenic and lytic pathways. Poly-lysogeny is also present in *Lb. plantarum* strain WCFS1 (Ventura et al. 2003). Its chromosome contains four prophage elements. Lp1 and Lp2 are two about 40-kb-long uninducible prophages that share closely related DNA packaging, head and tail genes, defining a second lineage of *pac*-site *Siphoviridae* in *Lb. plantarum*, distinct from *Lb. plantarum* Φ g1e, but related to *Bacillus* phage SPP1 and *Lactococcus* phage TP901-1. Comparative genomics identified candidate LCG, including genes with sequence similarities to putative LCG from *Strep. pyogenes* prophages and to a *Bacillus* plasmid. R-Lp3 and R-Lp4 represent short prophage remnants; R-Lp3 abuts Lp2 and displays sequence links to *cos*-site *Siphoviridae*.

The siphophage Φ adh is a prophage harboured by *Lb. acidophilus* ADH (Raya et al. 1989), which was later renamed *Lb. gasseri.* It could be induced by either MC or UV treatment, and a prophage-cured derivative of the strain was isolated from cells that survived UV exposure. It was demonstrated that Φ adh exhibit classic lytic and lysogenic cycles of replication. The prophage was morphologically, physically, and genetically characterized. Altermann et al. (1999) reported its complete genome sequence. It is a *cos*-type phage with 62 ORFs, of which many presented homology to *Lb. casei* phage A2 and, to a less extent, *Lb. plantarum* Φ JL-1. Phage Φ adh was used to mediate transduction of plasmid DNA (Raya et al. 1989). Another prophage was found on the genome of *Lb. gasseri* ATCC 33323 (Ventura et al. 2006). The predicted prophage LgaI shares the modular organization of the Sfi11-like *pac*-site *Siphoviridae* phages, but belongs to the family of *Myoviridae* phages (Ismail et al.

2009). The LgaI DNA is packed by the ''head-full'' mechanism and it is responsible for the strong autolytic phenotype observed for *Lb. gasseri* ATCC 33323. In addition, a copy of the LgaI resides in the chromosome of *Lb. gasseri* NCK102 (a prophage-cured derivative of *Lb. gasseri* strain ADH). LgaI was not inducible by MC in *Lb. gasseri* NCK102-adh (a re-lysogenised derivative of NCK102); however, it apparently contributed to the autolytic phenotype of this strain (Ismail et al. 2009). LgaI showed no amino acid sequence identity with the corresponding structural proteins from Φ adh, thus revealing two different lineages of structural genes in phages that infect *Lb. gasseri.* Notably, genome of prophage LgaI shows a considerable sequence similarity and conserved symmetry with the *Lb. johnsonii* Lj771 prophage genome. Moreover, the genetic organization of part of the prophage LgaI showed high similarity with *Strep. pneumoniae* prophage EJ-1 (Ventura et al. 2006).

Ventura et al. (2006) have also sequenced *Lb. salivarius* UCC 188 and, based on the presumptive prophage genome length, only prophages Sal1 y Sal2 appear to be complete, whereas Sal3 y Sal4 seem to be prophage remnants. Comparison of sequences of prophage remnants showed significant sequence similarity at the DNA level.

Little information is available on the occurrence of bacteriophages for *Lb. fermentum*. De Klerk et al. (1965) described the first temperate phage active against this species, which belongs to the *Siphovoridae* family. In 2001, Foschino et al. reported the isolation of six temperate siphophages after MC-induction of *Lb. fermentum* strains, two of which exhibited prolate heads, whereas five were able to form lysis plaques on indicators strains. Based on their results, the authors concluded that both prolate phages are identical viral particles harboured in two strains isolated from different specimens at different places and times. Analogously, two other temperate phages were found to be the same, even when harboured in different strains, though isolated from the same Grana cheese whey sample. Protein profiles, restriction analysis of DNA and Southern blot hybridization revealed a high degree of homology between four of the temperate phages and partial homologies were also detected among temperate and other virulent phages. Data on host ranges, using 31 *Lb. fermentum* strains, showed two separated clusters containing both temperate and virulent phages (Foschino et al. 2001). Some years later, Zhang et al. (2006) induced with MC another temperate phage (Φ PYB5) from a *Lb. fermentum* strain isolated form Chinese yoghurt. This *Siphoviridae* prophage was able to alter the growth behaviour of the lysogenic strain compared with the cured derivative. This phenotypic difference was coincident with characteristics of lysogenic conversion. Recently, the complete genomic sequence of Φ PYB5 was determined (Zhang et al. 2011). Its genome is highly modular with functionally related genes clustered together, and showed to be closely related to that of prophage 100-23 (of heterofermentative *Lb. reuteri*) and bacteriophage 10MC (of heterofermetative *Leuc. oenos*) in an evolutionary aspect.

Regarding to phages of *Lactobacillus* of the *casei* group, the sequence of phages A2 (García et al. 2003), Φ AT3 (Lo et al. 2005), Lca1 (Ventura et al. 2006) and Lcazh1 (Zhang et al. 2010) have been reported. An analysis of the genomic sequences of phages A2, Φ AT3, Lca1 and Lc-Nu permitted concluding that all of them are related and share a common ancestor. Accordingly, prophage Lrm1 from the industrial *Lb. rhamnosus* strain M1 possesses a genomic organization similar to those closely related bacteriophages (Durmaz et al. 2008).

The first studies on phages specific of *Lb. casei* were carried out on two diverse groups represented by phages PL-1 and Φ FSW, respectively (Sèchaud et al. 1988). Phage PL-1 lyses *Lb. casei* ATCC 27092, following a classical lytic cycle, but at the same time can lysogenize

Lb. casei ATCC 334 (Stetter 1977). Consequently, it should still be regarded as a temperate phage. The clearest demonstration of the relationship between lysogenic and lytic phages and the mechanism through which lysogenic phage acquire virulence was provided by Shimizu-Kadota and her colleagues who undertook a detailed analysis of phages which were either lysogenic or lytic for *Lb. casei* S-1 (Shimizu-Kadota et al. 1983, 1985, Shimizu-Kadota and Tsuchida 1984). Φ FSW is a temperate phage of *Lb. casei* ATCC 27139, but simultaneously the source of numerous virulent phages (generically named Φ FSV) that were independently isolated from different Japanese factories, which produced milk beverages fermented with *Lb. casei* S-1 in a closed vat system (Shimizu-Kadota et al. 1983). The virulent phages were generated by either point mutations or interaction with insertion sequences (IS). IS-like elements contribute to strain genetic instability. They are common and widely distributed in Gram-negative bacteria, whereas there are only a few reports in Gram-positive bacteria. IS elements have been found in *L. lactis* and in the genera *Lactobacillus* and *Leuconostoc* (Davidson et al. 1996). Among these, ISL*1*, a mobile genetic element present on the bacterial chromosome of some strains of *Lb. casei*, was the first IS-like element isolated, identified and sequenced in LAB (Shimizu-Kadota et al. 1988). ISL*1* can be inserted at different positions within prophage FSW, changing it into virulent phages Φ FSVs (Shimizu-Kadota et al. 1985). This was proposed to be due to either inactivation of operator(s) or presentation of a new promoter that was not affected by the prophage repressor, thus allowing the constitutive expression of genes essential for lytic growth of the lysogenic phage. Unexpectedly, ISL*1* was found to have a narrow host range; it seems to be restricted to the single species *Lb. casei*, as it was absent from 14 other *Lactobacillus* strains (9 species) and 15 unrelated strains (8 genera, 12 species). What is more, only 3 out of 19 *Lb. casei* strains tested were evidenced to harbour ISL*1* (Shimizu-Kadota et al. 1988). The authors first identified and characterized the integrase gene and the recombination sites of Φ FSW, and then constructed the plasmid pMSK742, capable of stably integrating into the chromosome of probiotic *Lb. casei* strain Shirota (transformed phenotype was not lost after 20 generations without selection pressure), which therefore would be a tool for desired genes to be expressed in this probiotic strain. Nevertheless, the utilization of genetically modified probiotic organisms is unfortunately not accepted.

Temperate phage A2 was isolated from the whey of a failed manufacture of an artisanal blue cheese made without use of external starters (Herrero et al. 1994), and it has been comprehensively studied because of both its temperate nature and the presence of cohesive ends in its genome, attributes that make it useful for the development of derived cloning vectors. Its temperate nature was established considering its ability to form stable lysogenic derivatives on both *Lb. casei* ATCC 393 and of *Lb. casei* ATCC 27092, which released phages after MC-induction. A true DNA integration process was observed by Southern hybridization, confirming the derived strains to be actually lysogenic (Herrero et al. 1994). Their genome has a length of 43,411 bp and a G+C content of 44.8%, close to the 46% reported for its host *Lb. casei*. No extensive homology with other sequenced *Lactobacillus* phages was observed (except for the replication module of *Lb. gasseri* temperate bacteriophage Φ adh) (Altermann et al. 1999). It was classified as a *cos*-site Sfi21-like *Siphoviridae* but the morphogenetic cluster was most related to the *S. aureus* phage PVL, which belongs to a different group. Hence, phage A2 was postulated to represent a bridge between two groups, reinforcing the assumption of a mosaic organization originated by the massive DNA exchange among dsDNA phages (García et al. 2003). Its genome contains 61 identified *orfs* and presents a peculiar attribute: *orf40* to *orf48*, contained within a region of DNA of about 10 kbp, seem to be non-essential for the phage according to results obtained using deletion mutants; it was postulated that they would have a regulatory function, serving for a correct DNA packaging into the phage head (García et al. 2003). Morphology and functional studies (such as host range, one-step growth curves, effect of temperature on phage propagation) of phage A2 were also investigated (Herrero et al. 1994).

Temperate phage PL-2 was MC-induced from *Lb. casei* ATCC 27092. Lysis was optimal at pH 5.5 to 7.1, on cells in the exponential growth phase at 30°C. Phage PL-2 is not cultivable, since there is no available host. The phage had an isometric head (45 nm in diameter) and a flexible, noncontractile tail (150 nm long and 10 nm wide) with transverse striations and a sharp tip. The morphology of the induced phage was different from *Lb. casei* phages PL-1, Φ FSW and Φ FSV, A2 and Lc-Nu. Phage DNA (35.2 kb) had complementary cohesive ends and its digestion patterns with 13 restricion endonucleases were different from those of the corresponding ones with other LAB phages including phage PL-1 (Nakashima et al. 1998).

Ventura et al. (2006) reported the existence of a prophage sequence (Lca1) within the genome of *Lb. casei* ATCC 334. The prophage belonged to the group of Sfi11-like *pac*-site *Siphoviridae* and its genome has 46,986 kb; a number of functional modules were identified by database matches of some of the identified ORFs. When the deduced protein products were compared to those encoded by phage A2, similarities were only observed between proteins specified by the tail and tail fiber modules, as well as by the lysogeny module for the integrase and cI repressor-encoding ORFs. This suggested that, although both phages infect the same bacterial host, they are genetically rather different.

Temperate *Siphoviridae* Φ AT3 (Lo et al. 2005) was MC-induced from *Lb. casei* ATCC 393 and resembles Sfi21-like phages. Its genome was completely sequenced and includes an IS element (IS*LC3*) in a gene that encodes a putative structural protein (ORF14). The functions of most predicted ORFs are unknown although they share sequence homology with comparable ORFs of *Lb. casei* phage A2 within the replication and downstream regions. It was seen that Φ AT3 integrates at the $tRNA^{Arg}$ gene locus of *Lb. rhamnosus* HN 001, similar to that observed in its native host *Lb. casei* ATCC 393.

The recently sequenced genome of *Lb. casei* BL23, a probiotic dairy strain, showed on most genes ca. 99% identical sequences to *Lb. casei* ATCC 334 and widely conserved synteny with its chromosome, which is 0.2 Mb shorter. Both strains share a core genome that represents 77% of BL23 genome, and a significant fraction of the rest is due to prophages and insertion elements (Mazé et al. 2010). As it was previously mentioned, *Lb. casei* Zhang has one defective prophage integrated into its chromosome, designated as Lcazh1 (Zhang et al. 2010).

The virulent *Siphoviridae* phage Lc-Nu (Tuohimaa et al. 2006) can infect an industrial probiotic strain of *Lb. rhamnosus.* Despite the absence of an *attP* and an integrase gene, the genome contains remnants of a repressor gene, suggesting a temperate origin for this phage. Phage Lc-Nu showed to be closely related to temperate *Lb. casei* phages Φ AT3 and A2, sharing with the former high DNA homology over the late transcribed genes, and with phage A2 over the genetic switch region. Moreover, high protein homology (over 90% identity) was found between phages Lc-Nu and Φ AT3 in several structural proteins. Lrm1, a prophage isolated by MC-induction from the industrial *Lb. rhamnosus* strain M1, was completely sequenced and characterized (Durmaz et al. 2008). Electron micrographs from the lysates revealed only a few intact viral particles, being this probably the cause of failure to find of indicator strains. Lrm1 showed similarities in the genomic organization and mosaic pattern with closely related *Lb. casei* temperate phages A2, Φ AT3 and Lca1, and with the virulent *Lb. rhamnosus* phage Lc-Nu. The authors suggested that prophage-encoded DNA methylase or endonuclease activity could provide a crucial function for *Lb. rhamnosus* M1 survival, resulting in the stability of the defective prophage in its lysogenic state.

6. TECHNOLOGICAL IMPLICATIONS IN DAIRY INDUSTRY

Lysogenic strains may serve as a reservoir of phages that are potentially able to attack strains in mixed-strain starter cultures. Therefore it is of major interest for starter cultures not to mix lysogenic strains potentially capable of releasing temperate phages with indicator strains sensitive to them (Klaenhammer 1984). Generally, indicator strains for temperate phages are not easily found (Séchaud et al. 1988). In view of the low success rate, it must be considered that many of the phages detected by electron microscopy may be defective and, hence, not able to be propagated on indicator strains. However, it is also possible that strains tested as indicator strains are not sensitive to the temperate phages. Besides, DNA homology among different temperate phages and total bacterial DNA was demonstrated (Jarvis 1989) suggesting that temperate phages carried by different strains may exhibit homoimmunity with one another and therefore, such strains would be unsuitable as indicator strains for other temperate phages.

Furthermore, not only the potentially spontaneous induction is a problem derived from the use of lysogenic strains, but also the ability of overcoming the superinfection immunity of certain temperate phages. Shimizu-Kadota et al. (1983) have demonstrated that a virulent phage appearing during milk fermentation with *Lb. casei* S-1 was originated from the temperate phage carried by this strain as a lysogen. The transition from temperate to virulent activity may accompany a mutation on phage genome capable to cause a catastrophic failure in a lysogenic, single-strain starter culture, since the virulent phage could infect the entire starter population (Davidson et al. 1990).

By using DNA hybridization, it was clear that at least some temperate and non-temperate phages share homology, suggesting that some non-temperate phages are variants of temperate ones, or genes may have been exchanged by recombination during the evolution of both phage types (Davidson et al. 1990). Homologous regions found in both lytic and temperates phages could serve as a pool to allow genetic recombination and production of new phages (Lodics and Steenson 1993). Therefore, prophages within the chromosome of a strain used in large scale industrial fermentations are potential sources, via DNA recombination events, of new phages infective for the host strain (Durmaz et al. 2008).

When a phage is able to lyse a starter strain, fermentation can be retarded or even fail, depending on the virulence and concentration of the phage. For LAB merely used for milk acidification by dairy industries, sensitive starter strains or strains for which homologous phages are detected during manufacture can be replaced with new strains in a relatively easy manner (Lodics and Steenson 1993). In contrast, probiotic strains are unique (probiotic claims are strain-specific), need a long time after isolation to be effectively considered probiotic (many well-documented studies are necessary to assure their properties) and are in consequence rather expensive cultures. Furthermore, the complete lysis of the strain produces the absence of its specific beneficial properties in the final product (Capra 2007). For all the exposed reasons, strain rotation for probiotic bacteria would be at first glance inappropriate. In consequence, the economic impact of viral infections on probiotic bacteria is even worse than that on LAB (Capra et al. 2009).

7. Lysogeny in Probiotics and Its Relationship with Functionality: Contribution or Destabilization?

There are some concerns that would let to think about a close relationship between lysogeny and functionality on probiotic lactobacilli. First of all, probiotic strains possess physiological and functional properties that are strain-specific. Thus, their performance cannot be yet made based on scientific evidence and can not be extrapolated even to highly genetically related strains (Sanders 1999). Secondly, lysogeny is widely distributed over the genus *Lactobacillus* (Séchaud et al. 1988, Davidson et al. 1990, Suárez et al. 2008, Capra et al. 2010, Mercanti et al. 2011), including most of probiotic strains currently used by the dairy industry. In this sense, it is highly probable that prophages are in fact more frequently present in the genomes of probiotic lactobacilli than in those of LAB in general. What is more, there are reports of virulent phages originated from prophages in probiotic lactobacilli, which became lytic for the same strain that release them (Shimizu-Kadota et al. 1983, Capra et al. 2010). Finally, as prophages and prophage remnants would help bacteria to adapt and survive in their specific ecological environment, we wonder if lysogeny is not directly related to probioticity, either by improving or counteracting it. If this was the case, probiotic properties of bacteria would be quite instable and therefore justify a continuous monitoring along their history of use as food additives. In the next section, the connection between prophage presence and the relative numbers of lactobacilli population in mucosa is discussed.

8. Bacteriophage Disbalance in Mucosa: Putative Role on Intestinal and Vaginal Disorders

Gut microbiota is highly unstable, and its composition changes according to several factors, either intrinsic or extrinsic, such as age, dietary changes, gastric acidity, diseases and use of antibiotics (Fooks and Gibson 2002, Marshall 1999). Within a given bacterial population, strains can interact and compete with each other, and phages are expected to have a significant role in driving the biodiversity of this complex ecosystem (Ventura et al. 2010). Owing to their impressive numbers, bacteriophages could have an important impact as well: in a recent study (Lepage et al. 2008) the population of bacteriophages specifically associated to the gut mucosa was estimated on 10^{10} per mm^3 of tissue. Alemayehu et al. (2009) stated that these elevated concentrations would significantly impact on the composition of the stable microbiota by causing an imbalance between protective and harmful bacteria. In this sense, Lepage et al. (2008) found perceptibly higher numbers of bacteriophages in the gut mucosa of Crohn's disease patients than in healthy individuals; this clearly indicates that populations in the gut are really influenced by phage infections. As it was previously explained, a spontaneous induction, though at low rate, can actually occur, and temperate phages can turn into virulent ones (Shimizu-Kadota et al. 1985, 2000). Moreover, a prophage released by a bacterial strain is considered not able to infect the same strain (and in most cases this is true), but frequently attacks related ones. However, recent evidence indicates that remarkably heat-resistant phages C_L1 and C_L2, isolated from noninfected lysed-cultures of *Lb. paracasei* A (a commercial dairy strain), were capable to lyse this strain (Capra et al. 2010). Thus, taking into consideration the high percentage of lysogenic probiotic strains currently added to dairy foods and their potential to release phages that could be virulent for the own or a different strain present in the gut, and the consequent impact on human health, more research should be focused on this issue in the near future.

The role of lactobacilli as probiotics is not limited to the gastrointestinal tract. In effect, they also play an active role in the maintaining of vaginal health by diverse mechanisms, which in general block the action of the anaerobic microflora: production of organic acids and other antimicrobial substances, competition for nutrients, blocking of pathogen adhesion by competition for receptors, steric exclusion and co-aggregation with pathogenic bacteria (Lepargneur and Rousseau 2002). More than 70% of the microorganisms isolated from vaginal exudates of healthy and fertile women are lactobacilli, and in many cases they are virtually exclusive (Eschenbach et al. 1989). However, anaerobic bacteria frequently outnumber vaginal lactobacilli, leading to a disorder called bacterial vaginosis (BV) associated with several health risks, for example preterm delivery, low birth weight, and neonatal mortality (Hillier et al. 1995). Several studies have correlated the alteration of this equilibrium with the presence of virulent phages of vaginal *Lactobacillus*, or furthermore, with prophages that are induced under certain circumstances causing lactobacilli depletion. Kiliç et al. (2001) studied the distribution, virulence, and types of vaginal *Lactobacillus* phages isolated from women of the United States and Turkey. They found a large prevalence of lysogeny in clinical isolates of *Lactobacillus* strains; the incidence of MC-inducible prophages was of 28% and 36% for vaginal strains isolated from American and Turkish women, respectively. It was also observed an unexpectedly high frequency of spontaneous phage release by vaginal lactobacilli (from 10^{-3} to 10^{-4} per cell), and some strains released phages with a broad host range, lytic against other vaginal lactobacilli regardless of their geographic origin. What is more, Kiliç et al. (2001) found that many lysogenic strains were superinfected not only by different phages, but also by the same phage, likewise previously mentioned phages C_L1 and C_L2, capable to infect the strain which originated them (Capra et al. 2010). More recently, Martín et al. (2009) suggested hydrogen peroxide to play a crucial role in the balance of *Lactobacillus* strains in the vaginal mucosa. They corroborated the elevated occurrence of lysogeny on vaginal *Lactobacillus*, but found that about a half of the prophages were not responsive to SOS activation. Hydrogen peroxide was produced by most of the strains tested, and as it acts as an inducer of the SOS response, it was postulated a regulatory role for this compound, by selecting strains that harbour SOS-insensitive, defective prophages which are therefore unable to promote vaginal lactobacilli phage-induced lysis. On the other hand, it has long been known that women who smoke are more likely to suffer BV (Hillier et al. 1995). In view of that, Pavlova and Tao (2000) observed that trace amounts of the chemical compound benzo[α]pyrene diol epoxide, present in vaginal secretion of women who smoke, significantly increased phage induction in lactobacilli, thus suggesting a phage-mediated mechanism of action for the development of BV.

CONCLUSION

Lysogeny is a very complex phenomenon occurring in the majority of bacterial cells, including LAB strains and, within this group, probiotic lactobacilli widely used by food industries, especially dairy companies. Prophages harboured by lysogenic strains represent a continuous source of virulent phages; indeed, they have been present within bacterial chromosomes all throughout their evolutionary history. Besides, a close relationship between temperate and virulent phages becomes every day more evident. Prophages could be considered enemies, on one hand, as they are potentially able to destroy an entire population after an induction event; this has enormous economical implicances on dairy industries, where the replacement of probiotic strains is not as simple as that of a merely acidificant LAB strain. On the other hand, however, prophages could be regarded as friends, contributing to the fitness of the strain by providing several advantages by means of LCG. LCG were first detected on pathogenic microorganisms, but later they were postulated to be useful for the adaptation on any type of bacteria to their ecological niche. In this sense, it is important to consider the particular situation of probiotics, as almost all commercially exploited strains contained prophage sequences and come from human mucosa environment, where phages either temperate or virulent possess a demonstrated regulatory function on the relative populations of lactobacilli in intestine and vagina. For all that has been exposed, we should reflect on the significance of prophages in probiotic lactobacilli not only from the point of view of their aptitude to kill an entire population during an industrial fermentation, but also analyzing their potential to either change the probiotic potential of the harbouring strain or alter the entire mucosa-associated ecosystem.

REFERENCES

Ackermann, H.-W., and DuBow, M. S. (1987). Viruses of Prokaryotes. In *General Properties of Bacteriophages Vol I* (pp. 202). Boca Raton, USA: CRC Press.

Alatossava, T., Forsman, P. and Ritzenthaler, P. (1995). Genome homology and superinfection immunity between temperate and virulent *Lactobacillus delbrueckii* bacteriophages. *Archives of Virology, 140(12)*, 2261-2268.

Alemayehu, D., Ross, R. P., O'Sullivan, O., Coffey, A., Stanton, C., Fitzgerald, G. F. and McAuliffe, O. (2009). Genome of a virulent bacteriophage Lb338-1 that lyses the probiotic *Lactobacillus paracasei* cheese strain. *Gene, 448,* 29-39.

Altermann, E., Klein, J. R. and Henrich, B. (1999). Primary structure and features of the genome of the *Lactobacillus gasseri* temperate bacteriophage Φ adh. *Gene, 236,* 333-346.

Arendt, E., Lonvaud, A. and Hamm, W. P. (1991a). Lysogeny in *Leuconostoc oenos*. *Journal of General Microbiology, 137*, 2135-2139.

Arendt, E. K., Neve, H. and Hammes, W. P. (1991b). Characterization of phage isolates from a phage-carrying culture of *Leuconostoc oenos* 58N. *Applied Microbiology and Biotechnology, 34(2),* 220-224.

Auad, L., De Ruiz Holgado, A. A. P., Forsman, P., Alatossava, T. and Raya, R. R. (1997). Isolation and characterization of a new *Lactobacillus delbrueckii* ssp. *bulgaricus* temperate bacteriophage. *Journal of Dairy Science, 80(11)*, 2706-2712.

Auad, L., Räisänen, L., Raya, R. R. and Alatossava, T. (1999). Physical mapping and partial genetic characterization of the *Lactobacillus delbrueckii* subsp. *bulgaricus* bacteriophage lb539. *Archives of Virology, 144(8),* 1503-1512.

Betley, M. J. and Mekalanos, J. J. (1985). Staphylococcal enterotoxin A is encoded by a phage. *Science, 229,* 185-187.

Botstein, D. (1980). A theory of modular evolution for bacteriophages. *Annals of the New York Academy of Sciences, 354,* 484-490.

Boyd, E. F. and Brüssow, H. (2002). Common themes among bacteriophage-encoded virulence factors and diversity among the bacteriophages involved. *Trends in Microbiology, 10,* 521-529.

Brüssow, H. (2001). Phages of dairy bacteria. *Annual Review of Microbiology, 55,* 283-303.

Brüssow, H. and Hendrix, R. W. (2002). Phage genomics: small is beautiful (minireview). *Cell, 108,* 13-16.

Brüssow, H., and Kutter, E. (2005). Phage ecology. In E. Kutter, and A. Sulakvelidze (Eds.), *Bacteriophages. Biology and applications* (pp. 129-163). Boca Raton, FL, USA: CRC Press.

Canchaya, C., Fournous, G., Chibani-Chennoufi, S., Dillmann, M. -L. and Brussow, H. (2003a). Phage as agents of lateral gene transfer. *Current Opinion in Microbiology, 6,* 417-424.

Canchaya, C., Proux, C., Fournous, G., Bruttin, A. and Brüssow, H. (2003b). Prophage genomics. *Microbiology and Molecular Biology Reviews, 67,* 238-276.

Capra, M. L. (2007). Bacteriofagos de *Lactobacillus casei / paracasei.* Caracterización y estudio de la fagorresistencia, Doctoral Thesis. Santa Fe, Argentina: Universidad Nacional del Litoral Facultad de Bioquímica y Ciencias Biológicas.

Capra, M. L., Quiberoni A., Ackermann, H.-W., Moineau, S. and Reinheimer, J. A. (2006). Characterization of a new virulent phage (MLC-A) of *Lactobacillus paracasei. Journal of Dairy Science, 89,* 2414-2423.

Capra, M. L., Binetti, A. G., Mercanti, D. J., Quiberoni, A. and Reinheimer, J. A. (2009). Diversity among *Lactobacillus paracasei* phages isolated from a probiotic dairy product plant. *Journal of Applied Microbiology, 107(4),* 1350-1357.

Capra, M. L., Mercanti, D. J., Reinheimer, J. A. and Quiberoni, A. L. (2010). Characterisation of three temperate phages released from the same *Lactobacillus paracasei* commercial strain. *International Journal of Dairy Technology, 63(3),* 396-405.

Carminati, D., and Giraffa, G. (1992). Evidence and characterization of temperate bacteriophage in *Streptococcus salivarius* subsp. *thermophilus* St18. *Journal of Dairy Research, 59,* 71-79.

Carminati, D., Mazzucotelli, L., Giraffa, G. and Neviani, E. (1997). Incidence of inducible bacteriophage in *Lactobacillus helveticus* strains isolated from natural whey starter cultures. *Journal of Dairy Science, 80,* 1505-1511.

Caso, J. L., De Los Reyes, C. G., Herrero, M., Montilla, A., Rodriguez, A. and Suarez, J. E. (1995). Isolation and characterization of temperate and virulent bacteriophages of *Lactobacillus plantarum. Journal of Dairy Science, 78,* 741-750.

Cavin, J. F., Drici, F. Z., Prevost, H. and Divies, C. (1991). Prophage curing in *Leuconostoc oenos* by mitomycin C induction. *American Journal of Enology and Viticulture, 42(3),* 163-166.

Chopin, A., Bolotin, A., Sorokin, A., Ehrlich, S. D. and Chopin, M. (2001). Analysis of six prophages in *Lactococcus lactis* IL1403: different genetic structure of temperate and virulent phage populations. *Nucleic Acids Research, 29,* 644-651.

Cluzel, P. J., Veaux, M., Rousseau, M., and Accolas, J. P. (1987a). Evidence for temperate bacteriophages in two strains of *Lactobacillus bulgaricus*. *Journal of Dairy Research, 54,* 397-405.

Cluzel, P.-J., Serio, J. and Accolas, J.-P. (1987b). Interactions of *Lactobacillus bulgaricus* temperate bacteriophage 0448 with host strains. *Applied and Environmental Microbiology, 53(8)*, 1850-1854.

Davidson, B. E., Powell, I. B. and Hillier, A. (1990). Temperate bacteriophages and lysogeny in lactic acid bacteria. *FEMS Microbiology Reviews, 87,* 79-90.

Davidson, B. E., Kordias, N., Dobos, M. and Hillier, A. J. (1996). Genomic organization of lactic acid bacteria. *Antonie van Leeuwenhoek, 70*, 161-183.

De Klerk, H. C., Coetzee, J. N. and Fourie, J. T. (1965). The fine structure of *Lactobacillus* bacteriophages. *Journal of General Microbiology, 38*, 35-38.

Desiere, F., Pridmore, R. D. and Brüssow, H. (2000). Comparative genomics of the late gene cluster from *Lactobacillus* phages. *Virology, 275,* 294-305.

Desiere, F., McShan, W. M., van Sinderen, D., Ferretti, J. J. and Brüssow, H. (2001). Comparative genomics reveals close genetic relationships between phages from dairy bacteria and pathogenic streptococci: evolutionary implications for prophage-host interactions. *Virology, 288,* 325-341.

Deutsch, S.-M., Neveu, A., Guezenec, S., Ritzenthaler, P. and Lorta, S. (2003). Early lysis of *Lactobacillus helveticus* CNRZ 303 in Swiss cheese is not prophage-related. *International Journal of Food Microbiology, 81(2)*, 147-157.

Durmaz, E., Miller, M. J., Azcarate-Peril, M. A., Toon, S. P. and Klaenhammer, T. R. (2008). Genome sequence and characteristics of Lrm1, a prophage from industrial *Lactobacillus rhamnosus* strain M1. *Applied and Environmental Microbiology, 74,* 4601-4609.

Ebrecht, A. C., Guglielmotti, D. M., Tremmel, G., Reinheimer, J. A. and Suárez, V. B. (2010). Temperate and virulent *Lactobacillus delbrueckii* bacteriophages: Comparison of their thermal and chemical resistance. *Food Microbiology, 27,* 515-520.

Eschenbach, D. A., Davick, P. R., Williams, B. L., Klebanoff, S. J., Young-Smith, K., Critchlow, C. M. and Holmes, K. K. (1989). Prevalence of hydrogen peroxide-producing *Lactobacillus* species in normal women and women with bacterial vaginosis. *Journal of Clinical Microbiology, 27,* 251-256.

Fayard, B., Haeflinger, M. and Accolas, J.-P. (1993). Interactions of temperate bacteriophages of *Streptococcus salivarius* subsp. *thermophilus* with lysogenic indicators affect phage DNA restriction pattern and host range. *Journal of Dairy Research, 60,* 385-399.

Ferretti, J. J., McShan, W. M., Ajdic, D., Savic, D. J., Savic, G., Lyon, K., Primeaux, C., Sezate, S., Suvorov, A. N., Kenton, S., Lai, H. S., Lin, S. P., Qian, Y., Jia, H. G., Najar, F. Z., Ren, Q., Zhu, H., Song, L., White, J., Yuan, X., Clifton, S. W., Roe, B. A. and McLaughlin, R. (2001). Complete genome sequence of an M1 strain of *Streptococcus pyogenes*. *Proceedings of the National Academy of Sciences of the United States of America, 98,* 4658-4663.

Fooks, L. J. and Gibson, G. R. (2002). Probiotics as modulators of the gut flora. *The British Journal of Nutrition, 88(Suppl. 1),* S39-S49.

Foschino, R., Picozzi, C. and Galli, A. (2001). Comparative study of nine *Lactobacillus fermentum* bacteriophages. *Journal of Applied Microbiology, 91,* 394-403.

García, P., Ladero, V. and Suárez, J. E. (2003). Analysis of the morphogenetic cluster and genome of the temperate *Lactobacillus casei* bacteriophage A2. *Archives of Virology, 48,* 1051-1070.

Goshorn, S. C. and Schlievert, P. M. (1989). Bacteriophage association of streptococcal pyrogenic exotoxin type C. *Journal of Bacteriology, 171,* 3068-3073.

Graham, D. M., Parmelee, C. E. and Nelson, F. E. (1952). The carrier state of lactic *Streptococcus* bacteriophage. *Journal of Dairy Science, 35(10),* 813-822.

Guttman, B., Raya, R., and Kutter, E. (2005). Basic phage biology. In E. Kutter, and A. Sulakvelidze (Eds.), *Bacteriophages. Biology and applications* (pp. 29-66). Boca Ratón, Florida, USA: CRC Press.

Hayashi, T., Makino, K., Ohnishi, M., Kurokawa, K., Ishii, K., Yokoyama, K., Han, C. G., Ohtsubo, E., Nakayama, K., Murata, T., Tanaka, M., Tobe, T., Iida, T., Takami, H., Honda, T., Sasakawa, C., Ogasawara, N., Yasunaga, T., Kuhara, S., Shiba, T., Hattori, M. and Shinagawa, H. (2001). Complete genome sequence of enterohemorrhagic *Escherichia coli* O157:H7 and genomic comparison with a laboratory strain K-12. *DNA Research, 8,* 11-22.

Herrero, M., de los Reyes-Gavilán, C. G., Caso, J. L. and Suárez, J. E. (1994). Characterization of Φ 393-A2, a bacteriophage that infects *Lactobacillus casei. Microbiology, 140,* 2585-2590.

Hillier, S. L., Nugent, R. P., Eschenbach, D. A., Krohn, M. A., Gibbs, R. S., Martin, D. H., Cotch, M. F., Edelman, R., Pastorek, J. G., Rao, A. V., McNellis, D., Regan, J. A., Carey, J. C. and Klebanoff, M. A. (1995). Association between bacterial vaginosis and preterm delivery of a low-birth-weight infant. The vaginal infections and prematurity study group. *The New England Journal of Medicine, 333,* 1737-1742.

Huang, A., Friesen, J. and Brunton, J. L. (1987). Characterization of a bacteriophage that carries the genes for production of Shiga-like toxin 1 in *Escherichia coli. Journal of Bacteriology, 169,* 4308-4312.

Ismail, E. A., Neve, H., Geis, A. and Heller, K. J. (2009). Characterization of temperate *Lactobacillus gasseri* phage LgaI and its impact as prophage on autolysis of its lysogenic host strains. *Current Microbiology, 58*, 648-653.

Jang, S. H., Hwang, M. H. and Chang, H. -I. (2010). Complete genome sequence of UMH1, a *Leuconostoc* temperate phage. *Archives of Virology, 155*, 1883-1885.

Jarvis, A. W. (1984). DNA-DNA homology between lactic streptococci and their temperate and lytic phages. *Applied and Environmental Microbiology, 47(5),* 1031-1038.

Jarvis, A. W. (1989). Bacteriophages of lactic acid bacteria. *Journal of Dairy Science, 72,* 3406-3428.

Josephsen, J., and Neve, H. (1998). Bacteriophages and lactic acid bacteria. In S. Salminen, and A. von Wright (Eds.), *Lactic Acid Bacteria. Microbiology and Functional Aspects* (pp. 385-436). New York, USA: Marcel Dekker Inc.

Kiliç, A. O., Pavlova, S. I., Alpay, S., Kiliç, S. S. and Tao, L. (2001). Comparative study of vaginal *Lactobacillus* phages isolated from women in the United States and Turkey: prevalence, morphology, host range, and DNA homology. *Clinical and Diagnostic Laboratory Immunology, 8,* 31-39.

Klaenhammer, T. R., and Fitzgerald, G. F. (1994). Bacteriophages and bacteriophage resistance. In M. J. Gasson, and W. de Vos (Eds.), *Genetics and biotechnology of lactic acid bacteria* (pp. 106-168). Glasgow, UK.: Blackie Academic and Professional.

Klaenhammer, T. R. (1984). Interactions of bacteriophages with lactic streptococci. *Advances in Applied Microbiology, 30,* 1-29.

Kleerebezem, M., Boekhorst, J., van Kranenburg, R., Molenaar, D., Kuipers, O. P., Leer, R., Tarchini, R., Peters, S. A., Sandbrink, H. M., Fiers, M. W. E. J., Stiekema, W., Lankhorst, R. M. K., Bron, P. A, Hoffer, S. M., Groot, M. N. N., Kerkhoven, R., de Vries, M., Ursing, B., de Vos, W. M. and Siezen, R. (2003). Complete genome sequence of *Lactobacillus plantarum* WCFS1. *Proccedings of the National Academy of Sciences of the United States of America, 100(4),* 1990-1995.

Kodaira, K. I., Oki, M., Kakikawa, M., Watanabe, N., Hirakawa, M., Yamada, K. and Taketo, A. (1997). Genome structure of the *Lactobacillus* temperate phage Φ g1e: the whole genome sequence and the putative promoter/repressor system. *Gene, 187(1),* 45-53.

Labrie, S. J. and Moineau, S. (2007). Abortive infection mechanisms and prophage sequences significantly influence the genetic make up of emerging lytic lactococcal phages. *Journal of Bacteriology, 189,* 1482-1487.

Lahbib-Mansais, Y., Mata, M., and Ritzenthaler, P. (1988). Molecular taxonomy of *Lactobacillus* phages. *Biochimie, 70,* 429-435.

Lawrence, J. G., Hendrix, R. W. and Casjens, S. (2001). Where are the pseudogenes in bacterial genomes? *Trends in Microbiology, 9,* 535-540.

Lee, C. Y., and Iandolo, J. J. (1986). Lysogenic conversion of staphylococcal lipase is caused by insertion of the bacteriophage L54a genome into the lipase structural gene. *Journal of Bacteriology, 166(2)*, 385-391.

Lepage, P., Colombet, J., Marteau, P., Sime-Ngando, T., Doré, J. and Leclerc, M. (2008). Dysbiosis in inflammatory bowel disease: a role for bacteriophages? *Gut, 57,* 424-425.

Lepargneur, J.-P. and Rousseau, V. (2002). Rôle protecteur de la flore de Doderleïn. *Journal de Gynécologie Obstétrique et Biologie de la Reproduction, 31,* 485-494.

Lo, T. C., Shih, T. C., Lin, C. F., Chen, H. W. and Lin, T. H. (2005). Complete genomic sequence of the temperate bacteriophage □AT3 isolated from *Lactobacillus casei* ATCC 393. *Virology, 339,* 42-55.

Lodics, T. A. and Steenson, L. R. (1993). Phage-host interactions in commercial mixed-strain dairy starter cultures: practical significance – A review. *Journal of Dairy Science, 76(8),* 2380-2391.

Loeffler, J. M. and Fischetti, V. A. (2006). Lysogeny of *Streptococcus pneumoniae* with MM1 phage: improved adherence and other phenotypic changes. *Infection and Immunity, 74(8),* 4486-4495.

Marshall, J. C. (1999). Gastrointestinal flora ad its alterations in critical illness. *Current Opinion in Clinical Nutrition and Metabolic Care, 2,* 405-411.

Martín, R., Soberón, N., Escobedo, S. and Suárez, J. E. (2009). Bacteriophage induction versus vaginal homeostasis: role of H_2O_2 in the selection of *Lactobacillus* defective prophages. *International Microbiology, 12,* 131-136.

Mata, M., Trautwetter, A., Luthaud, G. and Ritzenthaler, P. (1986). Thirteen virulent and temperate bacteriophages of *Lactobacillus bulgaricus* and *Lactobacillus lactis* belong to a single DNA homology group. *Applied and Environmental Microbiology, 52(4)*, 812-818.

Mazé, A., Boël, G., Zúñiga, M., Bourand, A., Loux, V., Yebra, M. J., Monedero, V., Correia, K., Jacques, N., Beaufils, S., Poncet, S., Joyet, P., Milohanic, E., Casarégola, S., Auffray, Y., Pérez-Martínez, G., Gibrat, J. F., Zagorec, M., Francke, C., Hartke, A. and Deutscher, J. (2010). Complete genome sequence of the probiotic *Lactobacillus casei* strain BL23. *Journal of Bacteriology, 192,* 2647-2648.

Mercanti, D. J., Carminati, D., Reinheimer, J. A. and Quiberoni, A. (2011). Widely distributed lysogeny in probiotic lactobacilli represents a potentially high risk for the fermentative dairy industry. *International Journal of Food Microbiology, 144,* 503-510.

Nakashima, Y., Hasuwa, H., Kakita, Y., Murata, K., Kuroiwa, A., Miake, F. and Watanabe, K. (1998). A temperate phage with cohesive ends induced by mitomycin C treatment of *Lactobacillus casei. Archives of Virology, 143,* 1621-1626.

Nakayama, K., Kanaya, S., Ohnishi, M., Terawaki, Y. and Hayashi, T. (1999). The complete nucleotide sequence of Φ CTX, a cytotoxin-converting phage of *Pseudomonas aeruginosa*: implications for phage evolution and horizontal gene transfer via bacteriophages. *Molecular Microbiology, 31,* 399-419.

Nechaev, S. and Severinov, K. (2008). The elusive object of desire - interactions of bacteriophages and their hosts. *Current Opinion in Microbiology, 11(2),* 186-193.

Pavlova, S. I. and Tao, L. (2000). Induction of vaginal *Lactobacillus* phages by the cigarette smoke chemical benzo[α]pyrene diol epoxide. *Mutation Research, 466,* 57-62.

Proux, C., van Sinderen, D., Suarez, J., Garcia, P., Ladero, V., Fitzgerald, G. F., Desiere, F. and Brüssow, H. (2002). The dilemma of phage taxonomy illustrated by comparative genomics of Sfi21-like *Siphoviridae* in lactic acid bacteria. *Journal of Bacteriology, 184,* 6026-6036.

Raya, R. R., Kleeman, E. G., Luchansky, J. B. and Klaenhammer, T. R. (1989). Characterization of the temperate bacteriophage Φ adh and plasmid transduction in *Lactobacillus acidophilus* ADH. *Applied and Environmental Microbiology, 55(9),* 2206-2213.

Sakurai, T., Takahashi, T. and Arai, H. (1970). The temperate phages of *Lactobacillus salivarius* and *Lactobacillus casei*. *Japanese Journal of Microbiology, 14(4)*, 333-336.

Sanders, M. E. (1999). Probiotics. *Food Technology, 53,* 67-76.

Séchaud, L., Cluzel, P. J., Rousseau, M., Baumgartner, A. and Accolas, J. P. (1988). Bacteriophages of lactobacilli. *Biochimie, 70*, 401-410.

Séchaud, L., Rousseau, M., Fayard, B., Callegari. M. L., Quénée, P., and Accolas, J.-P. (1992). Comparative study of 35 bacteriophages of *Lactobacillus helveticus*: morphology and host range. *Applied and Environmental Microbiology, 58(3),* 1011-1018.

Shimizu-Kadota, M. and Sakurai, T. (1982). Prophage curing in *Lactobacillus casei* by isolation of a thermo inducible mutant. *Applied and Environmental Microbiology, 43,* 1284-1287.

Shimizu-Kadota, M. and Tsuchida, N. (1984). Physical mapping of the virion and the prophage DNAs of a temperate *Lactobacillus phage* Φ FSW. *Journal of General Microbiology, 130,* 423-430.

Shimizu-Kadota, M., Sakurai, T. and Tsuchida, N. (1983). Prophage origin of a virulent phage appearing on fermentations of *Lactobacillus casei* S-1. *Applied and Environmental Microbiology, 45(2),* 669-674.

Shimizu-Kadota, M., Kiwaki, M., Hirokawa, H. and Tsuchida, N. (1985). ISL*1*: a new transposable element in *Lactobacillus casei*. *Molecular and General Genetics, 200,* 193-198.

Shimizu-Kadota, M., Flickinger, J. L. and Chassy, B. M. (1988). Evidence that *Lactobacillus casei* insertion element ISL*1* has a narrow host range. *Jouirnal of Bacteriology, 170,* 4976-4978.

Shimizu-Kadota, M., Kiwaki, M., Sawaki, S., Shirasawa, Y., Shibahara-Sone, H. and Sako, T. (2000). Insertion of bacteriophage Φ FSW into the chromosome of *Lactobacillus casei* strain Shirota (S-1): characterization of the attachment sites and the integrase gene. *Gene, 249,* 127-134.

Soberón, N., Martín, R. and Suárez, J. E. (2007). New method for evaluation of genotoxicity, based on the use of Real-Time PCR and lysogenic Gram-positive and Gram-negative bacteria. *Applied and Environmental Microbiology, 73,* 2815-2819.

Stetter, K. O. (1977). Evidence for frequent lysogeny in lactobacilli: temperate bacteriophages within the subgenus *Streptobacterium*. *Journal of Virology, 24,* 685-689.

Suárez, V., Zago, M., Quiberoni, A., Carminati, D., Giraffa, G. and Reinheimer, J. (2008). Lysogeny in *Lactobacillus delbrueckii* strains and characterization of two new temperate prolate-headed bacteriophages. *Journal of Applied Microbiology, 105,* 1402-1411.

Terzaghi, B. and Sandine, W. E. (1981). Bacteriophage production following exposure of lactic streptococci to ultra violet radiation. *Journal of General Microbiology, 122,* 305-311.

Tohyama, K., Sakurai, T., Arai, H. and Oda, A. (1972). Studies on temperate phages of *Lactobacillus salivarius* I. Morphological, biological, and serological properties of newly isolated temperate phages of *Lactobacillus salivarius*. *Japanese Journal of Microbiology, 16(5),* 385-395.

Tohyama, K. (1973). Studies on temperate phages of *Lactobacillus salivarius* II. Mode of action of defective phage 208 on nonlysogenic and homologous lysogenic *Lactobacillus saliavarius* strains. *Japanese Journal of Microbiology, 17(3),* 173-180.

Tuohimaa, A., Riipinen, K. A., Brandt, K. and Alatossava, T. (2006). The genome of the virulent phage Lc-Nu of probiotic *Lactobacillus rhamnosus*, and comparative genomics with *Lactobacillus casei* phages. *Archives of Virology, 151,* 947-965.

Vasala, A., Dupont, L., Baumann, M., Ritzenthaler, P. and Alatossava, T. (1993). Molecular comparison of the structural proteins encoding gene clusters of two related *Lactobacillus delbrueckii* bacteriophages. *Journal of Virology, 67(6),* 3061-3068.

Ventura, M., Canchaya, C., Kleerebezem, M., de Vos, W. M., Siezen, R. J., and Brüssow, H. (2003). The prophage sequences of *Lactobacillus plantarum* strain WCFS1. *Virology, 316(2),* 245-255.

Ventura, M., Canchaya, C., Pridmore, R. D. and Brüssow, H. (2004). The prophages of *Lactobacillus johnsonii* NCC 533: comparative genomics and transcription analysis. *Virology, 320,* 229-242.

Ventura, M., Canchaya, C., Bernini, V., Altermann, E., Barrangou, R., McGrath S., Claesson, M. J., Li, Y., Leahy,S., Walker, C. D., Zink, R., Neviani. E., Stelle. J., Broadbent. J., Klaenhammer, T. R., Fitzgerald, G. F., O'Toole, P. W. and van Sinderen, D. (2006). Comparative genomics and transcriptional analysis of prophages identified in the genomes of *Lactobacillus gasseri*, *Lactobacillus salivarius*, and *Lactobacillus casei*. *Applied and Environmental Microbiology, 72,* 3130-3146.

Ventura, M., Sozzi, T., Turroni, F., Matteuzzi, D. and van Sinderen, D. (2010). The impact of bacteriophages on probiotic bacteria and gut microbiota diversity. *Genes and Nutrition*, DOI 10.1007/s12263-010-0188-4.

Villion, M. and Moineau, S. (2009). Bacteriophages of *Lactobacillus*. *Frontiers in Bioscience, 14*, 1642-1660.

Weeks, C. R. and Ferretti, J. J. (1984). The gene for type A streptococcal exotoxin (erythrogenic toxin) is located in bacteriophage T12. *Infection and Immunity, 46,* 531-536.

Weinbauer, M. G. (2004). Ecology of prokaryotic viruses. *FEMS Microbiology Reviews, 28,* 127-181.

Wiederholt, K. M. and Steele, J. L. (1993). Prophage curing and partial characterization of temperate bacteriophages from thermolytic strains of *Lactococcus lactis* ssp. *cremoris*. *Journal of Dairy Science, 76,* 921-930.

Yokokura, T., Kodaira, S., Ishiwa, H. and Sakurai, T. (1974). Lysogeny in lactobacilli. *Journal of General Microbiology, 84,* 277-284.

Zhang, X., Kong, J., and Qu, Y. (2006). Isolation and characterization of a *Lactobacillus fermentum* temperate bacteriophage from Chinese yogurt. *Journal of Applied Microbiology,* 101, 857-863.

Zhang, W.-Y., Yu, D.-L., Sun, Z.-H., Chen, W., Hu, S.-N., Meng, H. and Zhang, H.-P. (2010). The comparative analysis of a prophage remnant Lcazh1 in relation to other *Lactobacillus* prophages, particularly Lp3. *International Journal of Dairy Technology, 63,* 413-417.

Zhang, X., Wanga, S., Guoa, T. and Kong, J. (2011). Genome analysis of *Lactobacillus fermentum* temperate bacteriophage Φ PYB5. *International Journal of Food Microbiology, 144(3)*, 400-405.

In: Bacteriophages in Dairy Processing
Editor: Andrea del Lujan Quiberoni et al.

ISBN 978-1-61324-517-0

Chapter 7

PHAGE RESISTANCE IN LACTIC ACID BACTERIA

Geneviève M. Rousseau, Hélène Deveau and Sylvain Moineau*
Département de Biochimie, de Microbiologie et de Bio-informatique,
Faculté des Sciences et de Génie, Groupe de Recherche en Écologie Buccale,
Faculté de Médecine Dentaire, Félix d'Hérelle Reference Center for Bacterial Viruses,
Université Laval, Québec City, Montreal, Canada

ABSTRACT

Lactic Acid Bacteria (LAB) are used in a variety of industrial fermentation processes because of their ability to convert a variety of substrates into complex products. Any technological process that relies on bacterial fermentation is vulnerable to bacteriophage infections. This chapter describes the relationship between bacteriophages and their LAB hosts in the context of the food fermentation industry. Specifically, we highlight the most significant antiviral mechanisms in LAB. The primary focus will be given to LAB used by the dairy industry because it has openly acknowledged the problem of phage infections and has teamed up with academia and starter cultures companies to develop natural and engineered phage resistance systems to curtail the propagation of diverse phages.

INTRODUCTION

Bacteria are among the most prolific organisms on earth, but bacteriophages (or phages) are estimated to be about 10-fold as numerous as bacteria (Whitman et al. 1998, Breitbart et al. 2005). Because of their rapid proliferation, with a progeny of up to hundreds per infected cell, in as little as tens of minutes to a day, phages could rapidly eradicate the bacterial population. Resistance mechanisms allow both populations to establish an equilibrium essential to their existence and co-evolution (Emond and Moineau 2007).

* E-mail address: sylvain.moineau@bcm.ulaval.ca. Tel.: + 1 418 656 3712; fax: +1 418 656 2861

The Problem

Virulent phages are often common in the majority of industrial processes using bacterial fermentation and the food fermentation industry is particularly susceptible to phage infections. Although phages don't represent a health hazard to consumers, phage infections can significantly slow down or even halt fermentation processes and alter product quality. Losses can be considerable and the risk associated with the phages cannot be neglected. It has been estimated that between 0.1 to 10% of dairy fermentation batches are compromised at various levels by phages (Moineau and Lévesque 2005, Bogosian et al. 2006). These phage infection events not only represent loss of revenues (due to failed or low quality products), but also a problem of additional costs due to lower productivity (Emond and Moineau 2007).

The Cause

Among lactic acid bacteria (LAB), fermentation failures or slowdowns have been reported more frequently when strains of *Lactococcus lactis* or *Streptococcus thermophilus* are used in processes. Some cases were also reported for strains of various *Lactobacillus* species (Brüssow and Súarez 2006, Villion and Moineau 2009). Since *L. lactis* is the most frequently used LAB on a daily basis, several studies have been published on the characterization of virulent lactococcal phages (Deveau et al. 2006b). These virulent phages are by far the main cause of milk fermentation failures. Interestingly, phages infecting *Lactobacillus* strains show the largest morphological and genetic diversity, possibly reflecting the large number of species in the *Lactobacillus* genus (Villion and Moineau 2009). Phages infecting *Streptococcus thermophilus* form a more homogenous group (LeMarrec et al. 1997, Guglielmotti et al. 2009). Although ten groups of lactococcal phages were reported (Deveau et al. 2006b), only three of them (936, c2, and P335 groups) are frequently isolated from failed industrial dairy fermentations. *Leuconostoc* phages have been isolated from sauerkraut, coffee and dairy fermentations (Boizet et al. 1992, Davey et al. 1995, Barrangou et al. 2002). In all cases, the isolated phages belong to the *Caudovirales* order (tailed bacteriophages with double stranded DNA genome located in the capsid). Usually, fermentation failures will be due to a clonal population of phages specifically infecting the LAB strains used. The use of other strains will likely dilute the dominant phages but when other LAB strains are used others phages will emerge. It should be noted that any given phage could stay within a factory for a prolonged period of time, despite the use of anti-viral hurdles (Rousseau and Moineau 2009).

The Consequences

The consequences of a phage infection include product loss, raw material spoilage, non-productive operation costs, and plant shutdown for decontamination purposes (Emond and Moineau 2007). In order to decrease the frequency of phage infections, many countermeasures are available and should be used. They include a good factory design, stringent sanitation procedures, efficient ventilation, phage monitoring protocol, LAB strains selections, and culture rotation schemes. These measures have been discussed elsewhere (Emond and Moineau 2007, Labrie and Moineau 2010). Nonetheless, these control strategies are adding to the costs of manufacturing fermented products.

THE DETECTION

To detect phage outbreaks, different types of assays can be used. Plaque assays with pure bacterial cultures is the most common method. The observation of clear phage plaques on plates confirms the presence of virulent phages in the fermentation vats. Another advantage of this microbiological assay is that it also informs on the level of contamination (PFU ml^{-1}) if the tested samples have been previously serially diluted. Finally, it also allows to rapidly identify LAB strains that are sensitive and insensitive to the predominant phages. However, whenever plaque assays cannot be done, tests assessing bacterial activity can be performed. The most common test of this type is the milk activity test. Briefly, the pH of milk containing starter culture and a sample to be tested is compared, after incubation, to the pH of a tube containing only the starter culture. If the pH is higher by at least 0.2 units, this may indicate reduced metabolic activity of the cells due phage infection (Moineau and Lévesque 2005).

A number of molecular methods have also been developed to detect the genomes of these phages (Labrie and Moineau 2000, Quiberoni et al. 2006, del Rio et al. 2007, Binetti et al. 2008, del Rio et al. 2008, Verreault et al. 2011). These PCR and qPCR systems offer to advantage of rapidly identifying the group of phages responsible for a fermentation failure. However, it does not determine which strains are sensitive to the phages.

THE TREATMENT

In the event of redundant phage outbreaks, the starter cultures containing the phage sensitive cells will be removed and new LAB strains will be used. Usually, these new LAB strains have the same technological properties. In the worse cases, a shutdown of production may be necessary to disinfect the entire facility (Emond and Moineau 2007). Phage-inactivating agents such as sodium hypochlorite and peracetic acid, which eliminate phage particles, are often recommended (Binetti and Reinheimer 2000, Lévesque and Moineau 2005). Although, the thorough efficacy of commercial sanitizers against dairy phages is often unavailable. The determination of the source of phage contamination is also a very important aspect. Phages can come from induced prophages (usually a virulent form of a prophage), from the factory environment (McIntyre et al. 1991, Moineau and Lévesque 2005) or even from the substrate, since phages can stay infectious after pasteurisation (Chopin 1980).

The latter is of particular interest if whey protein concentrates (or others ingredients) are added to milk prior to the fermentation to increase cheese yields. Recent studies have also shown a disturbing fact, namely the emergence of lactococcal phages with high thermal resistance (Madera et al. 2004, Atamer et al. 2009).

THE PREVENTION

Below, we will describe various microbiological methods and mechanisms to prevent phage outbreaks particularly in dairy plants. This section will deal with phage resistance mechanisms.

BACTERIOPHAGE-INSENSITIVE MUTANTS

As indicated above, whenever a bacterial strain "becomes" sensitive to bacteriophage infection (due the emergence of new phages), it should be replaced with a phage resistant strain. However, this type of strain is not always available (Moineau and Lévesque 2005). Ideally, isogenic variants should be used. This allows the conservation of the properties of the orginal bacterial strains. The isolation of spontaneous phage-resistant mutant selected after a classical phage challenge experiment is the easiest way to replace a sensitive strain (Forde and Fitzgerald 1999b, Coffey and Ross 2002). Mutants obtained from this method are called Bacteriophage-Insensitive-Mutants (BIMs). Since it is a random approach, mutations explaining the resistance phenotype are not always identified. Nonetheless, many BIMs are currently used successfully in industrial conditions. However, the efficiency of fermentation can be different from the original strain if the mutation affects an important cellular function (Emond and Moineau 2007, Labrie and Moineau 2010). It is also possible to find phage mutants infecting BIMs in manufacturing facilities (Emond and Moineau 2007).

The most effective BIMs are those obtained with a deletion in the phage receptor gene. This approach is an efficient method for controlling lactococcal phages of the c2 species. A spontaneous mutation in the phage infection protein (Pip), a membrane protein involved in the adsorption process of the c2-like phages, rendered the bacterial host resistant to c2-like phages (Valeyasevi et al. 1991). Replacing the wild-type chromosomal *pip* by a mutated allele also allowed bacterial strains to become resistant to phages without altering growth (Garbutt et al. 1997). The mutations brought to the *pip* gene and their subsequent industrial use as led to a decreasing rate of isolation of phages belonging to the c2 species in cheese plants worldwide (Labrie and Moineau 2010). Unfortunately, phage receptors are mostly unknown for LAB phages, c2 phages being the main exception.

INTRODUCTION TO THE NATURAL RESISTANT STRAINS

In addition to the design of a LAB strain rotation program, the screening of industrial strains for their phage sensitivity will inevitably highlight particularly resilient bacterial strains, especially if many phages are tested. The study of these LAB strains led to the isolation of a great variety of natural defence mechanisms, most frequently encoded on plasmids that are responsible for the phage resistance phenotype. These plasmids (and the resistance phenotype) can often be transferred into another (phage-sensitive) strain by conjugation. The resistance mechanisms can also be found on the bacterial chromosomes. In these cases, they are often in a hot spot of integration or in a prophage genomic sequence.

The efficiency of each mechanism can be evaluated by a relatively simple method named EOP (efficiency on plaquing), which is calculated by dividing the titers obtained on phage-resistant strains by titers obtained on sensitive strains (Sanders and Klaenhammer 1980). A barrier against LAB phages is considered very strong when the EOP is $<10^{-7}$. Mechanisms with an EOP value between 10^{-4} and 10^{-6} are considered as medium potency whereas an EOP between 10^{-1} and 10^{-3} indicates a weak system (Moineau 1999). Phage-resistance mechanisms can be separated into at least five groups: (1) adsorption blocking mechanisms, (2) DNA ejection blocking, (3) restriction/modification (R/M) systems, (4) abortive infection mechanisms (Abi), and (5) CRISPRs.

The vast majority of natural defense mechanisms have been isolated from *L. lactis* strains (Coffey and Ross 2002). Two reasons can explain this: (1) *L. lactis* is the most frequently used bacterium in dairy fermentations, and a large amount of studies were conducted on this species; (2) the majority of known mechanisms are encoded on plasmids and *L. lactis* strains harbour a large number of them. Consequently, more than 50 different phage resistance mechanisms from *L. lactis* strains have been reported.

Adsorption Blocking

The adsorption blocking mechanisms act at the first step of the lytic cycle: the phage/host recognition process (Figure 1). Two general situations resulting in the adsorption blocking mechanism have described in lactic acid bacteria (Klaenhammer 1994): the physical masking or the lack of receptors (Forde and Fitzgerald 1999b). First, the resistance phenotype can appear after the production by the cell of extracellular compounds such as exopolysaccharides (EPS), capsular polysaccharides (CPS), or cell wall proteins that mask the phage receptor or compete with the receptor for the phage adsorption (Forde and Fitzgerald 1999a, Looijesteijn et al. 2001, Coffey and Ross 2002). For example, the plasmid pSK11 from *L. lactis* subsp. *lactis* SK110 prevent phage adsorption by masking the phage receptor sites (de Vos et al. 1984). A galactose-containing lipoteichoic acid moiety was implicated in the shielding process (Sijtsma et al. 1988, 1990). Although some plasmids coding for the adsorption blocking mechanism have been reported, the characterization of the exact genetic determinant is often partial. However, phages infecting EPS or CPS producing LAB strains have already been isolated (Deveau et al. 2002, Broadbent et al. 2003) suggesting that the production of these polymers would not be a sufficient protection against phages. Another example is that some phages produce enzymes that degrade polysaccharide (Sutherland et al. 2004). Although, it still remains unclear whether or not LAB phages carry such polysaccharide degrading activities.

The adsorption step can also be blocked by a spontaneous mutation in the host chromosomal gene coding for phage receptor. For many BIMs, the resistance can be explained by this kind of mutation at the bacterial cell surface. Host receptors include structures such as carbohydrates, proteins, or lipoteichoic acids of the cell wall or membrane (Valyasevi et al. 1991, Dupont et al. 2004, Geller et al. 2005, Spinelli et al. 2006, Tremblay et al. 2006). For some lactococcal phages (c2 group), host adsorption has been shown to be a two-step phenomenon that requires Ca^{2+} ions. It comprises an initial reversible interaction with cell wall saccharides and a subsequent irreversible binding to the PIP protein (phage infection protein), that is located in the cytoplasmic membrane and exposes a large ectodomain to the exterior, as well as to another 32 kDa poorly characterized membrane protein (Valeyasevi et al. 1991, Geller et al. 1993, Monteville et al. 1994, Kraus and Geller 1998, Josephsen and Neve 2004).

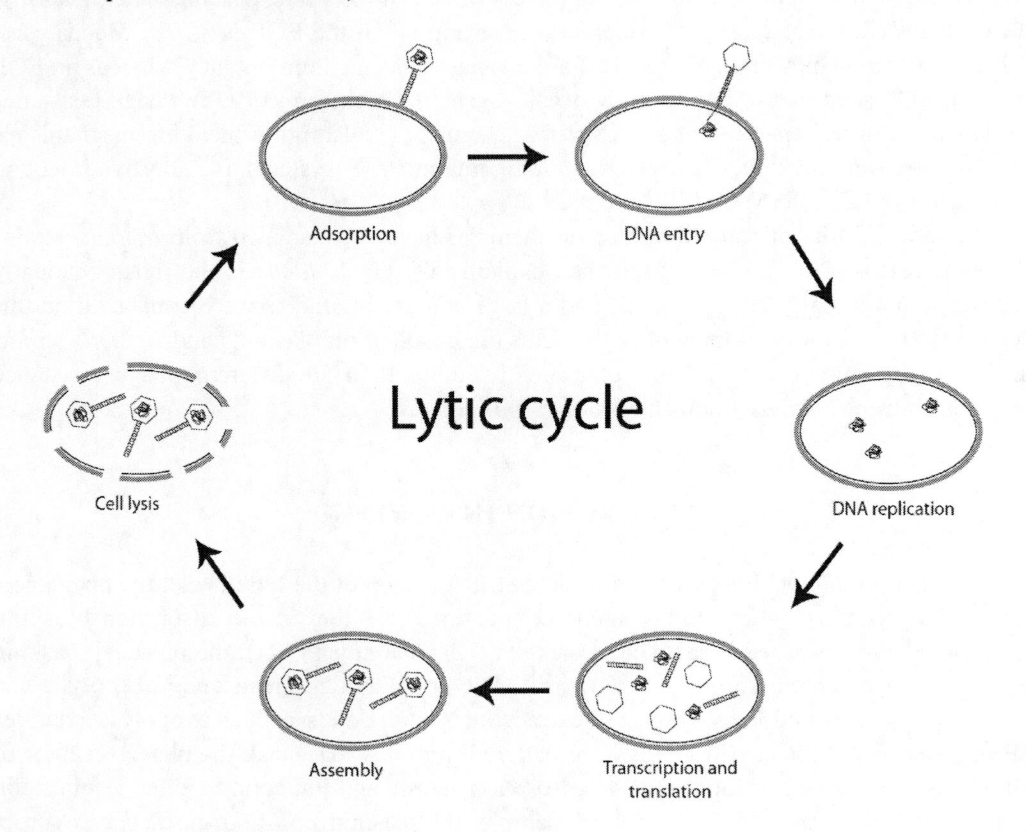

Figure 1. Lytic Cyle of Bacteriophages.

Inactivation of the *pip* gene leads to cells that are totally resistant to c2-like phages. Many lactococcal Pip-mutants strains are used commercially and are performing extremely well. However, Pip-mutants are still sensitive to lactococcal phages of the 936 and P335 groups. On the other hand, the exact host genes involved in adsorption of other LAB phages remains largely unknown. Moreover, it is not possible to state definitively the specific requirement(s) for adsorption since the process is likely to be phage-, strain- and receptor-dependent (Forde and Fitzgerald 1999b). Much research is needed in this area to further understand the nature of phage adsorption.

Blocking of DNA Entry

In this resistance mechanism, phages can still adsorb to the host cell, but viral DNA cannot enter into the cytosol. Not much is known about this mechanism in LABs. The best-known example is the superinfection exclusion system Sie_{2009} from the lactococcal phage Tuc2009 of the P335 species. The *sie2009* gene is located between the genes encoding the repressor and integrase, on the lysogeny module of the temperate lactococcal bacteriophage

Tuc2009. A protein associated with the cell membrane interferes with plasmid transduction and phage replication (McGrath et al. 2002a). The Sie_{2009} system confers immunity against lactococcal phages of the 936 group without disturbing phage adsorption, transfection and plasmid transformation (McGrath et al. 2002a). This function differs from phage immunity/repression systems (like λ-CI repressor protein) because it doesn't maintain lysogenic state and it is not specific to homoimmune phages (McGrath et al. 2002a). This system was also found in other prophages in the genomes of *Lactococcus lactis* strains (McGrath et al. 2002a, Labrie et al. 2010).

A Sie-like system was also found in the *Streptococcus thermophilus* prophage TP-J34. In this case, a signal-peptide on a 142-amino-acid lipoprotein (LTP) blocks the injection of phage DNA. Interestingly, this system provides resistance to some lactococcal phages once transformed into *L. lactis*. This system is similar to the Sie_{2009} because the *ltp* gene is also located between the repressor and the integrase within the lysogeny module of the phage TP-J34 (Sun et al. 2006). The Sie_{2009} and Ltp systems might operate by masking a component of the bacterial membrane that causes DNA release or by interacting with the phage particle, thereby preventing proper insertion of the phage tail tip into the cytoplasmic membrane (McGrath et al. 2002a, Sun et al. 2006, Mahony et al. 2008). Because many LAB strains harbor prophages or prophage remnants, it is likely that many strains harbor such systems. To our knowledge, their industrial efficacy has yet to be tested.

Other less characterized DNA blocking systems have been described in the literature. The first example of protection against phage DNA injection was shown with the phage 510 and his host *Lactococcus lactis* spp. l*actis* NCDO 510 mutant (Marshall and Berridge 1976, Forde and Fitzgerald 1999b). The blocking of phage DNA entry can also be plasmid-encoded as shown for pNP40. A protein product encoded by this plasmid could prevent phage DNA ejection and translocation of DNA across the membrane (Garvey et al. 1996). This type of resistance mechanism has also been seen for *Lactobacillus casei* YIT 9021 and the phage PL-1. In this case, the phage DNA could not be internalized in the cytoplasm, but was found to stay within the capsid (Watanabe et al. 1984, 1987). A membrane protein involved in the DNA infection process has also been identified in *Strep. thermophilus* by insertional mutagenesis of the vector pG(+)host9:ISS1 (challenged with phage Sfi19) (Lucchini et al. 2000).

RESTRICTION/MODIFICATION SYSTEMS

Restriction and modification (R/M) systems are arguably the most recognized intracellular defense mechanism. It is believed that most LAB strains possess at least one type of R/M systems (see below). The endonucleases are well known as they have been exploited commercially and are now part of the day-to-day experiments of molecular biology studies. From a phage resistance perspective, these mechanisms act early after the entry of DNA to protect the cell and prevent damage that would otherwise cause cell death (Emond and Moineau 2007). R/M systems are mostly made by two types of enzymes: endonuclease (Enase for endodeoxyribonuclease) and methylase (Mtase for DNA-methyltransferase) (Wilson and Murray 1991, Moineau and Lévesque 2005). First, the action of modification is to transfer a methyl group on an adenine or cytosine (Josephsen and Neve 2004). This way, bacteria protect their own DNA against the second enzymatic activity, namely the restriction. The latter action allows for the cutting of the invading DNA in or near a recognition site with a restriction endonuclease (Josephsen and Neve 2004). The recognition sequence usually comprises four to eight defined nucleotides. It can be continuous or not, symmetric or not and unique or degenerate. The «cognate enzymes» (from the same system), Enase and Mtase, recognize the same sequences (Wilson and Murray 1991).

There are at least, four types of R/M systems based on the co-factors needed, the subunit composition, the recognition sequence and the cutting site. Three of them are described here in more details for LAB (Voet and Voet 1998, O'Driscoll et al. 2004). R/M systems are the best understood and the most diverse anti-phage mechanisms. Despite of their high efficacy (because phage DNA is degraded), they are also the most fragile (leakiness) (Forde and Fitzgerald 1999b, Labrie and Moineau 2010). However, when combined with other defense mechanisms, they are even more powerful (Josephsen and Neve 2004). In general, they have a broad range of efficiency and can inactivate members of many diverse groups of phages. The efficacy of the R/M systems depend on the number of recognition sites in the phage double-stranded DNA genome. In general, EOP of phages decrease logarithmically as the number of phage recognition sites increase (Wilson and Murray 1991, Josephsen and Neve 2004).

The first group of R/M system is type I, which has three genes encoding for subunits needed for restriction, methylation, and specificity. This group cuts DNA far from the recognition site, at variable distances (> 1000 bp) (Voet and Voet 1998, Forde and Fitzgerald 1999b, O'Driscoll et al. 2004). It is the most biochemically and genetically complex system (Forde and Fitzgerald 1999b). This system needs a complex set of co-factors such as ATP, Mg^{2+} and AdoMet to operate. Type I restriction systems can also be classified into four subclasses or families: IA, IB, IC or ID. This classification is based on biochemical properties of the enzymes, sequence similarities and on the subunit complementation and transcriptional organisation of the constituent component (*hsd*R, *hsd*M and *hsd*S genes). *hsd* stands for host specificity for DNA. The subunits *hsd*M and *hsd*S are required to form the methyltransferase complex, while the three subunits together are needed for the activity of the endonuclease (Forde and Fitzgerald 1999b). In LAB, This system has been studied mostly in *Lactococcus lactis* (and to some limited extent in *Streptococcus thermophilus*) and some strains were shown to harbor plasmids with genes encoding only for the *hsd*S subunit. This subunit can sometimes be combined with chromosome- or plasmid-encoded type I R/M systems and alter their specificities. Furthermore, homologous recombinaisons can generate chimeric *hsd*S subunits with new specificities. This is why this system provides malleability (Shouler et al. 1998, Madsen et al. 2000, O'Sullivan et al. 2000, Seegers et al. 2000, Josephsen and Neve 2004).

Because their activities are easy to detect biochemically, the type II R/M systems are the most known in bacteria, including LABs. They necessitate simple co-factors and have simple organizations. Type II systems are usually composed of one endonuclease and one methylase. The endonucleases cleave at, or near, the site of recognition (Forde and Fitzgerald 1999b). Many type II R/M systems are unique. Some have two methylases such as the *Lactococcus* system LlaDCHI (Moineau et al. 1995) and ScrFI (Davis et al. 1993, Twomey et al. 1993, 1997). It has been suggested that the second MTase of the ScrFI system may have evolved to counteract the degeneracy of the restriction site (5'-CC/NGG-3') (Twomey et al. 1997, Forde and Fitzgerald 1999b). When IS elements are associated with phage resistance mechanisms, it is generally accepted that this may, on the first side, facilitate their dissemination to other strains, but on the other side, allow some phages to acquire some genes of the modification components of the R/M systems more easily (Forde and Fitzgerad 1999b). There is also type IIS referring «S» for shifted cleavage, up to 20 bp on one side of the target (Forde and Fitzgerald 1999b).

The last somewhat well described system for LABs is the type III. This system is multimeric as the type I, except that it only has two enzymes: *res* and *mod* for restriction and modification enzymes (Bickle 1993). Two inversely oriented asymetric recognition sites are required by the system. Cleavage occurs at approximately 25 nucleotides to one side of the recognition sequence. There is a distinctive feature for this type of system. The cleavage can occur only if neither strand is modified because methylation is done only on one DNA strand. The co-factors needed are Mg^{2+}, ATP and AdoMet (Forde and Fitzgerald 1999b). The first Gram+ type III R/M system was found on a 12 kb plasmid named pND801 in *L. lactis* subsp. *lactis* strain (named LL42-1). The name of this system is LlaFI (Su et al. 1999).

As indicated above, LAB strains carrying R/M systems will confer a significant protection against phages but it is leaky. Upon entry into the bacterial host, the phage genome is not modified, which gives the restriction enzyme a chance to cleave this DNA before modifying enzyme completes the methylation of the recognition sites. In this competition between modification and restriction, it happens, although at low frequency, that a phage successfully escapes restriction, because modification completes before the restriction enzyme had a chance to cut at recognition sites (Emond and Moineau 2007). If a phage genome contains a functional methylase gene, it may also limit the efficacy of a specific R/M system (Hill et al. 1991). A phage that escaped restriction becomes resistant to the R/M system and can pursue its lytic cycle unconstrained (Emond and Moineau 2007). Moreover, the genome of the phage progeny will also be modified, and be impervious to R/M in subsequent infection cycles. This « resistance » will be maintained for as long as the host use for propagation expresses the R/M system. Conversely, propagation of the modified phage on a strain that does not harbour the R/M system will produce a progeny that do not carry methylation; and the susceptibility to R/M will be restored. Thus, addition of R/M systems into LAB rotation programs is advisable to ensure that methylated phages will be purged upon LAB rotation, which will extend the useful lifespan of strains in industrial fermentations (Emond and Moineau 2007).

ABORTIVE INFECTION MECHANISMS

The abortive infection mechanisms refer to antiphage systems that inhibit phage multiplication between DNA entry and cell lysis (Hill et al. 1993, Haaber et al. 2010). The R/M and CRISPR/Cas systems (see below), which inhibit phage infection by cleaving incoming nucleic acids, are excluded from the Abi group. Abi mechanisms usually reduce phage burst size, phage efficiency of plaquing (EOP) and efficiency at which centers of phage infection form (ECOI). They are arguably considered the most powerful among the natural mechanisms because phage-infected Abi-containing cells will die, thereby limiting phage propagation (Sing and Klaenhammer 1990). Therefore, cells harboring Abi systems are like a trap that captures free phage particles and strongly limits their spread to other cells (altruistic suicide system) (Emond and Moineau 2007).

Abi systems have been identified in several bacterial species, but most of them have been studied in lactococci (Emond and Moineau 2007). At the time of writing this book chapter, 23 lactococcal Abi mechanisms have reported in the literature and at least one for *Strep. thermophilus* (Larbi et al. 1992). The exact modes of action of Abi systems are vastly unresolved because of our glaring lack of detailed understanding of LAB phage biology. Because of that, they are often categorized according to their ability of acting early (before or upon DNA replication) or late (after replication) (Josephsen and Neve 2004). There is also little or no sequence similarity between Abi systems (Emond and Moineau 2007). There are usually one or two genes per system (Boucher and Moineau 2001). Lactococcal Abi proteins have been generally tested against phages of the three main phage groups of lactococci. Because all lactococcal Abi systems are quite different, only the two most recent will be discussed here. All the other lactococcal Abi mechanisms are summarized in Table 1.

The plasmid pTR2030 harbors three different resistance mechanisms: the restriction/modification system LlaI, the abortive infection mechanism AbiA and the newly discovered AbiZ. AbiZ causes phage Φ31 to induce lysis of *Lactococcus lactis* strain NCK203 15 minutes earlier. AbiZ is probably the first lactococcal phage resistance mechanism to cause premature lysis of the infected cell. This system is effective only against phages of the P335 polythetic species (Durmaz and Klaenhammer 2007). This system is expressed constitutively and AbiZ appears to have some indirect interaction with the phage holin. Indeed, it has been argued that the AbiZ protein could interact, in some way, cooperatively with the phage holin or with a holin inhibitor to cause the premature cell lysis. During the late stage of phage infection, there is production of holin molecules, which accumulates in the cell membrane. Once a critical point is reached, the holin molecules create a hole within the cell membrane thereby providing access to the cell wall. Then, the phage lysin proceeds to degrade the cell wall, causing cell lysis (Gründling et al. 2001).

The most recently described lactococcal abortive infection mechanism is named AbiV. An insertion in a chromosomal region of the lactococcal strain MG1363 was found to confer resistance against phages of the 936 and c2 groups. A 201-amino acid deduced protein named AbiV prevented maturation of the replicated phage DNA of the 936 phage group (Haaber et al. 2008). The integration of pGhost9::IS*S1*, a system previously used to knock out factors involved in phage adsorption (Dupont et al. 2004) or DNA ejection (Lucchini et al. 2000), was found to activate AbiV. In fact, the promoter of the erythromycin resistance gene of pGhost9::IS*S1* was found to turn on this otherwise silent system in the strain MG1363 (Haaber et al. 2008). Also, the analysis of phage mutants insensitive to AbiV was used to further investigate the mode of action of the resistance mechanism. Sequencing of two whole-genomes of p2 mutants and homologous regions in the genomes of other AbiV-insensitive mutants from p2 and other lactococcal phages revealed mutations in an *orf* or a homologous *orf* sequence that is transcribed early. This *orf* was named *SaV* (for sensitivity to AbiV) (Haaber et al. 2009a).

Table 1. Abi Mechanisms in *Lactococcus lactis*

bi	# of genes	Size (aa)	Localizati on	Efficacy against phage groups 9 36	2	P 335	Effect on phage lytic cycle	References
	1	628	Plasmid	✓		✓	Interfere with DNA replication	Klaenhammer and Sanozky 1985, Hill et al. 1990, Coffey et al. 1991a, 1991b, Dinsmore and Klaenhammer 1994, 1997
	1	250	Chromoso me	✓			Affect RNA transcription	Cluzel et al. 1991, Parreira et al. 1996, Forde et al. 1999
	1	344	Plasmid	✓			Limit production of MCP	Durmaz. et al. 1992, Moineau et al. 1993
	1	366	Plasmid	✓				McLandsborough et al. 1995
	1	351	Plasmid	✓			Interfere with phage RuvC-like endonuclease	Anba et al. 1995, Bidnenko et al. 1995, 2002
1								
	2	286/297	Plasmid	✓			Affect phage packaging	Garvey et al. 1995
	1	341	Plasmid	✓			Interfere with DNA replication	Garvey et al. 1995
	2	249/397	Plasmid	✓			Affect RNA transcription	O'Connor et al. 1996
	1	346	Chromoso me	✓				Prevots et al. 1996
	1	331	Plasmid	✓			Affect phage packaging	Su et al. 1997
	1	282	Plasmid	✓				Deng et al. 1997
	1	599	Plasmid	✓		✓	Interfere with DNA replication	Émond et al. 1997, Boucher et al. 2000
	2	458/297	Plasmid	✓				Deng et al. 1999
	1	173	Prophage	✓				Prevots et al. 1998
	1	540	Plasmid	✓				Prevots and Ritzenthaler 1998
	1	244	Plasmid	✓			Interfere with DNA replication	Domingues et al. 2004
	1	183	Plasmid	✓			Affect DNA maturation and packaging	Émond et al. 1998
	4		Plasmid	✓				Twomey et al. 2000
	0	2 loci DNA structure	Plasmid	✓				Josephsen and Neve 2004
	2	127/213	Plasmid	✓		✓	Interfere with DNA replication	Bouchard et al. 2002
	2	589/341	Plasmid	✓		✓	Affect RNA transcription	Dai, et al. 2001
	1	201	Chromoso me	✓			Affect DNA maturation and packaging	Haaber et al. 2008, Haaber et al. 2009b
	1	432	Plasmid			✓	Cause premature lysis	Durmaz et al. 2007

According to Josephsen and Neve 2004, Deveau 2006a, Haaber et al. 2008.

Despite the fact that no putative function could be assigned to SaV, its fast-working toxic effect in *L. lactis* and *E. coli* suggest that it can shut down or redirect essential host functions (Haaber et al. 2009a). It has also been demonstrated that AbiV prevents phage protein synthesis and late gene transcription (Haarber et al. 2010). Preliminary biochemical analyses showed that AbiV and SaV proteins interact directly and form a complex consisting of 2 AbiV and 2 SaV molecules. It is probably the first demonstration of a direct interaction between an Abi protein and a phage protein (Haarber et al. 2010). Later, the report of the isolation of natural BIMs of *L. lactis* MG1363, which spontaneously express AbiV was made as well as the demonstration of conjugative transfer of *abiV* to other lactococcal strains. Such technological demonstration will avoid the use of genetic engineering technology as AbiV can now be naturally transferred into industrial starter strains (Haaber et al.2009b).

Many lactococcal Abis are currently in used (knowingly or not) within the dairy industry to limit phage proliferation. If an Abi+ LAB strain is used in an environment containing high levels of phages, the first fermentation cycle may somewhat suffer a slight delay because phage-infected Abi+ cells will die. However, the rate of acidification should resume to full speed in the second and subsequent fermentations if the Abi is effective. It should be noted that phages insensitive to Abi mechanisms have been documented and thus, despite their reported high efficiency, one should not solely rely on these Abi mechanisms.

CRISPR/CAS SYSTEM

A new adaptive bacterial immune system was recently described using strains of *Streptococcus thermophilus* and their phages as models. This system named CRISPR/Cas system, is present in the genome of 40% of bacteria (Grissa et al. 2007a) and in almost all archaebacteria (Godde and Bickerton 2006). CRISPR stands for Clustered Regularly Interspaced Short Palindromic Repeat and *Cas* for CRISPR-associated genes. This genetic structure was first described in 1987 in *Escherichia coli* (Ishino et al. 1987) but only recently (Barrangou et al. 2007) was a function attributed to the CRISPR/Cas systems. A number of excellent reviews have already been published on the CRISPR/Cas systems and the readers are directed to these articles for further information (Sorek et al. 2008, Koonin and Makarova 2009, Deveau et al. 2010, Horvath and Barrangou 2010, Karginov and Hannon 2010, Marraffini and Sontheimer 2010).

CRISPR loci are usually composed of 23 to 50 bp direct DNA repeats interspaced by nonrepetitive nucleotides of 26-72 bp called spacers. In a given locus, the size of the repeats and the spacers is conserved (Deveau et al. 2010). Almost half of the genomes that contain CRISPR/Cas system possess more than one locus (Godde and Bickerton 2006). Two to 588 repeats can be found per locus and they are usually highly conserved except for the last repeat (Jansen et al. 2002a, b, Horvath et al. 2008, Deveau et al. 2010). For the spacers, they have unique sequence within the same strain (Grissa et al. 2007b, Deveau et al. 2010, Horvath and Barrangou 2010). These spacers often have homology to phage and plasmid sequences. Upstream the CRISPR locus, there is a leader region of 20-534 bp that contains a promoter that transcribes the CRISPR locus (Jansen et al. 2002b, Tang et al. 2002, Tang et al. 2005, Lillestøl et al. 2006, Hale et al. 2008). *Cas* genes are quite always in proximity of the CRISPR region (Jansen et al. 2002, Deveau et al. 2010). Besides *cas* genes, other subtype-specific genes and genes encoding "repeat-associated mysterious proteins" (RAMP) have been also identified (Makarova et al. 2002, Haft et al. 2005, Makarova et al. 2006, Kunin et al. 2007, Horvath et al. 2009, Horvath and Barrangou 2010). As it will be discussed below, the CRISPR/Cas system can be separated into at least two functional stages: adaptation and interference. The adaptation stage is described as the acquisition of new spacers whereas the interference stage consists of targeting and cleaving invasive foreign nucleic acids.

When a phage-sensitive strain of *Strep. thermophilus* is challenged with a virulent phage in specific conditions, natural BIMs will eventually emerge. Analyses of these BIMs indicated that they have acquired at least one new repeat-spacer unit at the 5' end of a CRISPR locus. The newly added spacer (mostly 30 nucleotides in *Strep. thermophilus*) is 100% identical to a sequence (named proto-spacer) found in the genome of the infecting virulent phage (Barrangou et al.2007, Deveau et al. 2008). The molecular mechanism behind the acquisition of a repeat-spacer unit is still unknown (Labrie et al. 2010). Still, this new spacer must be identical to the corresponding phage protospacer to provide resistance (Deveau et al. 2010) and can be picked up from the sense or antisense strand of the phage genome (Deveau et al. 2008). Interestingly, different BIMs can acquire distinct spacers from the same phage genome and many spacers can be added. However, a motif seems to appear important for the acquisition of new spacers. It is the proto-spacer-associated motif (PAM) which is a short nucleotide sequence (e.g. NNAGAAW) flanking the proto-spacer at the 3' or at the 5' end (Deveau et al. 2010). The interference step starts with the transcription of the CRISPR locus. This CRISPR mRNA (pre-crRNA) is then processed (often in the repeat region) into small RNAs (crRNA) by Cas proteins (Hale et al. 2008, 2009), and those small genetic elements will target and lead to the cleavage of the invading foreign nucleic acids either RNA or DNA (Brouns et al. 2008, Marraffini et Sontheimer 2008, Hale et al. 2009, Garneau et al. 2010). In *Strep. thermophilus*, the target is phage DNA (Garneau et al. 2010).

The CRISPR-Cas system is still not infallible (Labrie et al. 2010). *Strep. thermophilus* phages can bypass the resistance provided by the newly acquired spacer(s) by a simple point mutation (or deletion) in the targeted proto-spacer. Of note, a mutation in the conserved PAM in the phage genome is also sufficient to dodge CRISPR. Interestingly, the iterative addition of new spacers targeting the emerging phage mutants is still possible.

The CRISPR loci can also be exploited for typing purposes as each unique LAB strain (carrying a CRISPR locus) harbor an unique set of spacers (Horvath and Barrangou 2010). In fact, the presence of orderly identical spacers in various strains, strongly suggest that these strains are related. Thus, the CRISPR loci can be used to select truly diverse strains for the development of strain rotation scheme. Clearly, the CRISPR/Cas system and the phage response illustrate the dynamic co-evolution between both groups of microorganisms (Labrie et al. 2010). The first phage-resistant *Strep thermophilus* strains based on the CRISPR/Cas system should be commercially available soon.

ENGINEERED MECHANISMS

Other phage resistance mechanisms using a biomimetic approach have been proposed. These systems are based on the knowledge of the phage constituents and on the identification of the factors essential for phage replication. A few LAB engineered phage-resistance systems are briefly described here. To our knowledge, none of these engineered phage-resistant strains are used commercially yet.

Origin-derived phage-encoded resistance (PER). During the early phage lytic development, once the phage DNA has entered the cell, some early phage proteins and host factors bind to the phage origin (*ori*) to initiate phage DNA replication. The formation of the initiation complex is dependent on the equilibrium between the concentration of *ori* sequences and on host- and phage-encoded initiation factors (Emond and Moineau 2007). The above features were exploited to construct phage-resistant strains.

Origin-derived phage-encoded resistance (PER) is a system in which multicopies of phage *ori* are provided in-*trans* on a plasmid carried by the host (Hill et al. 1990a, O'Sullivan et al. 1993). In order to obtain multicopies of the plasmid-encoded phage origin, the phage *ori* region can be cloned on a high-copy-number plasmid (O'Sullivan et al. 1993) and/or multicopies of the *ori* can be cloned on the plasmid (McGrath et al. 2001). Because the *ori*, located on the plasmid, will segregate most of the phage and host replication initiation factors, this will reduce the rate of formation of the initiation complex at the phage *ori*. A false phage genome replication signal delays phage development resulting in a reduction in the EOP and smaller plaques (Forde and Fitzgerald 1999, Emond and Moineau 2007). This system is only effective against phages that have simillar *ori*. This system has been engineered in at least *Lactococcus* (Hill et al. 1990a, O'Sullivan et al. 1993), *Lactobacillus* (Moscoso and Suarez 2000) and *Streptococcus thermophilus* (Foley et al. 1998, Stanley et al. 2000).

Gene silencing. Gene silencing or anti-sense RNAs is an artificial mechanism that uses molecules of anti-sense RNAs to prevent the translation of genes essential for phage development (Emond and Moineau 2007). This technology involves cloning of a target gene in the reverse orientation, relative to an active promoter (McGrath et al. 2002b). The mRNA produced will form a stable hybrid with the target mRNA. The translation will be inhibited because of the ineffective ribosome loading and/or because of the increased sensitivity to RNA degrading enzymes (Inouye 1988, McGrath et al. 2002b). This engineered resistance mechanism necessitates the knowledge of phage genomic nucleotide sequences and the characterization of single phage genes (Moineau and Lévesque 2005). The genes encoding proteins such as the putative subunit of DNA polymerase, single-stranded binding protein, putative transcription factor, major tail protein, terminase, major capsid protein and putative helicase have been targeted by antisense RNAs (Kim et al. 1992, Walker and Klaenhammer 2000).

Phage-triggered suicide systems. This system exploits a high-copy-number plasmid on which a toxic gene is cloned downstream from a phage inducible promoter (Djordjevic et al. 1997). This system is an engineered form of an abortive infection system. The lethal genes, made of restriction gene cassettes, kill the bacterial cell as well as the infecting phages once expressed. These cassettes are cloned without their corresponding methylase gene. This system works only with closely related phages that are able to induce the expression of the cloned promoter (Emond and Moineau 2007).

Subunit poisoning. This new type of engineered defense mechanism targets an essential phage enzyme that requires oligomerization to become activated. The plasmid-encoded mutant proteins must be structurally intact and able to form stable interactions with the native protein, thus inhibiting their activity (Sturino and Klaenhammer 2004). To do so, the mutated gene contains different amino acids in the catalytic domain of the enzyme. The mutant allele would associate with the wild-type subunits (through its binding domain which is usually different than the catalytic domain) to form dysfunctional heteropolymers that lost activity (Sturino and Klaenhammer 2004). This type of system appears to be very effective in *Strep. thermophilus* once combined with a PER system (Sturino and Klaenhammer 2004).

Phage structural proteins and antibodies. Another new type of designed anti-phage strategy is based on the use of a phage neutralizing heavy-chain antibody fragment obtained after an immunization of a llama with the lactococcal phage p2 of the 936 group. The phage display technology was then used to obtain p2 specific single-domain antibody fragments. A group of potent neutralizing antibodies that prevent phage infection was identified and the receptor-binding protein (RBP) was identified as the antigen (de Haard et al. 2005). Because the neutralizing heavy-chain antibody fragment could be produced in a food grade production organism and that a low amount is needed, the use of such molecules makes surprising possibility. However, the engineered mechanism is also effective against only a closely related group of phages (Emond and Moineau 2007).

Conclusion

Phages are found everywhere, where there are bacteria. Thus, it is not surprising to observe that phages are able to reach fermentation facilities. Dairy plants are man-made environments where the ecological balance is not the rule and where the eradication of bacteriophages is not possible. Phage infections still pose today a risk to bacterial fermentation processes mainly because of their rapid evolution. Many have tried to eliminate phages by taking control measures such as adapted factory designs and processses, cleaning and sanitation procedures and air control systems (Emond and Moineau 2007). Those control measures do not seem to be enough to protect some of the preferred strains. Some strategies have been very helpful for many years, to keep using some LAB strains with good characteristics in fermentation processes. Those strategies are the rotations of phage-unrelated strains, the use of BIMs and the selection of strains harboring phage-resistance mechanisms. The extensive knowledge about the physiology and the genetics of LAB as well as the biology of their phages have led to a better understanding of their co-evolution. As our understanding of LAB phages keep increasing and because new phage defense strategies are still needed, the next decade will certainly be fertile in new exiting developments for a better control of the phage populations in industrial processes.

References

Anba, J., Bidnenko, E., Millier, A., Ehrlich, D. and Chopin, M. C. (1995). Characterization of the lactococcal abiDl gene coding for phage abortive infection. *Journal of Bacteriology, 177*, 3818-3823.

Atamer, Z., Dietrich, J., Muller-Merbach, M., Neve, H., Heller, K. J. and Hinrichs, J. (2009). Screening for and characterization of *Lactococcus lactis* bacteriophages with high thermal resistance. *International Dairy Journal, 19,* 228-235.

Barrangou, R., Yoon S. S., Breidt, F. Jr, Fleming, H. P. and Klaenhammer, T. R. (2002). Characterization of six *Leuconostoc fallax* bacteriophages isolated from an industrial sauerkraut fermentation. *Applied and Environmental Microbiology, 68,* 5452-5458.

Barrangou, R., Fremaux, C., Deveau, H., Richards, M., Boyaval, P., Moineau, S., Romero, D. A. and Horvath, P. (2007). CRISPR provides acquired resistance against viruses in prokaryotes. *Science, 315,* 1709-1712.

Behnke, D. and Malke, H. (1978). Bacteriophage interference in *Streptococcus pyogenes.* I. Characterization of prophage-host systems interfering with the virulent phage A25. *Virology, 85,* 118-128.

Bickle, T. A. (1993). The ATP-dependent restriction enzymes. In: S. M. Linn, R. S. Lloyd and R. J. Roberts (Eds), *Nucleases* (2nd edition, pp. 89-109). Cold Spring Harbor, NY: Cold Spring Harbor Laboratory Press.

Bidnenko, E., Ehrlich, D. and Chopin, M.-C. (1995). Phage operon involved in sensitivity to the *Lactococcus lactis* abortive infection mechanism AbiD1. *Journal of Bacteriology, 177,* 3824-3829.

Bidnenko, E., Chopin, M.-C., Ehrlich, S. D. and Anba, J. (2002). *Lactococcus lactis* AbiD1 abortive infection effieciency is drastically increased by a phage protein. *FEMS Microbiology Letters, 214,* 283-287.

Binetti, A. G. and Reinheimer, J. A. (2000). Thermal and chemical inactivation of indigenous *Streptococcus thermophilus* bacteriophages isolated from Argentinian dairy plants. *Journal of Food Protection, 63,* 509-515.

Binetti, A. G., Capra, M. L., Alvarez, M. A. and Reinheimer, J. A. (2008). PCR method for detection and identification of *Lactobacillus casei/paracasei* bacteriophages in dairy products. *International Journal of Food Microbiology, 124,* 147-153.

Bogosian, G. (2006). Control of bacteriophage in commercial microbiology and fermentation facilities. In: R. Calendar (Eds), *The bacteriophages* (2nd edition, pp. 665-673). New York, NY : Oxrford University Press.

Boizet, B., Matal, M., Mignot, O., Ritzenthaler, P. and Tommazzo, S. (1992) Taxonomic characterization of *Leuconostoc mesenteroides* and *Leuconostoc oenos* bacteriophages. *FEMS Microbiology Letters, 90,* 211-216.

Bouchard, J. D., Dion. E., Bissonnette, F. and Moineau, S. (2002). Characterization of the two-component abortive phage infection mechanism AbiT from *Lactococcus lactis. Journal of Bacteriology, 184,* 6325-6332.

Boucher, I., Emond, E., Dion, E., Montpetit, D. and Moineau, S. (2000). Microbiological and molecular impacts of AbiK on the lytic cycle of *Lactococcus lactis* phages of the 936 and P335 species. *Microbiology, 146,* 445-453.

Boucher, I., Emond, E., Parrot, M. and Moineau, S. (2001). DNA sequence analysis of the three *Lactococcus lactis* plasmids encoding phage resistance mechanisms. *Journal of Dairy Science, 84,* 1610-1620.

Breitbart, M., Rohwer, F. and Abedon, S. T. (2005). Phage ecology and bacterial pathogenesis. In M. K. Waldor, D. I. Friedman and S. L. Adhya (Eds), *Phages: Their role in bacterial pathogenesis and biotechnology* (1st edition, pp. 66-91). Washington, DC: ASM Press.

Broadbent, J. R., McMahon, D. J., Welker, D. L. and Moineau, S. (2003). Biochemistry, genetics, and applications of exopolysaccharide production in *Streptococcus thermophilus*: a review. *Journal of Dairy Science, 86,* 407-423.

Brouns, S. J., Jore, M. M., Lundgren, M., Westra, E. R., Slijkhuis, R. J., Snijders, A. P., Dickman, M. J., Makarova, K. S., Koonin, E. V. and van der Oost, J. (2008). Small CRISPR RNAs guide antiviral defense in prokaryotes. *Science, 321,* 960-964.

Brüssow, H. and Súarez, J. E. (2006). *Lactobacillus* phages. In R. Calendar and S. T. Abedon (Eds), *The Bacteriophages,* (2e edition, pp. 653-663). Oxford, UK: Oxford University Press.

Chopin, M.-C. (1980). Resistance of 17 mesophilic lactic *Streptococcus* bacteriophages to pasteurization and spray-drying. *Journal of Dairy Research, 47,* 131-139.

Cluzel, P. J., Chopin, A., Ehrlich, S. D. and Chopin, M.-C. (1991). Phage abortive infection mechanism from *Lactococcus lactis* subsp. *lactis,* expression of which is mediated by an Iso-ISS 1 element. *Applied and Environmental Microbiology, 57,* 3547-3551.

Coffey, A. G., Fitzgerald, G. F. and Daly, C. (1991). Cloning and characterization of the determinant for abortive infection of bacteriophage from lactococcal plasmid pCI829. *Journal of General Microbiology, 137,* 1355-1362.

Coffey, A. and Ross, R. P. (2002). Bacteriophage-resistance systems in dairy stater strains: molecular analysis to application. *Antonie van Leeuwenhoek, 82,* 303-321.

Dai, G., Su, P., Allison, G. E., Geller, B. L., Zhu, P., Kim, W. S. and Dunn, N. W. (2001). Molecular characterization of a new abortive infection System (AbiU) from *Lactococcus lactis* LL51-1. *Applied and Environmental Microbiology, 67,* 5225-5232.

Davey, G. P., Ward, L. J. H. and Brown, J. C. S. (1995). Characterisation of four *Leuconostoc* bacteriophages isolated from dairy fermentations. *FEMS Microbiology Letters, 128,* 21-26.

Davis, R., van der Leile, D., Mercenier, A., Daly, C. and Fitzgerald, G. F. (1993). ScrFI restriction-modification system of *Lactococcus lactis* ssp. *cremoris* UC503: cloning and characterization of two ScrFI methylase genes. *Applied and Environmental Microbiology, 59,* 777-785.

de Haard, H. J., Bezemer, S., Ledeboer, A. M., Müler, W. H., Boender, P. J., Moineau, S., Coppemans, M. C., Verkleij, A. J., Frenken, L. G. and Verrips, C. T. (2005). Llama antibodies against a lactococcal protein located at the tip of the phage tail prevent phage infection. *Journal of Bacteriology, 187,* 4531-4541.

del Rio, B., Binetti, A. G., Martín, M. C., Fernández, M., Magadán, A. H. and Alvarez, M. A. (2007). Multiplex PCR for the detection and identification of dairy bacteriophages in milk. *Food Microbiology, 24,* 75-81.

del Rio, B., Martín, M. C., Martínez, N., Magadán, A. H. and Alvarez, M. A. (2008). Multiplex fast real-time PCR for quantitative detection and identification of cos- and pac-type *Streptococcus thermophilus* bacteriophages. *Applied and Environmental Microbiology, 74,* 4779-4781.

Deng, Y. M., Harvey, M. L., Liu, C. Q. and Dunn, N. W. (1997). A novel plasmid encoded phage abortive infection system from *Lactococcus lactis* biovar. *diacetylactis. FEMS Microbiology Letters, 146,* 149-154.

Deng, Y. M., Liu, C. Q. and Dunn, N. W. (1999). Genetic organization and functional analysis of a novel phage abortive infection system, AbiL, from *Lactococcus lactis. Journal of Biotechnology, 67,* 135-149.

Deveau, H. (2006a). Biodiversité des bactériophages infectant *Lactococcus lactis.* Doctoral Thesis, Université Laval, Québec, Canada.

Deveau, H., van Carsteren, M.-R. and Moineau, S. (2002). The effect of exopolysaccharides on phage-host interactions in *Lactococcus lactis*. *Applied and Environmental Microbiology, 68,* 4364-4369.

Deveau, H., Labrie S. J., Chopin M.-C. and Moineau, S. (2006b). Biodiversity and classification of lactococcal phages. *Applied and Environmental Microbiology, 72,* 4338-4346.

Deveau, H., Garneau, J. E. and Moineau, S. (2010). CRISPR/Cas system and its role in phage-bacteria interactions. *Annual Review of Microbiology, 64,* 475-493.

de Vos, W. M., Underwood, H. M., and Davies, F. L. (1984). Plasmid encoded bacteriophage resistance in *Streptococcus cremoris* SK11. *FEMS Microbiology Letters, 23,* 175-178.

Dinsmore, P. K. and Klaenhammer, T. R. (1994). Phenotypic consequences of altering the copy number of *abiA,* a gene responsible for aborting bacteriophage infections in *Lactococcus lactis*. *Applied and Environmental Microbiology, 60,* 1129-1136.

Dinsmore, P. K. and Klaenhammer, T. R. (1997). Molecular characterization of a genomic region in a *Lactococcus* bacteriophage that is involved in its sensitivity to the phage defense mechanism AbiA. *Journal of Bacteriology, 179,* 2949-2957.

Domingues, S., Chopin, A., Ehrlich, S. D. and Chopin, M.-C. (2004b). The lactococcal abortive phage infection system AbiP prevents both phage DNA replication and temporal transcription switch. *Journal of Bacteriology, 186,* 713-721.

Dupont, K., Janzen, T., Vogensen, F. K., Josephsen, J. and Stuer-Lauridsen, B. (2004). Identification of *Lactococcus lactis* genes required for bacteriophage adsorption. *Applied and Environmental Microbiology, 70,* 5825-5832.

Durmaz, E., Higgins, D. L. and Klaenhammer, T. R. (1992). Molecular characterization of a second abortive phage resistance gene present in *Lactococcus lactis* subsp. *lactis* ME2. *Journal of Bacteriology, 174,* 7463-7469.

Durmaz, E. and Klaenhammer, T. R. (2007). Abortive phage resistance mechanism AbiZ speeds the lysis clock to cause premature lysis of phage-infected *Lactococcus lactis*. *Journal of Bacteriology, 189,* 1417-1425.

Emond, E., Holler, B. J., Boucher, I., Vandenbergh, P. A., Vedamuthu, E. R., Kondo, J. K. and Moineau, S. (1997). Phenotypic and genetic characterization of the bacteriophage abortive infection mechanism AbiK from *Lactococcus lactis*. *Applied and Environmental Microbiology, 63,* 1274-1283.

Emond, E., Dion, E., Walker, S. A., Vedamuthu, E. R., Kondo, J. K. and Moineau, S. (1998). AbiQ, an abortive infection mechanism from *Lactococcus lactis*. *Applied and Environmental Microbiology, 64,* 4748-4756.

Emond, E. and Moineau, S. (2007). Bacteriophages and food fermentations. In: S. McGrath and D. van Sinderen (Eds.), *Bacteriophage: genetics and molecular biology* (1st edition, pp. 93-124). Norfolk, UK: Horizon Scientific Press/Caister Academic Press.

Foley, S., Lucchini, S., Zwahlen, M. C. and Brüssow, H. (1998). A short noncoding viral DNA element showing characteristics of a replication origin confers bacteriophage resistance to *Streptococcus thermophilus*. *Virology, 250,* 377-387.

Forde, A., Daly, C. and Fitzgerald, G. F. (1999). Identification of four phage resistance plasmids from *Lactococcus lactis* subsp. *cremoris* HO2. *Applied and Environmental Microbiology, 65,* 1540-1547.

Forde, A. and Fitzgerald, G. F. (1999a). Analysis of exopolysaccharide (EPS) production mediated by the bacteriophage adsorption blocking plasmid, pCI658, isolated from *Lactococcus lactis* ssp. *cremoris* HO2. *International Dairy Journal, 9,* 465-472.

Forde, A. and Fitzgerald, G. F. (1999b). Bacteriophage defence systems in lactic acid bacteria. *Antonie van Leeuwenhoek, 76,* 89-113.

Garbutt, K. C., Kraus, J. and Geller, B. L. (1997). Bacteriophage resistance in *Lactococcus lactis* engineered by replacement of a gene for a bacteriophage receptor. *Journal of Dairy Science, 80,* 1512-1519.

Garneau, J. E., Dupuis, M. E., Villion, M., Romero, D. A., Barrangou, R., Boyaval, P., Fremaux, C., Horvath, P., Magadán, A. H. and Moineau. S. (2010). The CRISPR/Cas bacterial immune system cleaves bacteriophage and plasmid DNA. *Nature 468,* 67-71.

Garvey, P., Fitzgerald G. F. and Hill, C. (1995). Cloning and DNA sequence analysis of two abortive infection phage resistance determinants from the lactococcal plasmid pNP40. *Applied and Environmental Microbiology, 61,* 4321-4328.

Garvey, P., Hill, C. and Fitzgerald, G. F. (1996). The lactococcal plasmid pNP40 encodes a third bacteriophage resistance mechanism, one which affects phage DNA penetration. *Applied and Environmental Microbiology, 62,* 676-679.

Geller, B. L., Ivey, R. G., Trempy, J. E. and Hettinger-Smith, B. (1993). Cloning of chromosomal gene required for phage infection of *Lactococcus lactis* subsp. *lactis* C2. *Journal of Bacteriology, 175,* 5510-5519.

Geller, B. L., Ngo, H. T., Mooney, D. T., Su, P. and Dunn, N. (2005). Lactococcal 936-species phage attachment to surface of *Lactococcus lactis. Journal of Dairy Science, 88,* 900-907.

Godde, J. S. and Bickerton, A. (2006). The repetitive DNA elements called CRISPRs and their associated genes: evidence of horizontal transfer among prokaryotes. *Journal of Molecular Evolution, 62,* 718-729.

Grissa, I., Vergnaud, G. and Pourcel, C. (2007a). CRISPRFinder: a web tool to identify clustered regularly interspaced short palindromic repeats. *Nucleic Acids Research, 35,* W52-57.

Grissa, I., Vergnaud, G. and Pourcel, C. (2007b). The CRISPRdb database and tools to display CRISPRs and to generate dictionaries of spacers and repeats. *BCM Bioinformatics, 8,* 172.

Gründling, A., Manson, M. D. and Young, R. (2001). Holins kill without warning. *Proceedings of the National Academy of Sciences of the United States of America, 98,* 9348-9352.

Guglielmotti, D. M., Deveau, H., Binetti, A. G., Reinheimer, J. A., Moineau, S. and Quiberoni, A. (2009). Genome analysis of two virulent *Streptococcus thermophilus* phages isolated in Argentina. *International Journal of Food Microbiology, 136,* 101-109.

Haaber, J., Moineau, S., Fortier, L.-C. and Hammer, K. (2008). AbiV, a novel antiphage abortive infection mechanism on the chromosome of *Lactococcus lactis* subsp. *cremoris* MG1363. *Applied and Environmental Microbiology, 74,* 6528-6537.

Haaber, J., Rousseau, G. M., Hammer, K. and Moineau, S. (2009). Identification and characterization of the phage gene *sav* involved in sensitivity to the lactococcal abortive infection mechanism AbiV. *Applied and Environmental Microbiology, 75,* 2484-2494.

Haaber, J., Moineau, S. and Hammer, K. (2009b). Activation and transfer of the chromosomal phage resistance mechanism AbiV in *Lactococcus lactis*. *Applied and Environmental Microbiology, 75,* 3358-3361.

Haaber, J., Samson, J. E., Labrie, S. J., Campanacci, V., Cambillau, C., Moineau, S. and Hammer, K. (2010). Lactococcal abortove infection protein AbiV interacts directly with the phage protein SaV and prevents translation of phage proteins. *Applied and Environmental Microbiology, 76,* 7085-7092.

Hale, C., Kleppe, K., Terns, R. M. and Terns, M. P. (2008). Prokaryotic silencing (psi) RNAs in *Pyrococcus furiosus*. *RNA, 14,* 2572-2579.

Hale, C. R., Zhao, P., Olson, S., Duff, M. O., Graveley, B. R., Wells, L., Terns, R. M. and Terns, M. P. (2009). RNA-guided RNA cleavage by a CRISPR RNA-Cas protein complex. *Cell, 139,* 945-956.

Hershey, A. D. and Chase, M. (1952). Independent functions of viral protein and nucleic acid in growth of bacteriophage. *Journal of General Physiology, 36,* 39-56.

Hill, C., Miller, L. A. and Klaenhammer, T. R. (1990). Cloning, expression, and sequence determination of a bacteriophage fragment encoding bacteriophage resistance in *Lactococcus lactis*. *Journal of Bacteriology, 172,* 6419-6426.

Hill, C., Miller, L. A. and Klaenhammer, T. R. (1990). Nucleotide sequence and distribution of the pTR2030 resistance determinant (hsp) which arborts bacteriophage infection in lactococci. *Applied and Environmental Microbiology, 56,* 2255-2258.

Hill, C., Miller, L. A. and Klaenhammer, T.R. (1991). In vivo genetic exchange of a functional domain from a type II A methylase between lactococcal plasmid pTR2030 and a virulent bacteriophage. *Journal of Bacteriology 173*, 4363-4370.

Hill, C. (1993). Bacteriophage and bacteriophage resistance in lactic acid bacteria. *FEMS Microbiology Letters, 12,* 87-108.

Horvath, P., Romero, D. A., Coûté-Monvoisin, A. C., Richards, M., Deveau, H., Moineau, S., Boyaval, P., Fremaux, C. and Barrangou, R. (2008). Diversity, activity, and evolution of CRISPR loci in *Streptococcus thermophilus*. *Journal of Bacteriology, 190,* 1401-1412.

Haft, D. H., Selengut, J., Mongodin, E. F. and Nelson, K. E. (2005). A guild of 45 CRISPR-associated (Cas) protein families and multiple CRISPR/Cas subtypes exist in prokaryotic genomes. *Public Library of Science Computational Biology, 1,* e60.

Horvath, P., Coûté-Monvoisin, A. C., Romero, D. A., Boyaval, P., Fremaux, C. and Barrangou, R. (2009). Comparative analysis of CRISPR loci in lactic acid bacteria genomes. *International Journal of Food Microbiology, 131,* 62-70.

Horvath, P. and Barrangou, R. (2010). CRISPR/Cas, the immune system of bacteria and Archaea. *Science, 327,* 167-170.

Inouye, M. (1988). Antisense RNA: Its functions and applications in gene regulation – a review. *Gene, 72,* 25-34.

Ishino, Y., Shinagawa, H., Makino, K., Amemura, M. and Nakata, A. (1987). Nucleotide sequence of the *iap* gene, responsible for alkaline phosphatase isozyme conversion in *Escherichia coli*, and identification of the gene product. *Journal of Bacteriology, 169,* 5429-5433.

Jansen, R., Embden, J. D., Gaastra, W. and Schouls, L. M. (2002a). Identification of a novel family of sequence repeats among prokaryotes. *OMICS, 6,* 23-33.

Jansen, R., Embden, J. D., Gaastra, W. and Schouls, L. M. (2002b). Identification of genes that are associated with DNA repeats in prokaryotes. *Molecular Microbiology, 43,*1565-1575.

Josephsen, J. and Neve, H. (2004). Bacteriophage and antiphage mechanisms of lactic acid bacteria. In S. Salminen, A. von Wright and A. Ouwehand (Eds.), *Lactic acid bacteria: Microbiological and functional aspects* (third édition, revised and expanded., pp. 295-350). New York, NY: Marcel Dekker, Inc.

Karginov, F. V. and Hannon, G. J. (2010). The CRISPR system: small RNA-guided defense in bacteria and archaea. *Molecular Cell, 37,* 7-19.

Kim, J. H., Kim, S. G., Chung, D. K., Bor, Y. C. and Batt, C. A. (1992). Use of antisense RNA to confer bacteriophage resistance in dairy starter cultures. *Journal of Industrial Microbiology, 10,* 71-78.

Klaenhammer, T. R. and Sanozky, R. B. (1985). Congugal transfer from *Streptococcus lactis* ME2 of plasmids encoding phage resistance, nisin resistance and lactose-fermenting ability- evidence for a high-frequency conjugative plasmid responsible for abortive infection of virulent bacteriophage. *Journal of General Microbiology, 131,* 1531-1541.

Koonin, E. V. and Makarova, K. S. (2009). CRISPR-Cas: an adaptive immunity system in prokaryotes. *F1000 Biology Reports, 1,* 95.

Kraus, J. and Geller, B. L. (1998). Membrane receptor for prolate phages is not required for infection of *Lactococcus lactis* by small or large isometric phages. *Journal of Dairy Science, 81*, 2329-2335.

Kunin, V., Sorek, R. and Hugenholtz, P. (2007). Evolutionary conservation of sequence and secondary structures in CRISPR repeats. *Genome Biology, 8,* R61.

Labrie, S. and Moineau, S. (2000). Multiplex PCR for detection and identification of lactococcal bacteriophages. *Applied and Environmental Microbiology, 66,* 987-994.

Labrie, S. J. and Moineau, S. (2010). Bacteriophages in industrial food processing : incidence and control in industrial fermentation. In: *Bacteriophages*, (1st edition, pp. 199-216). Washington, DC: ASM Press.

Labrie, S. J., Samson, J. E. and Moineau, S. (2010). Bacteriophage resistance mechanisms. *Nature Reviews Microbiology, 8,* 317-327.

Larbi, D., Decaris, B. and Simonet, J. M. (1992). Different bacteriophage resistance mechanisms in *Streptococcus salivarius* subsp. *thermophilus. Journal of Dairy Research, 59.* 349-357.

Le Marrec, C., van Sinderen, D., Walsh, L., Stanley, E., Vlegels, E., Moineau, S., Heinze, P., Fitzgerald, G. and Fayard, B. (1997). *Streptococcus thermophilus* bacteriophages can be divided into two distinct groups based on mode of packaging and structural protein composition. *Applied and Environmental Microbiology, 63,* 3246-3253.

Lillestøl, R. K., Redder, P., Garrett, R. A. and Brügger, K. (2006). A putative viral defense mechanism in archaeal cells. *Archaea, 2,* 59-72.

Looijesteijn, P. J., Trapet, L., de Vries, E., Abee, T. and Hugenholtz, J. (2001). Physiological function of exopolysacchides produced by *Lactococcus lactis. International Journal of Food Science, 64,* 71-80.

Lucchini, S., Sidoti, J. and Brüssow, H. (2000). Broad-range bacteriophage resistance in *Streptococcus thermophilus* by insertional mutagenesis. *Virology, 27,* 267-277.

Madera, C., Monjardín, C. and Suárez, J. E. (2004). Milk contamination and resistance to processing conditions determine the fate of *Lactococcus lactis* bacteriophages in dairies. *Applied and Environmental Microbiology, 70,* 7365-7371.

Madsen, A., Westphal, C. and Josephsen, J. (2000). Characterization of a novel plasmid-encoded HsdS subunit, S. *Lla*W12I, from *Lactococcus lactis. Plasmid, 44,* 196-200.

Mahony, J., McGrath, S., Fitzgerald, G. F. and van Sinderen, D. (2008). Identification and characterization of lactococcal-prophage-carried superinfection exclusion genes. *Applied and Environmental Microbiology, 74,* 6206-6215.

Makarova, K. S., Aravind, L., Grishin, N. V., Rogozin, I. B. and Koonin, E. V. (2002). A DNA repair system specific for thermophilic Archaea and bacteria predicted by genomic context analysis. *Nucleic Acids Research, 30,* 482-496.

Makarova, K. S., Grishin, N. V., Shabalina, S. A, Wolf, Y. I. and Koonin, E. V. (2006). A putative RNA-interference-based immune system in prokaryotes: computational analysis of the predicted enzymatic machinery, functional analogies with eukaryotic RNAi, and hypothetical mechanisms of action. *Biology Direct, 1,* 7.

Marraffini, L. A. and Sontheimer, E. J. (2008).CRISPR interference limits horizontal gene transfer in staphylococci by targeting DNA. *Science, 322,* 1843-1845.

Marraffini, L. A. and Sontheimer, E. J. (2010). CRISPR interference: RNA-directed adaptive immunity in bacteria and archaea. *Nature Reviews Genetics, 11,*181-190.

Marshall, R. J. and Berridge, N. J. (1976). Selection and some properties of phage-resistant starters for cheese-making. *Journal of Dairy Research, 43,* 449-458.

McIntyre, K., Heap, H. A., Davey, G. P. and Limsowtin, G. K. Y. (1991). The distribution of lactococcal bacteriophage in the environment of a cheese manufacturing plant. *International Dairy Journal, 1,* 183-197.

McGrath, S., Fitzgerald, G. F. and van Sinderen, D. (2001). Improvement and optimization of two engineered phage resistance mechanisms in *Lactococcus lactis. Applied and Environmental Microbiology, 67,* 608-616.

McGrath, S., Fitzgerald, G. F. and van Sinderen, D. (2002). Identification and characterization of phage-resistance genes in temperate lactococcal bacteriophages. *Molecular Microbiology, 43,* 509-520.

McGrath, S., van Sinderen, D. and Fitzgerald, G. F. (2002). Bacteriophage-derived genetic tools for use in lactic acid bacteria. *International Dairy Journal, 12,* 3-15.

McLandsborough, L. A., Kolaetis, K. M., Requena, T. and McKay, L. L. (1995). Cloning and characterization of the abortive infection genetic determinant abiD isolated from pBF61 of *Lactococcus lactis* subsp. *lactis* KR5. *Applied and Environmental Microbiology, 61,* 2023-2026.

Moineau, S., Durmaz, E., Pandian, S. and Klaenhammer, T. R. (1993). Differentiation of two abortive mechanisms by using monoclonal antibodies directed toward lactococcal bacteriophage capsid proteins. *Applied and Environmental Microbiology, 59,* 208-212.

Moineau, S., Walker, S. A., Vedamuthu, E. R. and Vandenbergh, P. A. (1995). Cloning and sequencing of LlaDCHI restriction/modification genes from *Lactococcus lactis* and relatedness of this system to the *Streptococcus pneumoniae* DpnII system. *Applied and Environmental Microbiology, 61,* 2193-2202.

Moineau, S. (1999). Applications of phage resistance in lactic acid bacteria. *Antonie van Leeuwenhoek, 76,* 377-382.

Moineau, S. and C. Lévesque. (2005). Control of bacteriophages in industrial fermentations. In: *Bacteriophages: Biology and Applications.* E. Kutter, and A. Sulakvelidze (Eds). CRC Press., p. 285-296.

Monteville, M. R., Ardestani, B. and Geller, B. L. (1994). Lactococcal bacteriophages require a host cell wall carbohydrate and a plasma membrane protein for adsorption and ejection of DNA. *Applied and Environmental Microbiology, 60,* 3204-3211.

Moscoso, M. and Suárez, J. E. (2000). Characterization of the DNA replication module of bacteriophage A2 and use of its origin of replication as a defense against infection during milk fermentation by *Lactobacillus casei. Virology, 273,* 101-111.

O'Connor, L., Coffey, A., Daly, C. and Fitzgerald, G. F. (1996). AbiG, a genotypically novel abortive infection mechanism encoded by plasmid pCI750 of *Lactococcus lactis* subsp. *cremoris* UC653. *Applied and Environmental Microbiology,* 62, 3075-3082.

O' Driscoll, J., Glynn, F., Cahalane, O., O'Connell-Motherway, M., Fitzgerald, G. F. and van Sinderen, D. (2004). Lactococcal plasmid pNP40 encodes a novel, temperature-sensitive restriction-modification system. *Applied and Environmental Microbiology, 70,* 5546-5556.

O'Sullivan, D. J., Hill, C. and Klaenhammer, T. R. (1993). Effect of increasing the copy number of bacteriophage origins of replication, in trans, on incoming-phage proliferation. *Applied and Environmental Microbiology, 59,* 2449-2456.

O'Sullivan, D., Twomey, D. P., Coffey, A., Hill, C., Fitzgerald, G. F. and Ross, R. P. (2000). Novel type I restriction specificities through domain shuffling of HsdS subunits in *Lactococcus lactis. Molecular Microbiology, 36,* 866-875.

Parreira, R., Ehrlich, S. D. and Chopin, M. C. (1996). Dramatic decay of phage transcripts in lactococcal cells carrying the abortive infection determinant AbiB. *Molecular Microbiology 19,* 221-230.

Prévôts, F., Daloyau, M., Bonin, O., Dumont, X. and Tolou, S. (1996). Cloning and sequencing of the novel abortive infection gène abiH of *Lactococcus lactis* ssp. *lactis* biovar. *diacetylactis* S94. *FEMS Microbiology Letters, 142,* 295-299.

Prévôts, F. and Rizenthaler, P. (1998). Complete sequence of the new lactococcal abortive phage resistance gène abiO. *Journal of Dairy Science, 81,* 1483-1485.

Prévôts, F., Tolou, S., Delpech, B., Kaghad, M. and Daloyau, M. (1998). Nucleotide sequence and analysis of the new chromosomal abortive infection gène abiN of *Lactococcus lactis* subsp. *cremoris* SI 14. *FEMS Microbiology Letters, 159,* 331-336.

Quiberoni, A., Tremblay, D., Ackermann, H.-W., Moineau, S. and Reinheimer, J. A. (2006). Diversity of *Streptococcus thermophilus* phages in a large production cheese factory in Argentina. *Journal of Dairy Science, 89,* 3791-3799.

Rousseau, C., Nicolas, J. and Gonnet, M. (2009). CRISPI: a CRISPR interactive database. *Bioinformatics, 25,* 3317-3318.

Sanders, M. E. and Klaenhammer, T. R. (1980). Restriction and modification in group N Streptococci: effect of heat on development of modified lytic bacteriophage. *Applied and Environmental Microbiology, 40,* 500-5006.

Schouler, C., Gautier, M., Ehrlich, S. D. and Chopin, M.-C. (1998). Combinational variation of restriction-modification specificities in *Lactococcus lactis*. *Molecular Microbiology, 28,* 169-178.

Seegers, J. F. M. L., van Sinderen, D. and Fitzgerald, G. F. (2000). Molecular characterization of the lactococcal plasmid pCIS3 : natural stacking of specificity subunits of a type I restriction/modification system in a single lactococcal strain. *Microbiology, 146,* 435-443.

Sijtsma, L., Sterkenburg, A. and Wouters, J. T. (1988). Properties of the cell walls of *Lactococcus lactis* subsp. *cremoris* SK110 and SK112 and their relation to bacteriophage resistance. *Applied and Environmental Microbiology, 54,* 2808-2811.

Sijtsma, L., Jansen, N., Hazeleger, W. C., Wouters, J. T. M. and Hellingwerf, K. J. (1990). Cell surface characteristics of bacteriophage-resistant *Lactococcus lactis* subsp. *cremoris* SK110 and its bacteriophage-sensitive variant SK112. *Applied and Environmental Microbiology, 56,* 3230-3233.

Sing, W. D. and Klaenhammer, T. R. (1990). Plasmid-induced abortive infection in lactococci: A review. *Journal of Dairy Science, 73,* 2239-2251.

Sorek, R., Kunin, V. and Hugenholtz, P. (2008). CRISPR - a widespread system that provides acquired resistance against phages in bacteria and archaea. *Nature Reviews in Microbiology, 6,* 181-186.

Spinelli, S., Campanacci, V., Blangy, S., Moineau, S., Tegoni, M. and Cambillau, C. (2006). Modular structure of the receptor binding proteins of *Lactococcus lactis* phages. The RBP structure of the temperate phage TP901-1. *Journal of Biological Chemistry, 281,* 14256-14262.

Stanley, E., Walsh, L., van der Zwet, A., Fitzgerald, G. F. and van Sinderen, D. (2000). Identification of four loci isolated from two *Streptococcus thermophilus* phage genomes responsible for mediating bacteriophage resistance. *FEMS Microbiology Letters, 182,* 271-277.

Sturino, J. M. and Klaenhammer, T. R. (2004). Bacteriophage defense systems and stategies for lactic acid bacteria. *Advances in Applied Microbiology, 56,* 331-378.

Su, P., Harvey, M., Im, H. J. and Dunn, N. W. (1997). Isolation, cloning and characterisation of the abil gene from *Lactococcus lactis* subsp. *lactis* Ml38 encoding abortive phage infection. *Journal of Biotechnology, 54,* 95-104.

Su, P., Im, H., Hsieh, H. Kang'A, S. and Dunn, N.W. (1999). LlaFI, a type III restriction and modification system in *Lactococcus lactis*. *Applied and Environmental Microbiology, 65,* 686-693.

Sun, X., Göhler, A., Heller, K. J. and Neve, H. (2006). The ltp gene of temperate *Streptococcus thermophilus* phage TP-J34 confers superinfection exclusion to *Streptococcus thermophilus* and *Lactococcus lactis*. *Virology, 350,* 146-157.

Sutherland, I. W., Hugues, K. A., Skillman, L. C. and Tait, K. (2004). The interaction of phage and biofilms. *FEMS Microbiology Letters, 232,* 1-6.

Tang, T. H., Bachellerie, J. P., Rozhdestvensky, T., Bortolin, M. L., Huber, H., Drungowski, M., Elge, T., Brosius, J. and Hüttenhofer, A. (2002). Identification of 86 candidates for small nonmessager RNAs from the archeon *Archaeoglobus fulgidus. Proceedings of the National Academy of Sciences of the United States of America, 99,* 7536-7541.

Tang, T. H., Polacek, N., Zywicki, M., Huber, H., Brugger, K., Garrett, R., Bachellerie, J. P. and Hüttenhofer, A. (2005). Identification of novel noncoding RNAs as potential antisense regulators in the archaeon *Sulfolobus solfataricus. Molecular Microbiology, 55,* 469-481.

Tangney, M. and Fitzgerald, G. F. (2002). Effectiveness of the lactococcal abortive infection systems AbiA, AbiEm AbiF and AbiG against P335 type phages. *FEMS Microbiology Letters, 210,* 67-72.

Tremblay, D. M., Tegoni, M., Spinelli, S., Campanacci, V., Blangy, S., Huyghe, C., Desmyter, A, Labrie, S., Moineau, S. and Cambillau, C. (2006). Receptor-binding protein of *Lactococcus lactis* phages: identification and characterization of the saccharide receptor-binding site. *Journal of Bacteriology, 188,* 2400-2410.

Twomey, D. P., Davis, R., Daly, C. and Fitzgerald, G. F. (1993). Sequence of the gene encoding a second ScrFI m5C methyltransferase of *Lactococcus lactis*. *Gene, 136,* 205-209.

Twomey, D. P., Gabillet, N., Daly, C. and Fitzgerald, G. F. (1997). Molecular characterization of the restriction endonuclease gene (scrFIR) associated with the ScrFI restriction-modification system from *Lactococcus lactis* UC503. *Microbiology, 143,* 2277-2286.

Twomey, D. P., De Urraza, P. J., McKay, L. L. and O'Sullivan, D. J. (2000). Characterization of AbiR, a novel multicomponent abortive infection mechanism encoded by plasmid pKR223 of *Lactococcus lactis* subsp. *lactis* KR2. *Applied and Environmental Microbiology, 66,* 2647-51.

Valyasevi, R., Sandine, W. E. and Geller, B. L. (1991). A membrane protein is required for bacteriophage c2 infection of *Lactococcus lactis* subsp. *lactis* C2. *Journal of Bacteriology, 173,* 6095-6100.

van der Oost, J., Jore, M. M., Westra, E. R., Lundgren, M. and Brouns, S. J. (2009). CRISPR-based adaptive and heritable immunity in prokaryotes. *Trends in Biochemical Sciences, 34,* 401-7.

Verreault, D., Gendron, L., Rousseau, G.M., Veillette, M., Massé, D., Lindsley, W.G., Moineau, S. and Duchaine, C. (2011). Detection of airborne lactococcal bacteriophages. *Applied and Environmental Microbiology, 77,* 491-497.

Villion, M. and Moineau, S. (2009). Bacteriophages of *Lactobacillus*. *Front. Biosci., 14,* 1661-1683.

Voet, D. and Voet, J. G. (1998). *Biochimie*. Bruxelles, Belgium: De Boeck University.

Walker, S. A. and Klaenhammer, T. R. (2000). An explosive antisense RNA strategy for inhibition of a lactococcal bacteriophage. *Applied and Environmental Microbiology, 66,* 310-319.

Watanabe, K. and Takesue, S. (1972). The requirement for calcium in infection with *Lactobacillus* phage. *Journal of General Virology, 17,* 19-30.

Watanabe, K., Ishibashi, K., Nakashima, Y. and Sakurai, T. (1984). A phage-resistant mutant of *Lactobacillus casei* which permits phage adsorption but not genome injection. *Journal of General Virology, 65,* 981-986.

Watanabe, K., Ishibashi, K., Iki, K., Nakashima, Y., Hayashida, M. and Amako, K. (1987). Characteristics of some phage-resistant strains of *Lactobacillus casei*. *Journal of Applied Microbiology, 63,* 197-200.

Whitman, W. B., Colemann, D. C. and Wiebe, W. J. (1998). Prokaryotes: The unseen majority. *Proceedings of the National Academy of Sciences of the United States of America, 95,* 6578-6583.

Wilson, G. G. and Murray, N. M. (1991). Restriction and modification systems. *Annual Review of Genetics, 25,* 585-627.

In: Bacteriophages in Dairy Processing
Editor: Andrea del Lujan Quiberoni et al.
ISBN 978-1-61324-517-0

Chapter 8

BIOCIDES FOR DAIRY BACTERIOPHAGE INACTIVATION

Viviana Suárez ***and Daniela Guglielmotti***
Instituto de Lactología Industrial (UNL-CONICET),
Facultad de Ingeniería Química,
Universidad Nacional del Litoral, Santiago del Estero
Santa Fe, Argentina

ABSTRACT

One of the main problems encountered in dairy fermentations is phage infection, which can alter the quality of fermented products or delay manufacturing processes. Several control strategies such as sanitizers, thermal and chemical treatments, strain rotation, spontaneous mutant phage resistant strains and phage inhibitory media are used in dairy industries to minimize this problem. Disinfectants are used in food manufacture to eliminate microbial populations on surfaces and equipments, preventing the spread of contaminations. Besides having a broad spectrum of activity and a rapid efficiency, a biocide for industrial hygiene must have a low toxicological risk, be easy to manipulate and rinse off, biodegradable, not corrosive and economical.

Biocides used in dairy industry and tested against dairy bacteriophages could be classified into those used in laboratory and those used in plant. Alcohols (isopropanol and ethanol), oxidant compounds (sodium hypochlorite and potassium monopersulfate) and quaternary ammonium compounds are extensively used in laboratory, while peracids (peracetic acid, mix of peracetic / peroctanoic acids), chloride-based compounds (alkaline chloride foam and p-toluensulfonchloroamide sodium salt) and acid products (ethoxylated nonylphenol / phosphoric acid) are those preferred in dairy plants. In general, alcohols are not very effective against bacteriophages. Low concentrations of sodium hypochlorite (100 ppm – 400 ppm) have usually a good inactivation effect on LAB phages, but for some phages higher concentration are needed (800 – 1,200 ppm) in order to completely inactive them. Biocides with potassium monopersufate (1%), as main

* E-mail address: vivisuar@fiq.unl.edu.ar. Tel.: +54 342 4530302; fax: +54 342 4571162

active agent, have excellent destruction capacity of phages and viruses in general, even though they have not been tested on LAB phages. The information about the antiseptic action of quaternary ammonium compounds against dairy bacteriophages is very limited. One report revealed that a quaternary ammonium chloride-based compound was very efficient against *Lactobacillus delbrueckii* phages. Peracetic acid is widely used in food industry including dairy plants because degradation products are innocuous and biodegradable. It was assayed (0.15% v/v, 40ºC) against several phages of LAB used in dairy industry and it was able to completely inactivate them in a short time. Chlorine-based compounds revealed different effectiveness against *Lb. delbrueckii* phages. Alkaline chloride foam produced a rapid inactivation of *Lb. delbrueckii* phages possibly due to its high pH value, while a p-toluensulfonchloroamide solution showed a low efficiency to destroy phage particles. A biocide composed by ethoxylated nonylphenol and phosphoric acid showed great inactivation activity against the same phages, probably due to its very low pH.

INTRODUCTION

One of the main problems encountered in dairy fermentations is the ubiquitous presence of virulent bacteriophages, which can alter the quality of fermented products or delay manufacturing processes. Even though a plethora of phage control measures have been introduced since the discovery of bacteriophages as the major cause of fermentation failures, phages remain a high risk for the dairy industry (Moineau and Lévesque 2005, Emond and Moineau 2007, Villion and Moineau 2009).

The lysogenic state in lactic acid bacteria (LAB) is considered the principal source of bacteriophages in dairy industrial environments. Temperate phages can disturb the normal process of fermentations when, by mutation, they turn into virulent phages capable of overcoming superinfection immunity. Another problem derived from the use of lysogenic strains is their potentially spontaneous induction, releasing phage particles that could infect sensitive strains of starter cultures (Davidson et al. 1990). The consequences of this fact are evident when commercial starters are used, because they are composed of a low number of strains.

Considerable efforts have been aimed to minimize this problem. Several control strategies such as sanitizers, thermal and chemical treatments, strain rotation regimes, spontaneous mutant phage resistant strains and phage inhibitory media (PIM), are used in dairy industries (Suárez et al. 2007). In addition, genes that encode natural resistance mechanisms can be introduced into starter strains (Madera et al. 2004). However, phages represent a real and persistent threat and the authors are aware of recent unpublished cases where phage infection actually limited the fermentation process and/or caused product downgrading (McGrath et al. 2004).

Sanitization is the use of antimicrobial products to kill microorganisms. The aim of disinfection is to reduce the surface population of viable cells left after cleaning and prevent microbial growth on surfaces before production restart (Simões et al. 2010). Disinfectants and sanitizers have a myriad of uses from farm to fork including the cleaning of handling equipment, processing equipment, floors, processing water and wash water. Disinfectants used in food processing or food production may be included to eliminate microbial populations in process water, prevent the spread of contamination from one surface to another and reduce the numbers on the surface of produce, such as by disinfectants used in the produce industry (Hirneisen et al. 2010).

1. Biocide Selection Criteria

The applicability of disinfectants is selected according to the spectrum of activity based on the chemical characterization of the agent and the suitability to different food production processes. Besides a broad spectrum of activity and a rapid efficiency, a modern disinfectant for industrial hygiene must have the following additional properties for an efficacious application process:

- low toxicological risk
- ease of manipulation
- easily to rinse off without any residual problems
- low ecological risk, classification as readily biodegradable, concern to water hazard classes
- compatibility with different technological surface material, no corrosion problems
- active in the presence of organic matter
- low cost application

2. Factors Affecting Biocide Effectiveness

The ideal disinfectant is not yet available. The level of disinfection achieved will depend on many factors such as contact time, temperature, type and concentration of the active ingredient, the presence of organic matter, pH of the medium, the type and quantum of microbial load, between others. The chemical disinfectants at working concentrations rapidly lose their strength on standing.

2.1. Physical Factors

- Surface characteristics: the surface can be cleaned before use of sanitizers due to the effectiveness of sanitization requires a direct contact with the microorganisms and virus.
- Exposure time: generally, the longer time a sanitizer chemical is in contact with the equipment surface, the more effective the sanitization effect; intimate contact is as important as prolonged contact.
- Temperature: this factor is also important to microbial kill by a chemical sanitizer; temperatures up of 55°C are not recommendable because the corrosive nature of most chemical sanitizers.

- Concentration: generally, the activity of biocides increases with increased concentration; however, the use of sanitizers at levels above recommendations does not sanitize better and can be corrosive.
- Soil: the presence of organic material can minimize the activity of sanitizers, which could not sanitize an unclear surface.

2.2. Chemical Factors

- pH: many sanitizers can be affected by pH solution values, and some of them are only active at certain pH conditions (acidic or alkaline).
- Water properties: impurities of water used in solution can affect markedly the action of some biocides.
- Inactivators: some inactivators present in detergent residue can react with sanitizers giving rise to non-biocide products. It is necessary to rinse surfaces before their disinfection.

2.3. Biological Factors

The type of microorganism to inactivate is important to properly select the biocide agent. In general, spores are, more resistant than vegetative cells. Some biocides are more effective against Gram positive than Gram negative bacteria (or vice versa) and can vary in their effectiveness against yeast, moulds, fungi and virus.

3. Biocides against Virus

3.1. Aldehydes

Aldehydes act through alkylation of amino-, carboxyl- or hydroxyl group, and probably damages nucleic acids. They kill all microorganisms, including spores. Glutaraldehyde and formaldehyde are the most used aldehyde compounds.

Some disadvantages of aldehydes are that vapors are irritating (must be neutralized by ammonia), they have poor penetration, they leave non-volatile residue and that their activity is reduced in the presence of proteins. Gluteraldehyde requires alkaline pH and only those articles that are wettable can be sterilized. The main characteristics of glutaraldehyde and formaldehyde are described below.

a) Glutaraldehyde

Glutaraldehyde is an important dialdehyde used as a disinfectant and sterilant, in particular for low-temperature disinfection and sterilization of endoscopes and surgical equipment. Glutaraldehyde has a broad spectrum of activity against bacteria and their spores, fungi and viruses.

The mechanism of action of glutaraldehyde involves a strong association with the outer layers of bacterial cells, specifically with unprotonated amines on the cell surface, possibly representing the reactive sites (Bruck 1991). Such an effect could explain its inhibitory action on transport and on enzyme systems. Glutaraldehyde is more active at alkaline than at acidic pH values. As the external pH is altered from acidic to alkaline, more reactive sites will be formed at the cell surface, leading to a more rapid bactericidal effect.

Glutaraldehyde is a potent virucidal agent (Kobayashi et al. 1984). It reduces the activity of hepatitis B surface antigen (HBsAg) and especially hepatitis B core antigen (HBcAg in hepatitis B virus HBV) (Adler-Storthz et al. 1983) and interacts with lysine residues on the surface of hepatitis A virus (HAV) (Passagot et al. 1982).

Bacteriophages were recently studied to obtain information about mechanisms of virucidal action. Many glutaraldehyde-treated *Pseudomonas aeruginosa* F116 phage particles had empty heads, implying that the phage genome had been ejected. The aldehyde was possibly bound to F116 double-stranded DNA but without affecting the molecule; glutaraldehyde also interacted with proteins of phage F116, which were postulated to be involved in the ejection of the nucleic acid (Maillard et al. 1994, 1996a, 1996b, Maillard and Russell 1997).

b) Formaldehyde

Formaldehyde (methanal, CH_2O) is a monoaldehyde that exists as a freely water-soluble gas. Formaldehyde solution or "formalin" is an aqueous solution containing 34 to 38% (w/w) CH_2O with methanol to delay polymerization. Its clinical use is generally as a disinfectant and sterilant in liquid or in combination with low-temperature steam. Formaldehyde is an extremely reactive chemical that interacts with protein, DNA and RNA in vitro, showing to be bactericidal, sporicidal and virucidal, but it works slower than glutaraldehyde (Power 1995). Its action against virus includes alteration of HbsAg and HBcAg of HBV (Adler-Storthz et al. 1983) and extensively reaction with nucleic acid (e.g., the DNA of phage T2) (Grossman et al. 1961, Stewart et al. 1991). This biocide forms protein-DNA cross-links in SV40 (simian virus 40), thereby inhibiting DNA synthesis (Permana and Snapka 1994).

As a desinfectant, it is generally used as 40% formaldehyde (formalin) solution for surface disinfection and fumigation of rooms, chambers, operation theatres, biological safety cabinets, wards, sick rooms etc. Fumigation is achieved by boiling formalin, heating paraformaldehyde or treating formalin with potassium permanganate. It also sterilizes bedding, furniture and books. A solution of formalin (10%) and tetraborate (0.5%) sterilizes clean metal instruments. Gluteraldehyde (2%) is used to sterilize thermometers, cystoscopes, bronchoscopes, centrifuges, anasethetic equipments, etc. An exposure of at least 3 h at alkaline pH is required for action by gluteraldehyde. Formaldehyde (2%) at 40°C for 20 min is used to disinfect wool, while a 0.25% solution at 60°C for 6 h it is used to disinfect animal hair and bristles.

3.2. Ozone

Ozone is an emergent chemical agent employed in environmental decontamination, used for decades as a primary disinfectant for drinking water in many European countries. It has received more interest in other countries after the discovery of potentially harmful by-products of chlorine disinfection. Ozone is a very strong oxidant and an effective disinfectant against most waterborne pathogens (Shin and Sobsey 2003). For viruses, early research on ozone applications focused mainly in water (Tyrrell et al. 1995, Lazarova et al. 1998, Shin and Sobsey 2003, Thurston-Enriquez et al. 2005). Some studies report that a relatively low ozone concentration (less than 1 mg l^{-1}) and a short contact time (1 min) are sufficient to inactivate 99% virus, such as rotaviruses, parvoviruses, feline calicivirus, and hepatitis A virus (Shin and Sobsey 2003, Thurston-Enriquez et al. 2005). These previous studies revealed that the susceptibility of viruses was highly related to ozone concentrations, pH value, water temperature, residence time, mixing degree, and organic compounds. Tseng and Li (2006) reported that the survival fraction of airborne viruses (specific of *Escherichia coli* and *Pseudomonas*) decreased exponentially with increasing ozone dose. Virus required ozone doses of 0.34-1.98 and 0.80-4.19 min-mg/m^3 for 90% and 99% inactivation, respectively. In the airborne phase with a short contact time, virus susceptibility to ozone could be related to the type of virus capsid architecture, and with or without envelope. Regarding the relative humidity (RH) effects, the susceptibility for viruses was higher at 85% RH than that at 55% RH. According to the authors, this might be related to the generation of more radicals from ozone which reacted with more water vapour at higher RH. Currently, there are no studies focusing on the ozone effect on specific dairy bacteriophages.

3.3. Hydrogen Peroxide Vapour (HPV)

Another relatively new biocide used in environmental control of virus is the hydrogen peroxide vapour (HPV). Systems for vaporising hydrogen peroxide have been used in the pharmaceutical clean rooms for disinfection, microbiology laboratories to disinfect both safety cabinets and entire laboratories (to replace formaldehyde fumigation) and are currently being introduced into UK hospitals (Pottage et al. 2010). Hydrogen peroxide, a potent oxidant, acts against bacteria and viruses by forming HO, free radicals which react with thiol groups in proteins, lipids and nucleic acids (Mc Donnell and Russell 1999). In viruses, these reactions prevent the correct functioning of proteins and nucleic acids, thus inhibiting the infection and replication process. Hydrogen peroxide reacts with organic material that reduces its efficacy. When planning a gaseous disinfection it is important to consider the level of contamination present and the presence of a potentially protective soiling. A pre-cleaning of surfaces with appropriate disinfectants prior to the deployment of gaseous disinfection is recommended. Gaseous disinfection with hydrogen peroxide systems alone should not be regarded as a combined cleaning / disinfection procedure. There is no information about the efficiency of this biocide against dairy phages.

3.4. Chlorine Dioxide (ClO_2)

Chlorine dioxide (ClO_2) is a biocide with an inactivation capacity that exceeds chlorine and its derivatives. Chloride dioxide presents great biocide efficiency in a wide pH range (between 3 and 10). Its action against virus includes adsorption and penetration in proteins of viral capsid and reaction with RNA or DNA of virus. As result, ClO_2 damages the genetic capacity of virus.

Chlorine dioxide (CD) is mainly used in water disinfection. In relation to this application, advantages of ClO_2 disinfection include (i) oxidation of iron and manganese (reduces discoloration of finished water), ii) no trihalomethane formation, (iii) no reaction with ammonia, (iv) less sensitive than chlorine to pH conditions typical of drinking water, (v) a relatively persistent residual and (vi) reduction of tastes and odors caused by organic and sulphurous compounds.

Disadvantages of ClO_2 disinfection include (i) formation of organic halides, (ii) formation of chlorite and chlorate and (iii) production of taste and odors at concentrations higher than 0.5 mg l^{-1} (Thurston-Enriquez et al. 2005).

Resistance of animal and human virus to the biocide action of chlorine dioxide is fairly well reported. Recently, Sanekata et al. (2010) evaluated the antiviral activity of a chlorine dioxide gas solution (CD) and sodium hypochlorite (SH) against feline calicivirus, human influenza virus, measles virus, canine distemper virus, human herpesvirus, human adenovirus, canine adenovirus and canine parvovirus. CD at concentrations ranging from 1 to 100 ppm produced potent antiviral activity, inactivating > 99.9% of the viruses with a 15 s treatment for sensitization. The antiviral activity of CD was approximately 10 times higher than that of SH. In addition, Zoni et al. (2007) reported that for complete inactivation of HAV and feline calicivirus, concentrations from 0.6 mg l^{-1} were required. This observation is accurate for coxsackie B5 as well, but this virus has shown a good sensitivity at all concentration tested according to regression analysis results. For feline calicivirus and HAV, at low concentrations of disinfectant, prolonged contact times were needed to obtain a 99.99% reduction of viral titres (about 16 and 20 min, respectively). Lukasik et al. (2003) studied the action of chloride dioxide against viral and bacterial pathogens in inoculated strawberries.

Poliovirus 1 and the bacteriophages PRD1, X-174, and MS2 were used as model and surrogate organisms. Significant reductions (higher than 98%) in numbers of viruses were obtained with 200 ppm of stabilized chlorine dioxide washes.

On the other hand, little is known about whether gaseous ClO_2 has antiviral action. The gaseous agents show features of excellent diffusibility and penetrability, making it possible to access sites that are difficult to disinfect with conventional liquids agents. Morino et al. (2009) reported that low concentration of ClO_2 gas (mean 0.08 ppm) could inactivate FCV (feline calcivirus) in the wet state with 0.5 % fetal bovine serum (FBS) within 6 h in moderate RH (> 3 log_{10} reductions) and FCV in dry state with 2 % FBS within 10 h in high RH (> 3 log_{10} reductions) at 20°C. They proposed that low concentrations of ClO_2 gas (lower than 0.3 ppm) may be suitable as a disinfectant to prevent NV (norovirus) infection.

This sanitizer has become more widely used in the dairy industry, predominately in the sanitizing of surfaces of equipment, floor drains, and other areas to greatly reduce the microbial load in these area. However, until now no results were reported about the biocide action specifically against dairy bacteriophages.

4. Biocides against Dairy Phages

Biocides used in dairy industry and proved against dairy bacteriophages could be classified into those used in laboratory and those that are used in plant. Within those used in the laboratory, the most common are alcohols (isopropanol and ethanol), oxidant compounds (sodium hypochlorite and potassium monopersulfate) and quaternary ammonium compounds. In plant, however, other compounds are routinely used, as well as peracids (peracetic acid, mix of peracetic / peroctanoic acids), chloride-based compounds (alkaline chloride foam and p-toluensulfonchloroamide sodium salt) and acid products (ethoxylated nonylphenol / phosphoric acid) (Ebrecht et al. 2010).

4.1. Laboratory Environment

a) Alcohols

Isopropanol and ethanol are the most common alcohols used in disinfection of surfaces. In general, it is well known that isopropanol is not effective against lactic acid bacteria phages. As example, isopropanol (used as pure) showed to be extremely harmlessness and not efficient to inactivate the Argentinian temperate bacteriophages Cb1/204 and Cb1/342, specific of *Lb. delbrueckii* species (Ebrecht et al. 2010). For these phages, T_{99} values (defined as the time necessary to inactivate the 99% of initial phage particles) were higher than 35 min (Figure 1). Similar results were reported for other species of dairy bacteriophages according to Quiberoni et al. (1999, 2003), Binetti and Reinheimer (2000), Suárez and Reinheimer (2002), Capra et al. (2005) and Briggiler Marcó et al. (2009).

On the other hand, ethanol appears as a biocide with variable capacity of phage inactivation, which depends on the concentration used and the phage characteristics as well. In general, it was observed that 75% and 100% ethanol were the most efficient concentrations allowing in some cases, the total destruction of phage particles between 5 and 45 min. (Suárez et al. 2002). Table 1 shows T_{99} values obtained for lactococci bacteriophages. These results revealed that the effect is highly dependent of the phage treated.

Table 1. T_{99} values obtained for virulent phages of lactococci strains treated with ethanol at diverse concentrations

Concentration (% v/v) [a]	T_{99} (min) [b]			
	001	046	QF12	QP4
10	> 45	> 45	> 45	> 45
50	> 45	> 45	> 45	> 45
75	18.5	> 45	4.9	1.3
100	1.9	24.3	3.6	3.8

[a]: Diluted in destilled water.

[b]: Time (min) to inactivate 99% of phage particles.

It is clearly seen that first-order model is not suitable to describe neither thermal nor chemical phage inactivation kinetics. A mathematical model to describe the inactivation kinetic by ethanol and isopropanol agents on lactococci phages was proposed by Buzrul et al. (2007). They stated that this modelling approach could be also used for other biocides. However, as important consideration, this should be done with sufficient data points at appropriate time intervals.

b) Sodium Hypochlorite

Sodium hypochlorite is the most widely used chlorine-releasing disinfecting agent, and is active against a broad spectrum of microorganisms. Solutions of this desinfectant are widely used for hard-surface disinfection (household bleach) and can be used for disinfecting spillages of blood containing human immunodeficiency virus or HBV (McDonnell and Russell 1999).

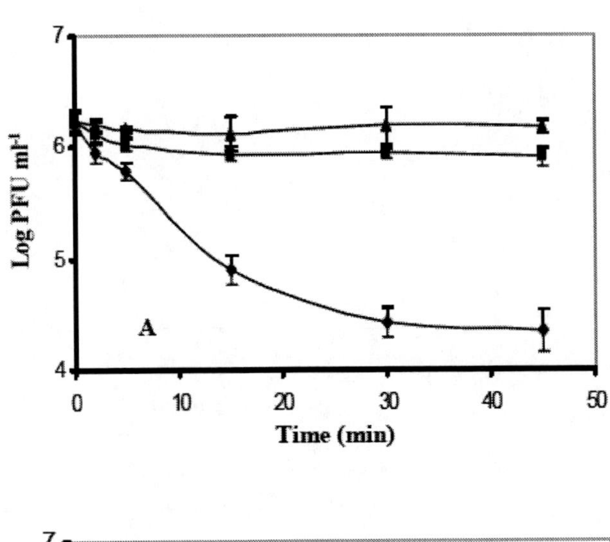

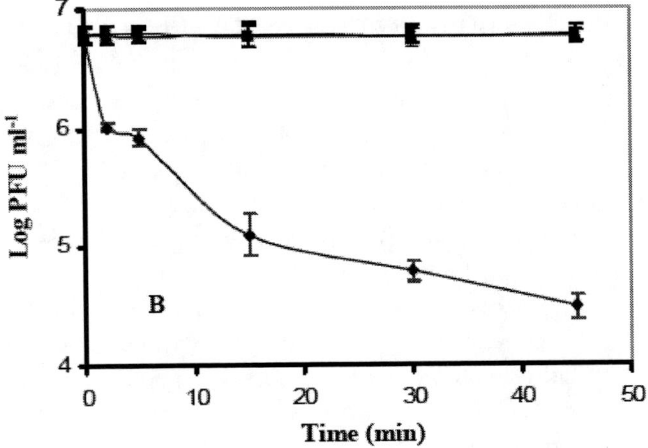

Figure 1. Chemical inactivation kinetics of Argentinian temperate phages Cb1/204 (A) and Cb1/342 (B) treated with 10% (▲), 50% (■) and 100% (◆) of isopropanol (Data adapted from Ebrecht et al. 2010).

In LAB, there are several reports that inform the activity of this disinfectant on specific bacteriophages, virulent and temperate, of diverse LAB species. A concentration of 100 ppm (active chlorine) was effective to inactivate temperate *Lb. delbrueckii* phages Cb1/204 and Cb1/342, according to Ebrecht et al. (2010). There were necessary 20 min and 30 min of treatment, respectively, to inactivate all phage particles (at high initial concentration) (Figure 2). This concentration was reported to be efficient against bacteriophages of *Lactobacillus helveticus* (Quiberoni et al 1999), *Streptococcus thermophilus* (Binetti and Reinheimer 2000) and *Lactococcus lactis* (Suárez and Reinheimer 2002). However, this last study also demonstrated that, for some phages (046 and QP4) a minimum concentration of 300 ppm of residual-free chlorine was necessary to completely inactivate them.

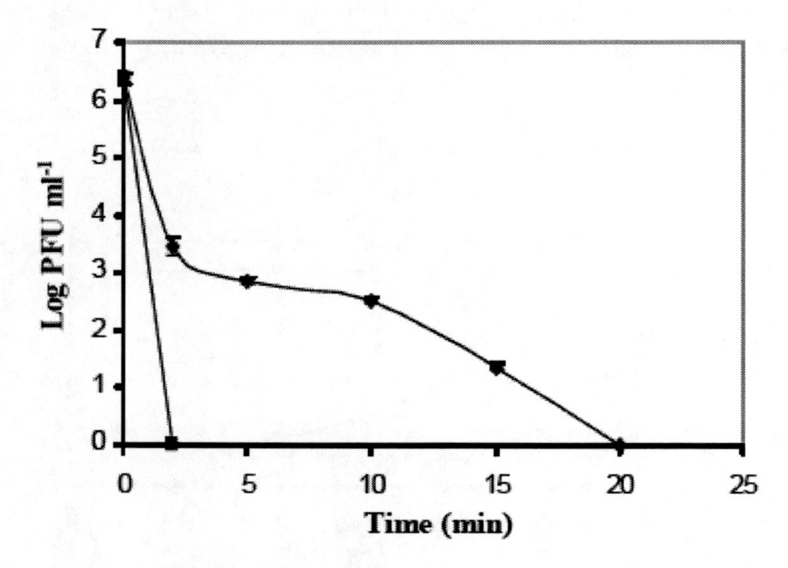

Figure 2. Chemical inactivation kinetics of Argentinian temperate phage Cb1/204 treated with 100 ppm (◆) and 200 ppm (■) of sodium hypochlorite (Data adapted from Ebrecht et al. 2010).

Hunter and Whitehead (1940) and Bennet and Nelson (1954) have demonstrated that concentrations higher than 500 ppm of sodium hypochlorite were necessary to inactivate phages of *Lactococcus lactis* subsp. *lactis*, but these tests were carried out using milk and whey as suspension media. Moreover, Avsaroglu et al. (2007) reported that, for lactococcal phages isolated from Turkish dairy plants, concentrations between 2,000 ppm and 5,000 ppm were required to reduce initial concentration of phage particles (30 min and 20 min, respectively). Even though some *Lb. delbrueckii* temperate phages were sensitive to low concentrations of sodium hypochlorite (Ebretch et al. 2010), there were found some lytic (or virulent) phages with higher resistance to this biocide. Regarding this, concentrations of 300 ppm (phage LL-H), 400 ppm (phage BYM), 400 ppm (phage YAB) and 1,200 ppm (phage Ib_3) (Figure 3) of residual-free chlorine were necessary to obtain a total inactivation of phage particles within 15, 5, 45 and 45 min, respectively (Quiberoni et al 2003). As shown by the results, *Lb. delbrueckii* phage Ib_3 showed a remarkably resistance to high concentrations of sodium hypochlorite. Capra et al. (2004) reported that phages of *Lactobacillus casei / paracasei* supported concentrations until 800 ppm of active chlorine.

In general, sodium hypochlorite is a good disinfectant agent to eliminate dairy bacteriophages, except for special cases of very resistant ones. However, it should be convenient to remember that its use is limited because of its corrosive effect, its inactivation by organic materials and its instability. In addition, the inhalation of chlorine gas causes irritation of respiratory tract and can produce cough, dyspnea, clinical pulmonary edema and pneumonitis.

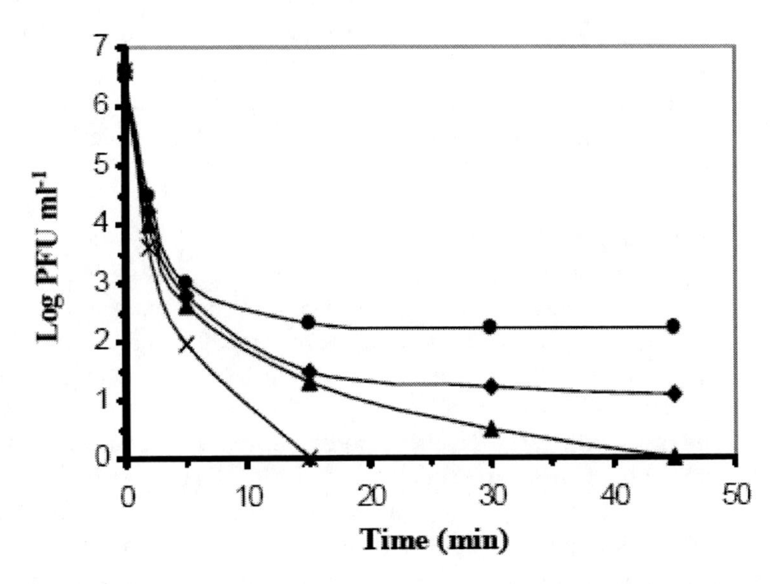

Figure 3. Chemical inactivation kinetics of Argentinian phage Ib_3 treated with 800 ppm (●), 1,000 ppm (◆), 1,200 ppm (▲) and 1,400 ppm (×) of residual free-chlorine (Data adapted from Quiberoni et al. 2003).

c) Potassium Monopersulfate

This compound is commercialized as a balanced and stabilized combination of peroxide compounds, organic acids and an inorganic stabilizator system, with potassium monopersufate as main active agent. Oxidation of bacterial structures leads to cellular death. Acid power together with oxidizing power (due to the components of the formulation), and the ability of incorporated surfactant, get a superiority in the destruction of the biofilm, which results in greater bactericidal activity. The recommended use concentration is 1% and as used, does not irritate skin or eyes. In addition, its use is compatible with diverse materials as steel, aluminium and plastic.

Until now, there are no studies about the effect of this biocide against dairy bacteriophages. Anyway, Halfhide et al. (2008) reported the inhibitory effect of this biocide against 6 phages belonging to *Myoviridae* and *Siphoviridae* families, but none of them were dairy bacteriophages. However, many lactic acid bacteria phages belong to one of these two families and always to the *Caudovirales* order. Results obtained in this study demonstrated that all phages were inactivated rapidly when potassium monopersufate was used. Because it inactivated the phages so quickly it was not possible, from the data, to assess whether some strains were more susceptible than others.

Solomon et al. (2009) compared the susceptibilities of hepatitis A, virus and feline calicivirus with MS2 and X-174 phages when they all were treated with this biocide. This group reported that MS2 phage particles were reduced in 1.26 log after the treatment with 0.1%, 3.24 log using 0.25%, 4.44 log using 0.5% and below the detection limit after the treatment using 1% of the biocide. Phage X-174 was more sensitive than MS2 when treated with this disinfectant. HAV was resistant to 0.1%, 0.25% and 0.5% of the biocide, although it was reduced to below the detectable limit after treatments with 1%. Comparing between phages and viruses susceptibility studies, it seems that the response of phage MS2 to this biocide is very similar to that found for FCV.

These studies supported the affirmation that biocides with potassium monopersufate as main active agent used in the concentration recommended (1%), have excellent destruction capacity of phages and viruses in general.

d) Quaternary Ammonium Compounds

The quaternary ammonium compounds (QAC) are antiseptics with a low toxicity. Their action is based on the inactivation of energy-producer enzymes, protein denaturalization and disruption of cellular membranes. Holah and co-workers (2002) showed that QAC are by far the most commonly used disinfectant in the food industry in the United Kingdom. This is primarily due to their adequate performance as biocide (though not as effective as hypochlorite) combined with their excellent non-tainting, non-toxic and non-corrosive (to human skin as well as hard surfaces) nature, which makes them the ideal choice for manual application in the food industry. The common methods of application, namely hand spraying, high pressure spraying and manual application, apply relatively small volumes of product to the equipment surface, where the product remains until it drains off or evaporates.

Disinfectants based on QAC are also widely used in the hospital environment. Optimal concentration of use depends on their composition. Jiménez and Chiang (2006) proved the viricidal activity of R-82 compound (commercial combination of diakyl dimethyl ammonium chloride and alkyl dimethyl benzyl ammonium chloride) against feline calcivirus, a surrogate of norovirus. They demonstrated that 10 min with a 1:256 dilution of R-82 inactivated completely a high concentration of virus (reduction of log NMP ml^{-1} approx. 6.6). In particular, this QAC was effective due to its dual composition and a combination along with the high alkaline pH, which gave a synergistic effect.

The QACs have an effect on lipid, enveloped (including human immunodeficiency virus and HBV) but not nonenveloped viruses (Springthorpe and Satter 1990). QAC-based products induced disintegration and morphological changes of human HBV, resulting in loss of infectivity. In studies with different phages (298–301, 303–305, 307), CPC significantly inhibited transduction by phage F116 and inactivated the phage particles. Furthermore, CPC altered the protein bands of phage F116 but did not affect the DNA within the capsid.

On the other hand, Sickbert-Bennett et al. (2005) evaluated the effect of hand hygiene agents against the phage MS2, a surrogate of clinically important non-enveloped viruses. They demonstrated that the benzethonium chloride, as hand-washing compound, was a good disinfectant on this bacteriophage. However, they also demonstrated that others sanitizers showed better inhibitory activity than quaternary ammonium compounds.

Besides potassium monopersufate-based biocides, Solomon et al. (2009) studied the effect of QAC against hepatitis A virus (HAV), feline calicivirus (FC), phages MS2 and X-174. The QAC disinfectant was tested at 1, 5 and 10 times the recommended concentration

(848, 4,240 and 8,480 ppm active quaternary ammonium). Phage X-174 was resistant to all concentrations of QAC. The number of phage MS2 particles was reduced in approximately 0.63 log after the treatment with 1× QAC, 1.86 log after using 5× QAC, and in 3.34 log when 10× QAC was used. Surprisingly, treatment with 1× QAC resulted in greater reductions of FCV than treatment with higher levels of QAC (10×).

The information about the antiseptic action of QAC against dairy bacteriophages is very limited. Ebrecht et al. (2010) informed that a quaternary ammonium chloride-based compound was very efficient against three virulent and two temperate *Lb. delbrueckii* phages. In all cases, 1.0 % (v/v) of QAC destroyed 99% of total phage particles within 2 min. Even the pH value of this solution was 9.3 all phages maintained their viability as shown by the pH control.

Table 2 shows T_{99} values when biocide A was tested against these phages. These results demonstrate that the use of QAC in surface disinfection should be a good alternative to eliminate dairy phages.

Table 2. T_{99} values for temperate (Cb1/342 and Cb1/204) and virulent (BYM, YAB and Ib_3) *Lactobacillus delbrueckii* phages treated with biocide A at diverse concentrations

Concentration (% v/v) / pH	T_{99} (min) [a]				
	Cb1/342	Cb1/204	BYM	YAB	Ib_3
3.0 / 10.5	< 2.0	< 2.0	< 2.0	< 2.0	< 2.0
2.5 / 9.9	< 2.0	< 2.0	< 2.0	< 2.0	< 2.0
2.0 / 9.7	< 2.0	< 2.0	< 2.0	< 2.0	< 2.0
1.5 / 9.6	< 2.0	< 2.0	< 2.0	< 2.0	< 2.0
1.0 / 9.3	< 2.0	< 2.0	< 2.0	< 2.0	< 2.0

[a]: Time to inactivate 99% of phage particles.

4.2. Equipment, Utensils and Plant Environment

a) Compounds Based on Peracetic Acid

Peracetic or peroxiacetic acid does not exist commercially as pure compound. The available products are mix in equilibrium of peracetic acid, acetic acid, hydrogen peroxide and water in variable proportions, according to the predetermined amounts of acetic acid and hydrogen peroxide used as initial reaction in the manufacturing process. Peracetic acid is an effective bactericidal, sporicidal, fungicidal and even virucidal agent. Its biocide action is due to the oxidant capacity of peracetic acid and peroxide hydrogen, which produce the oxidation of bacteria, fungi and viral proteins. Thus, hydrogen peroxide may also contribute to the disinfection action and bacteriostatic effects of peracetic acid product (Liberti and Notarnicola 1999). Hydrogen peroxide has not been widely used as a sole disinfectant for water or wastewater treatment, mainly due to its slow disinfection action and low disinfection efficiency (Liberti et al. 2000, Wagner et al. 2002).

The recommended concentration of peracetic acid biocides ranges from 0.02% to 2%, depending on the supplier's recommendations. This product is widely used in food industry including dairy plants because the degradation products are innocuous and biodegradable. It is recommended in sanitization of equipments, pipes, tanks, pasteurizers and surfaces in general.

Peracetic acid was assayed against several phages of LAB that are used in dairy industry. In all cases, at a concentration of 0.15% and 40°C, the peracetic acid was able to completely inactivate phage suspensions of high titre (10^6 PFU ml^{-1}) in short time (< 5 min). This fact was demonstrated for *Lb. helveticus* (Quiberoni et al. 1999), *Strep. thermophilus* (Binetti and Reinheimer 2000), *L. lactis* (Suárez and Reinheimer 2002), *Lb. delbrueckii* (Quiberoni et al. 2003, Ebrecht et al. 2010), *Lb. casei / paracasei* (Capra et al. 2004) and *Lactobacillus plantarum* (Briggiler Marcó et al. 2009) bacteriophages. These results show that this biocide is highly effective against phages of lactic acid bacteria, making it an excellent choice to use in the dairy industry.

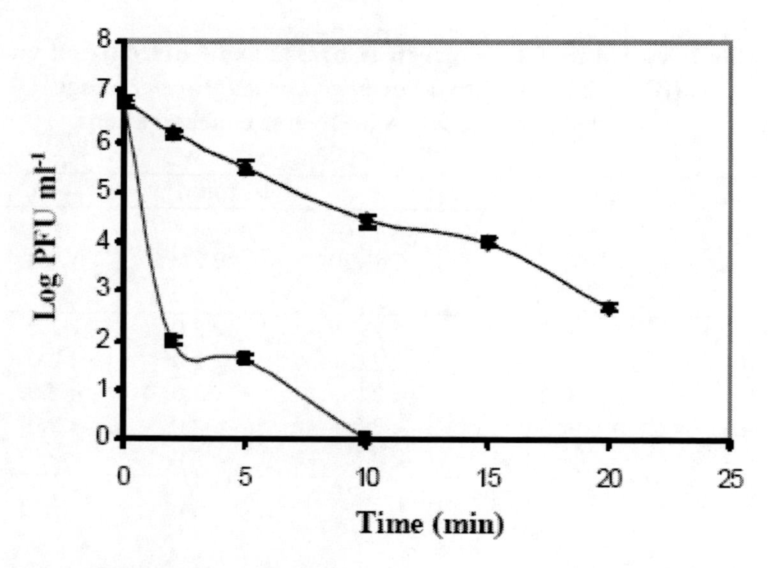

Figure 4. Chemical inactivation kinetics of virulent phage Ib_3 treated with 0.13% (v/v) (■) and 0.26% (v/v) (◆) of biocide based in peracid / peroctanoic acid combination (Data adapted from Ebrecht et al. 2010).

Another peracid compound generally used is a mix of peracetic and peroctanoic acids. This commercial product is a combination of peracids that, in theory, should be more efficient than the peracetic acid alone. This product is used simillarly to peracetic acid and it has a comparable action.

Studies of this biocide against virulent and temperate *Lb. delbrueckii* phages were performed by Ebrecht et al. (2010). These authors demonstrated that this biocide was efficient to rapidly destroy all phages studied, except for phage Ib_3. Figure 4 shows inactivation kinetics of phage Ib_3 when 0.13% and 0.26% (v/v) of this biocide were used. These solutions showed pH values of 3.45. However, when the tested phages were incubated at this low pH (without biocide) for 2 min, none of them was inactivated. On the other hand, phage Ib_3 was incubated at this pH for 10 min and 20 min, because it showed to be very resistant to this biocide (0.26% and 0.13%, respectively). These phage particles were not affected by the solution at this pH value (pH control). According to these results, peracid / peroctanoic acid-based biocide was less efficient than the use of peracetic acid alone, which is in contradiction to the affirmations stated by the suppliers.

b) Chlorine-Based Compounds

Alkaline chloride foam. This product is advised to be used in concentrations between 2.5% and 7.5% (v/v) and, applied with machine, forms a persistent foam layer that adheres well to any surface, both horizontal and vertical. It is used in cleaning of equipments with difficult access for others degreasers. It combines a degreasing action with a biocide capacity based on oxidant power of hypochlorite. Due to this product is strongly alkaline must be handled with great caution. To manipulate it, it is recommended the use of personal protective elements, such as rubber gloves, safety goggles, rubber boots and plastic apron.

There is not information about the action of this biocide against bacteriophages in general. The only study about this subject was published by Ebrecht et al. (2010). In this work, the biocidal effect of this compound against temperate and virulent *Lb. delbrueckii* phages was tested. The concentration used was 2.5 % (v/v) and the solution pH value was 12.4. At this extreme pH, the authors reported that phage particles were completely destroyed after 2 min of exposure, even in control without biocide. These results suggest that the high pH value may have a great contribution on phage inactivation.

p-toluensulfonchloroamide (sodium salt). The product is marketed at a concentration of 1% (w/v) p-toluensulfonchloroamide in hydro alcoholic solution. In dairy industry, it is used in disinfection of milking equipment, pasteurizers, tanks and pipes. Specifications assure that this compound destroy Gram negative and positive bacteria, fungi, spores and virus. Its action was tested against temperate and virulent *Lb. delbrueckii* phages by Ebrecht et al. (2010). In this work, the biocide was used pure (1%, w/v) and the pH value of the solution was 6.1. This biocide showed a low efficiency to destroy the particles of assayed phages, being Ib_3 the most resistant (Figure 5). Only phage Cb1/204 showed a T_{99} value lower than 45 min, while the others overcame this time.

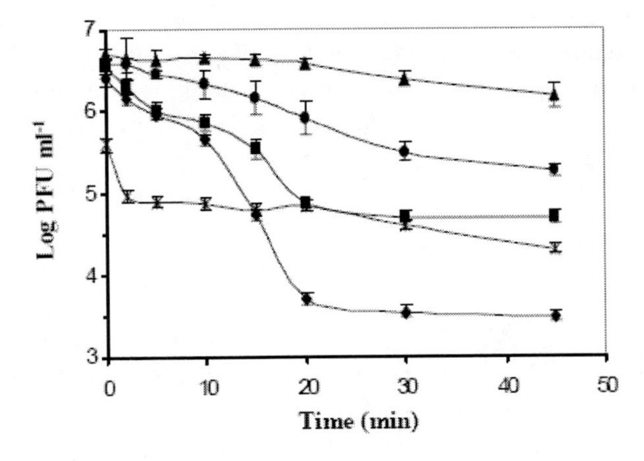

Figure 5. Chemical inactivation kinetics of temperate Cb1/204 (◆), Cb1/342 (■), and virulent phages Ib_3 (▲), YAB (•) and BYM (*) of *Lactobacillus delbrueckii* phages treated with p-toluensulfonchloroamide (1%) (Data adapted from Ebrecht et al. 2010).

c) Acid Compounds

Ethoxylated nonylphenol / phosphoric acid. This biocide contains a mixture of surfactants, dispersants, phosphoric acid and a new inhibitor for aluminum. Combination of cleaning agents removes especially fats and proteins existing in the dairy industry, also presenting a bactericidal activity comparable to normal disinfectants; phosphoric acid content ensures perfect elimination of lime precipitation and stone milk; the product contains an inhibitor to prevent corrosion of aluminum. This product is used in cleaning of tankers, storage tanks and pipes, packers, butter (continuous system). Due to its dual capacity as cleaner and sanitizer, it is perfect to be used in CIP (cleaning-in-place) cleaning processes. The recommended concentration is 0.8% and the pH value of this solution is 2.0.

Ebrecht et al. (2010) presented results for virulent and temperate *Lb. delbrueckii* phages. T_{99} values were, in all cases, less than 2 min. pH inactivation controls were performed by subjecting high concentrations of phage to the pH value of the biocide solution (pH 2). Extreme pH values of the biocide produced total destruction of phage particles after 2 min. These results allow us to hypothesize that this extreme pH value could have a great contribution on phage inactivation (acid inactivation).

Conclusion

Bacteriophages are natural constituents of most microbiota and are found in all ecosystems where bacteria flourish. Because of their ubiquity in nature, phages eventually find their way into fermentation facilities and threaten the industrial process used to manufacture fermented foods. Over the years, fermentation industries have developed a series of control measures to keep phage in check, including adapted factory design and processes, cleaning and sanitation procedures, and tailored air control systems. Strategies based upon culture rotation of phage-unrelated strains or strains harboring natural phage-resistance mechanisms tremendously helped in the containment of phages in these industries. More recently, genetic engineering has enabled the design of new strains that combine excellent industrial performance with enhanced phage resistance (Genetically Modified Organisms); however their application should remain marginal until they receive general acceptance in the public. Although complete eradication of bacteriophages in industrial fermentation settings is not possible, the implementation of an integrated multihurdle program and appropriate staff education should provide the means to contain phage contamination in the food fermentation industry.

References

Adler-Storthz, K., Sehulster, L. M., Dreesman, G. R., Hollinger, F. B and Melnick, J. L. (1983). Effect of alkaline glutaraldehyde on hepatitis B virus antigens. *European Journal of Clinical Microbiology 2*, 316-320.

Avsaroglu, M. D., Buzrul, S., Alpas, H. and Akcelik, M. (2007). Hypochlorite inactivation kinetics of lactococcal. *LWT – Food Science and Technology, 40*, 1369-1365.

Bennet, F. and Nelson, F. (1954). Action of certain viricidal agents on lactic *Streptococcus* bacteriophage in liquids. *Journal of Dairy Science, 36,* 847-856.

Binetti, A. G. and Reinheimer, J. A. (2000). Thermal and chemical inactivation of indigenous *Streptococcus thermophilus* bacteriophages isolated from Argentinian dairy plants. *Journal of Food Protection, 63 (4),* 509-515.

Briggiler Marcó, M., De Antoni, G. L., Reinheimer, J. A. and Quiberoni, A. (2009). Thermal, chemical, and photocatalytic inactivation of *Lactobacillus plantarum* bacteriophages. *Journal of Food Protection, 72 (5),* 1012-1019.

Bruck, C. W. (1991). *Role of glutaraldehyde and other liquid chemical sterilants in the processing of new medical devices*. In R. F. Morrissey and Y. I. Prokopenko (Eds), *Sterilization of medical products*, vol. V. (pp. 376-396). Morin Heights, Canadá: Polyscience Publications.

Buzrul, S., Ozturk, P., Alpas, H. and Akcelik, M. (2007). Thermal and chemical inactivation of lactococcal bacteriophages. *LWT - Food Science and Technology, 40*, 1671–1677.

Capra, M. L., Quiberoni, A. and Reinheimer, J. A. (2004). Thermal and chemical resistance of *Lactobacillus casei* and *Lactobacillus paracasei* bacteriophages. *Letters in Applied Microbiology, 38*, 499-504.

Davidson, B., Powell, I. B. and Hillier, A. J. (1990). Temperate bacteriophages and lysogeny in lactic acid bacteria. *FEMS Microbiology Reviews, 87*, 79-90.

Ebrecht, A., Guglielmotti, D., Tremmel, G., Reinheimer, J. and Suárez, V. (2010). Temperate and virulent *Lactobacillus delbrueckii* bacteriophages: Comparison of their thermal and chemical resistance. *Food Microbiology, 27*, 515-520.

Emond, E. and Moineau, S. (2007). *Bacteriophages and food fermentations*. In: McGrath, S. and van Sinderen, D. (Eds.), *Bacteriophage: Genetics and MolecularBiology* (pp. 93-124). New York, USA: Horizon Scientific Press/Caister Academic Press.

Grossman, L., Levine, S. S. and Allison, W.S. (1961). The reaction of formaldehyde with nucleotides and T2 bacteriophage DNA. *Journal of Molecular Biology 3*, 47–60.

Halfhide D., Gannon, B. W., Hayes, C. M. and Roe, J. M. (2008). Wide variation in effectiveness of laboratory disinfectants against bacteriophages. *Letters in Applied Microbiology, 47,* 608–612

Hirneisen, K. A., Black, E. P., Cascarino, J. L., Fino, V. R., Hoover, D. G. and Kniel, K. E. (2010). Viral Inactivation in Foods: A Review of Traditional and Novel Food-Processing Technologies. *Comprehensive Reviews in Food Science and Food Safety, 9,* 3-20.

Holah, J. T., Taylor J. H., Dawson, D. J. and Hall, K. E. (2002). Biocide use in the food industry and the disinfectant resistance of persistent strains of *Listeria monocytogenes* and *Escherichia coli*. *Journal of Applied Microbiology Symposium Supplement, 92*, 11S-120S.

Hunter, G. and Whitehead, H. (1940). The action of chemical disinfectants on bacteriophages or the lactic streptococci. *Journal of Dairy Research, 11*, 62-66.

Jimenez, L. and Chiang, M. (2006). Virucidal activity of a quaternary ammonium compound disinfectant against feline calicivirus: a surrogate for norovirus. *American Journal of Infection Control, 34 (5)*, 269–273.

Kobayashi, H., Tsuzuki, M., Koshimizu, K., Toyama, H., Yoshihara, N., Shikata, T., Abe, Mizuno, K., Otomo, N. and Oda, T. (1984). Susceptibility of hepatitis B virus to disinfectants or heat. *Journal of Clinical Microbiology, 20*, 214–216.

Lazarova, V., Janex, M. L., Fiksdal, L., Oberg, C., Barcina, I. and Pommepuy, M. (1998). Advanced wastewater disinfection technologies: short and long term efficiency. *Water Science and Technology, 38 (12),* 109–117.

Liberti, L. and Notarnicola, M. (1999). Advanced treatment and disinfection for municipal wastewater reuse in agriculture. *Water Science and Technology, 40 (4–5)*, 235–245.

Liberti, L., Lopez, A., Notarnicola, M., Barnea, N., Pedahzur, R. and Fattal, B. (2000). Comparison of advanced disinfection methods for municipal wastewater reuse in agriculture. *Water Science and Technology, 42 (1–2)*, 215–220.

Lukasik, J., Bradley, M. L., Scott, T. M., Dea, M., Koo, A., Hsu, W.-Y., Bartz, J. A. and Farrah, S. R. (2003). Reduction of poliovirus 1, bacteriophages, *Salmonella Montevideo* and *Escherichia coli* 0157:H7 on strawberries by physical and disinfectant washes. *Journal of Food Protection, 66*, 188-193.

Madera, C., Monjardín, C. and Suárez, J. E. (2004). Milk contamination and resistance to processing conditions determine the fate of *Lactococcus lactis* bacteriophages in dairies. *Applied and Environmental Microbiology, 70 (12)*, 7365-7371.

Maillard, J.-Y. and Russell, A. D. (1997). Viricidal activity and mechanisms|of action of biocides. *Science Progress, 80*, 287–315.

Maillard, J.-Y., Beggs, T. S., Day, M. J., Hudson, R. A. and Russell, A. D. (1994). Effect of biocides on MS2 and K coliphages. *Applied and Environmental Microbiology, 60*, 2205–2206.

Maillard, J.-Y., Beggs, T. S., Day, M. J., Hudson, R. A. and Russell, A. D. (1996a). Damage to *Pseudomonas aeruginosa* PAO1 bacteriophage F116 DNA by biocides. *Journal of Applied Bacteriology, 80*, 540–554.

Maillard, J.-Y., Beggs, T. S., Day, M. J., Hudson, R. A. and Russell, A. D. (1996b). The effect of biocides on proteins of *Pseudomonas aeruginosa* PAO bacteriophage F116. *Journal of Applied Bacteriology, 80*, 291–295.

McDonnell, G. and Russell, D. A. (1999). Antiseptics and Disinfectants: Activity, Action, and Resistance. *Clinical Microbiology Reviews, 12 (1)*, 147-179.

McGrath, S., Fitzgerald, G. and van Sinderen, D. (2004). *Starter cultures: bacteriophages*. In: Fox, P., Mc Sweeney, P. and Cogan, T. (Eds.), *Cheese: Chemistry, Physics and Microbiology*, vol. 1. (pp. 163-189). London, United Kingdom: Elsevier.

Moineau, S. and Lévesque, C. (2005). Control of bacteriophages in industrial fermentations. In: Kutter, E. and Sulakvelidze, A. (Eds.), *Bacteriophages: Biology and Applications* (pp. 285-296). Boca Raton, FL, United States: CRC Press.

Morino, H., Fukuda, T., Miura, T., Lee, C., Shibata, T. and Sanekata, T. (2009). Inactivation of feline calicivirus, a norovirus surrogate, by chlorine dioxide gas. *Biocontrol Science, 14 (4),* 147-153.

Passagot, J., Crance, J., Biziagos, E., Laveran, H., Agbalika, F. and Deloince, R. (1987). Effect of glutaraldehyde on the antigenicity and infectivity of hepatitis A virus. *Journal of Virology Methods, 16*, 21–28.

Permana, P. A. and Snapka. R. M. (1994). Aldehyde-induced protein-DNA crosslinks disrupt specific stages of SV40 DNA replication. *Carcinogenesis, 15*, 1031–1036.

Pottage, T., Richardson, C., Parks, S., Walker, J. T. and Bennett. A. M. (2010). Evaluation of hydrogen peroxide gaseous disinfection systems to decontaminate viruses. *Journal of Hospital Infection, 74,* 55-61.

Power, E. G. M. (1995). Aldehydes as biocides. *Progress in Medicinal Chemistry, 34*, 149–201.

Quiberoni, A., Guglielmotti, D. M. and Reinheimer, J. A. (2003). Inactivation of *Lactobacillus delbrueckii* bacteriophages by heat and biocides. *International Journal of Food Microbiology, 84*, 51-62.

Quiberoni, A., Suárez, V. B. and Reinheimer, J. A. (1999). Inactivation of *Lactobacillus helveticus* bacteriophages by thermal and chemical treatments. *Journal of Food Protection, 62 (8)*, 894-898.

Sanekata, T., Fukuda, T., Miura, T., Morino, H., Lee, C., Maeda, K., Araki, K., Otake, T., Kawahata, T. and Shibata, T. (2010). Evaluation of the Antiviral Activity of Chlorine Dioxide and Sodium Hypochlorite against Feline Calicivirus, Human Influenza Virus, Measles Virus, Canine Distemper Virus, Human Herpesvirus, Human Adenovirus, Canine Adenovirus and Canine Parvovirus. *Biocontrol Science, 15 (2),* 45-49.

Shin, G. and Sobsey. M. D. (2003). Reduction of Norwalk Virus, poliovirus 1 and coliphage MS2 by ozone disinfection of water. *Applied and Environmental Microbiology, 69 (7)*, 3975-3978.

Sickbert-Bennett, E., Weber, D. J., Gergen, M. F., Sobsey, M. D., and Rutala, W. A. (2005). Comparative efficacy of hand hygiene agents in the reduction of bacteria and viruses. *American Journal of Infection Control, 33 (2)*, 67-77.

Simões, L. C., Simões, M. and Vieira, M. J. (2010). Influence of the diversity of bacterial isolates from drinking water on resistance of biofilms to disinfection. *Applied and Environmental Microbiology, 76 (19)*, 6673-6679.

Solomon, E. B., Fino, V., Wei, J. and Kniel, K. (2009). Comparative susceptibilities of hepatitis A virus, feline calicivirus, bacteriophage MS2 and bacteriophage X-174 to inactivation by quaternary ammonium and oxidative disinfectants. *International Journal of Antimicrobial Agents, 33*, 288-289.

Springthorpe, V. S. and Satter, S. A. (1990). Chemical disinfection of viruscontaminated surfaces. *Critical Reviews in Environmental Control, 20*, 169–229.

Stewart, G., Jassim, S. and Denyer, S. P. (1991). Mechanisms of action and rapid biocide testing. *Society for Applied Bacteriology Technical Series, 27*, 319–329.

Suárez, V. and Reinheimer, J. A. (2002). Effectiveness of thermal treatments and biocides in the inactivation of Argentinian *Lactococcus lactis* phages. *Journal of Food Protection, 65 (11)*, 1756-1759.

Suárez, V., Capra, M., Rivera, M. and Reinheimer, J. (2007). Inactivation of Calcium – Dependent Lactic Acid Bacteria Phages by Phosphates. *Journal of Food Protection, 70 (6)*, 1518-1522.

Thurston-Enriquez, J. A., Haasb, C. N., Jacangelo, J. and Gerbad, C. P. (2005). Inactivation of enteric adenovirus and feline calicivirus by ozone. *Water Research , 39 (15)*, 3650-3656.

Tseng, C.-C. and Li, C.-S. (2006). Ozone for Inactivation of Aerosolized Bacteriophages. *Aerosol Science and Technology, 40 (9)*, 683 – 689.

Tyrrell1, S. A., Rippey, S. R. and Watkins, W. D. (1995). Inactivation of bacterial and viral indicators in secondary sewage effluents, using chlorine and ozone. *Water Research, 29 (11)*, 2483-2490.

Villion, M. and Moineau, S. (2009) Bacteriophages of *Lactobacillus. Frontiers in Bioscience 14*, 1661-1683.

Wagner, M., Brumelis, D. and Gehr, R. (2002). Disinfection of wastewater by hydrogen peroxide or peracetic acid: development of procedures for measurement of residual disinfectant and application to a physicochemically treated municipal effluent. *Water Environment Research, 74 (1)*, 33–50.

Zoni, R., Zanelli, R., Riboldi, E., Bigliardi, L. and Sansebastiano, G. (2007). Investigation of virucidal activity of chlorine dioxide, experimental data on feline calicivirus, HAV and Coxsackie B5. *Journal of Preventive Medicine and Hygiene, 48*, 91-95.

In: Bacteriophages in Dairy Processing
Editor: Andrea del Lujan Quiberoni et al.
ISBN 978-1-61324-517-0

Chapter 9

THERMAL RESISTANCE OF BACTERIOPHAGES IN THE DAIRY INDUSTRY

Zeynep Atamer[1,*], Horst Neve[2], Knut J. Heller[2] and Jörg Hinrichs[1]

[1]University of Hohenheim, Department of Food Science and Biotechnology, Garbenstrabe, Stuttgart, Germany

[2]Max Rubner-Institut, Federal Research Institute of Nutrition and Food, Department of Microbiology and Biotechnology, Hermann-Weigmann-Strabe, Kiel, Germany

ABSTRACT

The manufacture of fermented dairy products is dependent on the availability of active starter cultures, which are selected on basis of their bacteriophage insensitivity. Phages of lactic acid bacteria, which can originate either from raw milk or from lysogenic starter cultures, are still a major cause of fermentation failure in the dairy industry. These phages can affect the activity of both acid-producing and flavor-producing starter bacteria.

Knowledge on the genetic diversity of phages in dairy environment and their thermal resistance enables the industry to design processes that eliminate phages more efficiently. Heat resistant phages are suitable test phages for the evaluation of heating processes. Therefore, thermal inactivation of thermo-resistant phages is investigated to improve process safety. This chapter focuses on the characterization and inactivation of dairy phages. Firstly, in Section 1 the characterization of bacteriophages in the dairy industry and their occurrence as well as their heat resistances are presented. Hereafter, the attention in Section 2 turns to the thermal inactivation of phages and inactivation kinetics. The influence of heating media on thermal inactivation of phages is also illustrated.

* E-mail address: zeynep.atamer@uni-hohenheim.de. Tel.: + 49 711 45924208; fax: + 49 711 45923617

INTRODUCTION

The manufacture of fermented dairy products is dependent on the availability of suitable starter cultures, which are selected based on their bacteriophage insensitivity. Phages of lactic acid bacteria, which can originate both from raw milk and from lysogenic starter cultures, are still a major cause of fermentation failure in the dairy industry. Examples of phage infection of *Streptococcus thermophilus* starter cells are illustrated in Figure 1. Cells before and after infection by a lytic phage are shown (Figure 1A and 1D, respectively), as well as adsorption of a phage particle to the surface of a *Strep. thermophilus* cell before (Figure 1B) and after ejection of the phage DNA into the infected cell (Figure 1C).

Lactic acid bacteria phages can affect the activity of both acid-producing (e.g., *Lactococcus lactis*, *Strep. thermophilus*, *Lactobacillus* sp.) and flavor-producing (e.g., *Leuconostoc* sp., *L. lactis* subsp. *lactis* biovar *diacetylactis*) starter bacteria. Any given dairy factory is faced with fermentation problems caused by phages in up to 10% of their milk fermentations (Moineau and Levesque 2005).

Phages that attack *L. lactis* bacteria are commonly isolated from failed fermentations. However, there is a lack of information on the occurrence of *Leuconostoc* phages in dairies and on problems caused by this group of phages. Knowledge on the genetic diversity of phages in dairy environment and their thermal resistance enables the industry to design processes that eliminate phages more efficiently. Heat resistant phages are suitable test phages for the evaluation of heating processes. Therefore, thermal inactivation of thermo-resistant phages has been investigated with the aim to improve process safety. This chapter focuses on the characterization and inactivation of dairy phages. Firstly, in Section 1, characterization of dairy bacteriophages and their occurrence in the dairy industry as well as their heat resistance are presented.

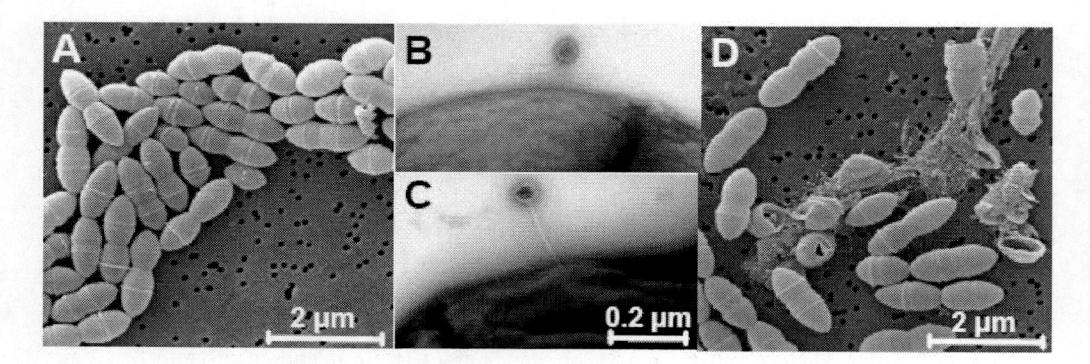

Figure 1. Illustration of the beginning of phage-induced lysis of a *Strep. thermophilus* culture. The culture is shown in the scanning electron micrographs before (A) and after (D) infection by a lytic phage. A number of cells have already lysed and released their cell contents containing a new phage progeny (not visible at this low magnification). Adsorption of a phage particle to the surface of a *Strep. thermophilus* cell is shown by the transmission electron micrographs before (B) and after (C) ejection of phage DNA into the infected cell.

Hereafter, the attention in Section 2 turns to thermal inactivation of phages and inactivation kinetics. The influence of heating media on thermal inactivation of phages is also discussed.

1. Characterization and Thermal Stability of Dairy Bacteriophages

1.1. Characterization of Phages Infecting Dairy Starter Cultures

In principle, all types of dairy starter cultures can be prone to infection by bacteriophages, since these bacterial viruses are widely disseminated in dairy environments. This is certainly true for lactic acid producing cultures with a growth preference either at lower temperatures (i.e., mesophilic cultures with *L. lactis* strains) and at higher temperatures (i.e., thermophilic cultures with *Strep. thermophilus* strains). As will be discussed later, mesophilic flavour producing cultures are also contributing to milk fermentations (i.e., dairy *Leuconostoc* cultures) which can also be affected negatively by lytic bacteriophages.

Dairy phage populations reveal a notably high degree of biodiversity. Particularly, lactococcal starter bacteria can be attacked by a broad spectrum of different phages which have been divided into 10 distinct phage species (Samson and Moineau 2010). In Figure 2, representative phages are illustrated for 8 lactococcal phage species. Except for phages of the P335 species, phages are strictly lytic.

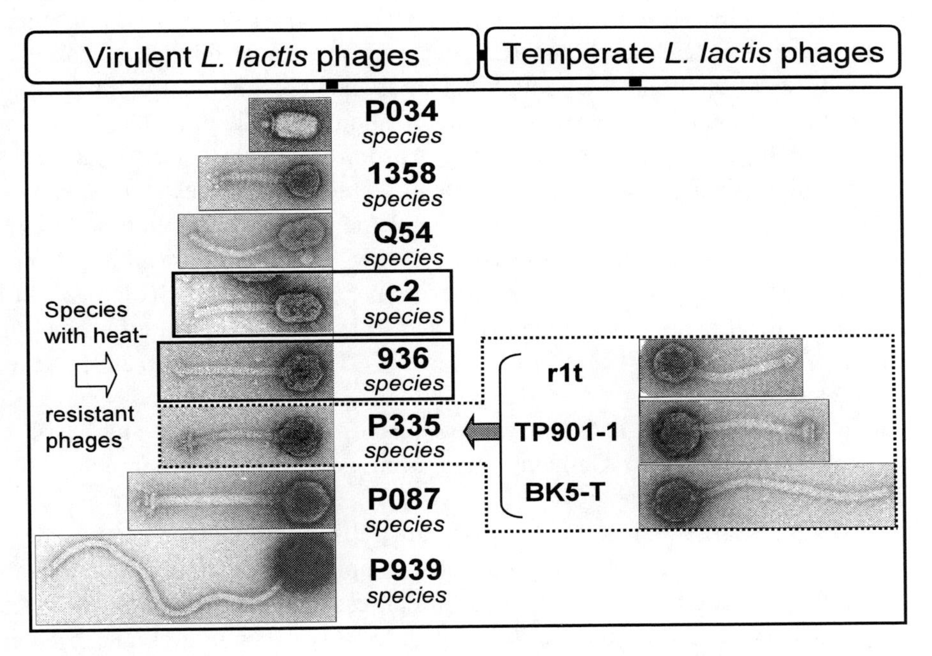

Figure 2. Current taxonomy of lactococcal phages by classification into genetically distinct phage species. The 936-, c2-phages (solid frames) and P335-phages (dashed frame) are disseminated in dairies world-wide. The P335 phage species is polythetic and comprises different subgroups of virulent and temperate phages as shown in the dashed frame and by the grey arrow. Two lactococcal phage species represented by only one phage isolate (phage KSY1, phage 1706) are not shown in this graph. Lactococcal phages with maximum heat tolerance belong to the 936 phages (see white arrow).

The P335 phage species is also addressed as a "polythetic" phage species, since phages of different morphotypes can be differentiated. P335 phages can also enter a lysogenic life cycle, in which they integrate their DNA as prophage DNA into the bacterial chromosome. It is shown later in this chapter, that lactoccocal phages with a high thermo-resistance phenotype belong to the 936 phage species. The 936-type phages are wide-spread in dairies throughout the world. In addition, c2- and – to a lesser degree - P335 phages are also ubiquitous in dairy environments (Moineau and Levesque 2005). Lactococcal phages are members of 2 phage families, i.e. *Siphoviridae* (with long non-contactile tails) and *Podoviridae* (with very short tails, see phage P034 in Figure 2). Only *Siphoviridae* phages are known today for dairy *Leuconostoc* (with short tails, see Figure 2) and *Strep. thermophilus* strains (with long tails, see Figure 1) (Quiberoni et al. 2010, Atamer et al. 2011). For a detailed description of the characteristics of dairy phages, readers are referred to Chapters 1, 3, and 5.

1.2. Occurence of Thermo-Resistant Phages in the Dairy Environment

Raw milk is the main source for phages emerging in the dairies, which is proven in different studies by phage isolation from raw milk samples (McIntyre et al. 1991, Bruttin et al. 1997, Madera et al. 2004). Madera et al. (2004) showed that in almost 10% of the analyzed 900 raw milk samples collected from different dairies, viable lactococcal phages were detected. The concentration of phages infecting lactic acid bacteria in raw milk ranges between 10^1 and 10^4 phages per ml (McIntyre et al. 1991, Neve et al. 2008). In raw milk, phages can multiply in the lactic acid bacteria, which are present in milk as contaminants originating from different environmental sources such as plants or material of plant origin. The phage population in raw milk may survive pasteurization (e.g., Chopin 1980, Suárez and Reinheimer 2002, Müller-Merbach et al. 2005, Abedon 2008, Atamer et al. 2009). Consequently, thermo-resistant phages can frequently enter the process via raw milk. Therefore, rigid phage control strategies have to be implemented in dairies. In case of a new phage contamination, phages can accumulate rapidly during fermentation, resulting in phage concentrations up to 10^9 phages per ml cheese whey or per g milk product (McIntyre et al. 1991, Brüssow 1994, Bruttin et al. 1997, Atamer et al. 2011), up to 10^8 plaque-forming units (PFU) ml^{-1} in brine, and up to 10^8 PFU m^{-3} in air (Neve et al. 1994, Neve et al. 2005, Neve et al. 2008, Atamer et al. 2009).

Phage surveys for lactococcal and *Leuconostoc* phages in large dairies and small and medium-sized enterprise dairies in Germany have been performed (Atamer et al. 2009, Atamer et al. 2011).

The phage status in large and in small dairies are illustrated in Figure 3, whereas in Figure 4 maximal phage titers detected in the collected samples are given. It can be seen that phages are still an important problem in both types of dairies. However, contamination of the samples with phages was observed mostly in small dairies, which indicates phage control measures are more effective in large plants. In samples of whey, cheese, brine, butter milk and butter cream highest phage titers was detected. Heat resistances of the detected phages are illustrated in Figure 4 that cause a recontamination risk due to recycling of whey cream and whey protein into cheese milk.

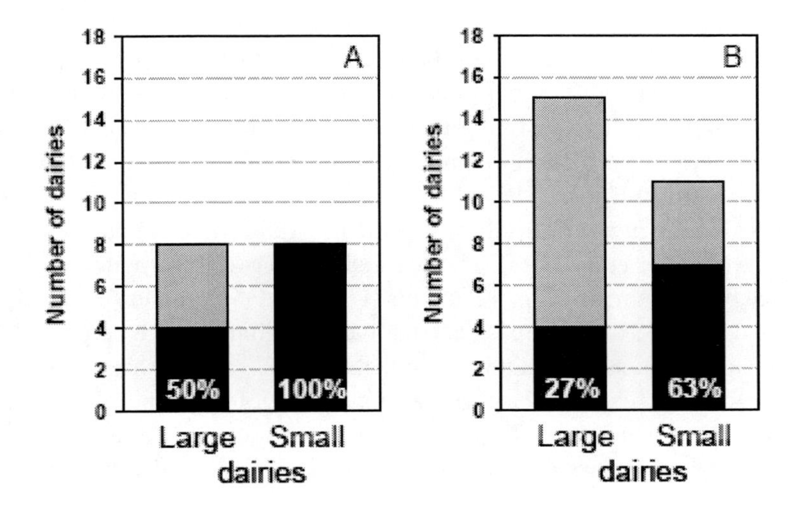

Figure 3. Lactococcal (A, 16 dairies) and *Leuconostoc* (B, 26 dairies) phage status in large and in small dairies. Black bars: Dairies with phage contamination. Grey bars: Dairies without phage contamination (Data adapted from Atamer et al. 2009 and Atamer et al. 2011).

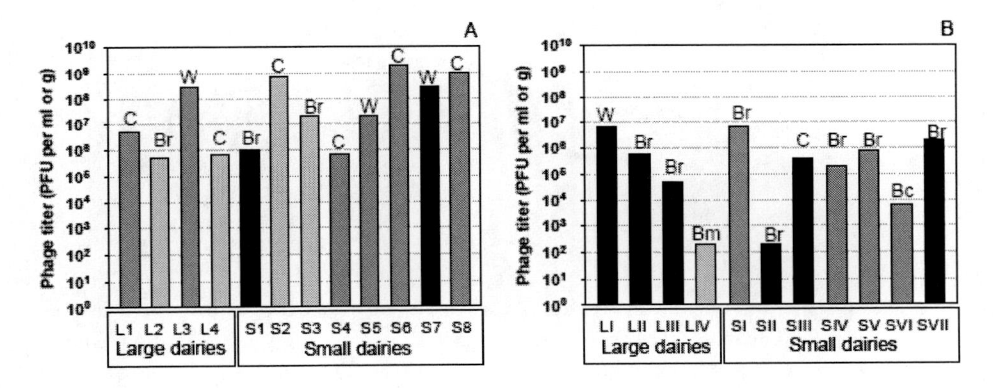

Figure 4. Maximal lactococcal (A) and *Leuconostoc* (B) phage titers detected in samples of large (L1-4, LI-IV) and small (S1-8, SI-VII) German dairies. Highest titers were detected in whey (W), cheese (C), brine (Br), butter milk (Bm) or in butter cream (Bc) as indicated. The graphs also illustrate the heat tolerance of the various phage populations. Bars are labeled in light grey, when all phages of the individual dairy phage populations did not survive an 80°C heat treatment in skim milk either for 5 min (lactococcal phages, Fig. 4A) or for 1 min (*Leuconostoc* phages , Fig. 4B). Dark grey bars: Some or all phage isolates survived this heat treatment at 80°C. Black bars: Some or all phage isolates survived higher treatments at 85°C and 90°C (lactococcal phages, Fig. 4A) or 85°C (*Leuconostoc* phages, Fig. 4B) (Data adapted from Atamer et al. 2009 and Atamer et al. 2011).

1.3. Screening of Dairy Phages Based on Their Thermal Resistance

Heat resistances of dairy phages have been investigated by isolating phages from dairy environments followed by heat inactivation studies (Atamer et al. 2009, Atamer et al. 2011). Phages were isolated from samples of whey, brine, yoghurt, sour cream, quarg, various cheese types, butter milk, and butter cream from different dairies and starter culture manufacturers. In samples collected from dairies, high phage titres, partly even far beyond the critical threshold of around 10^7 phages per ml (Figure 4), were detected, at which activity of the infected starters was reported to decrease significantly. Neve et al. (1995), however, reported that the limit of detection of starter growth failure in a miniaturized starter activity test was as low as 10^2 PFU ml^{-1}. The collected lactococcal phages (56 phages) were tested for their thermo-resistance in a series of heat treatments, which were conducted at 80, 85, 90 and 95°C for 5 min (Atamer et al. 2009). Thermo-resistant phages surviving 80°C for 5 min heat treatment were detected in significant numbers (Figure 5A) in samples obtained from all three sources (large dairies: 33% of the phages; small and medium-sized enterprise dairies: 44% of the phages; phages from starter culture manufacturer: 46% of the phages). A limited set of 4 phage isolates (i.e., approx. 7%) survived heating at 95°C for 5 min. The results showed that thermo-resistant phages are part of the natural phage populations in all types of dairies and they can be isolated from dairy products, from whey samples and also from brine.

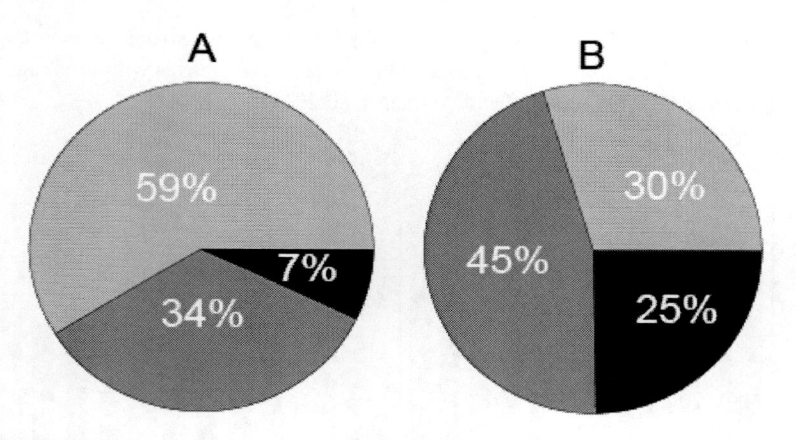

Figure 5. Variation of thermal tolerance (%) within a set of 56 dairy lactococcal phages (A) and 77 *Leuconostoc* phages (B). Lactococcal phages were heated in skim milk for 5 min (Fig. 5A), whereas *Leuconostoc* phages were heated in skim milk for 1 min (Fig. 5B). Light grey: percentage of phages not surviving an 80°C treatment, dark grey: percentage of phages surviving an 80°C treatment, black: percentage of phages surviving 85°C and 90°C (A) or 85°C only (B).

Compared to *L. lactis* phages infecting acid-producing starter culture bacteria, *Leuconostoc* phages attacking flavor-producing starters were found to be less thermo-resistant. Therefore, heating temperatures of 65, 70, 75, 80 and 85°C (for 1 min) for the screening of *Leuconostoc* phages (77 phages) were applied (Atamer et al. 2011). Hosts of the *Leuconostoc* phages belonged either to species *Leuconostoc pseudomesenteroides* or *Leuconostoc mesenteroides*. Although *Leuconostoc* phages showed lower thermal resistance than lactococcal phages, approximately 45% (Figure 5B) of the *Leuconostoc* phages tested survived 1-min heating at 85°C. It can be concluded that the high temperature short time pasteurization of raw milk (75°C for 30 s) is not sufficient to ensure a complete inactivation of both lactococcal and *Leuconostoc* phages. For comparison, heat resistance phenotypes of all collected lactococcal and *Leuconostoc* phages are presented in Figure 6A and Figure 6B, respectively. Micrographs of the most heat-resistant *L. lactis* and *Leuc. pseudomesenteroides* phages (labeled by arrows in Figure 6A and Figure 6B) are shown in Figure 7.

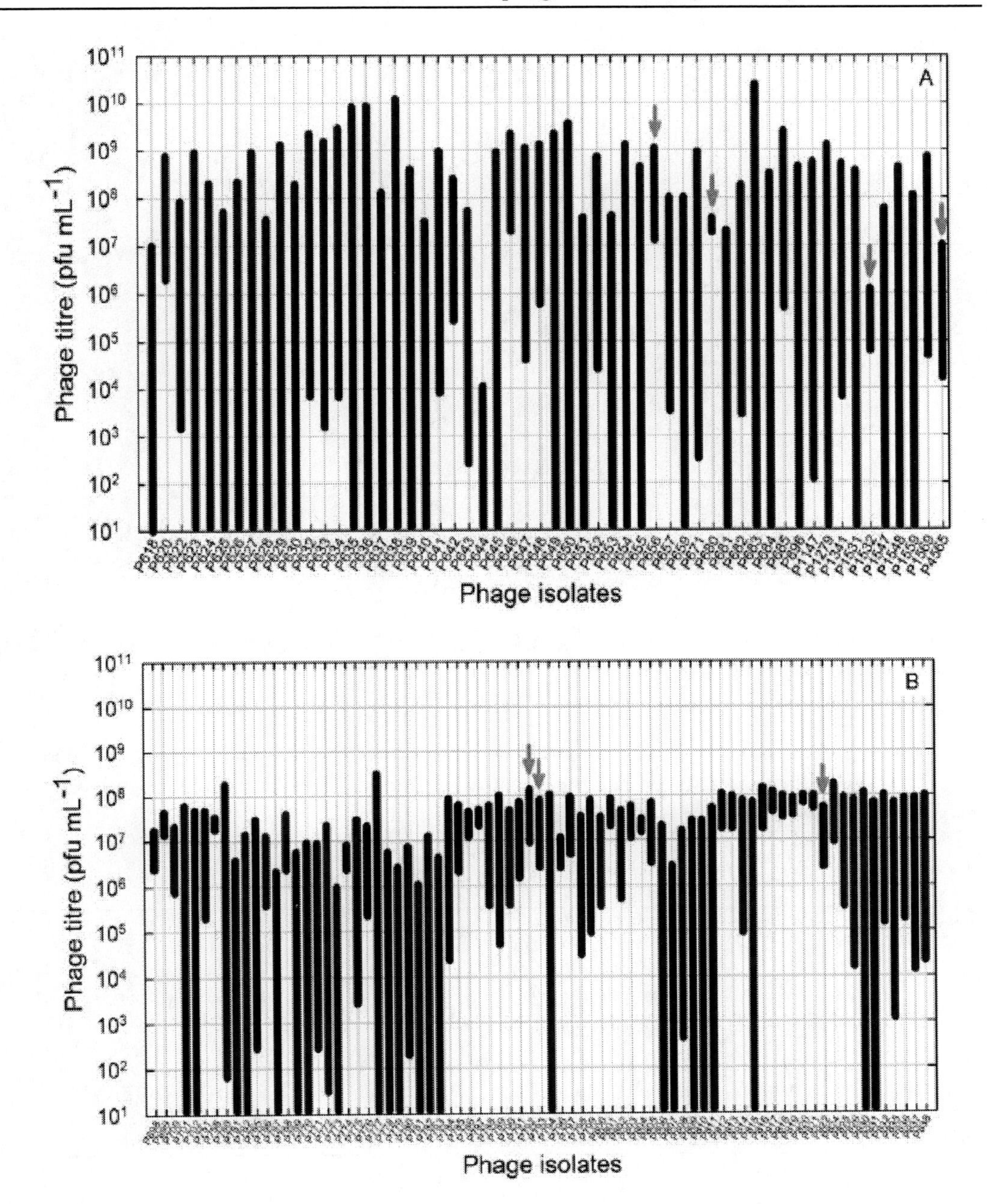

Figure 6. Inactivation of 56 lactococcal phages by a 5-min heat treatment at 80°C in milk (A) and inactivation of 77 *Leuconostoc* phages by a 1-min heat treatment at 80°C in milk (B). The bars indicate the initial (top) and the residual (bottom) phage titres after heat treatment. Phages P656, P680, P1532 and P4565 in Fig. 6A and phages P792, P793 and P822 in Fig. 6B (labeled with arrows) survived with highest titres in milk at temperatures above 85°C for 5 min (lactococcal phages) and at 85°C for 1 min (for *Leuconostoc* phages), representing the thermo-resistant phages (Data adapted from Atamer et al. 2009 and Atamer et al. 2011).

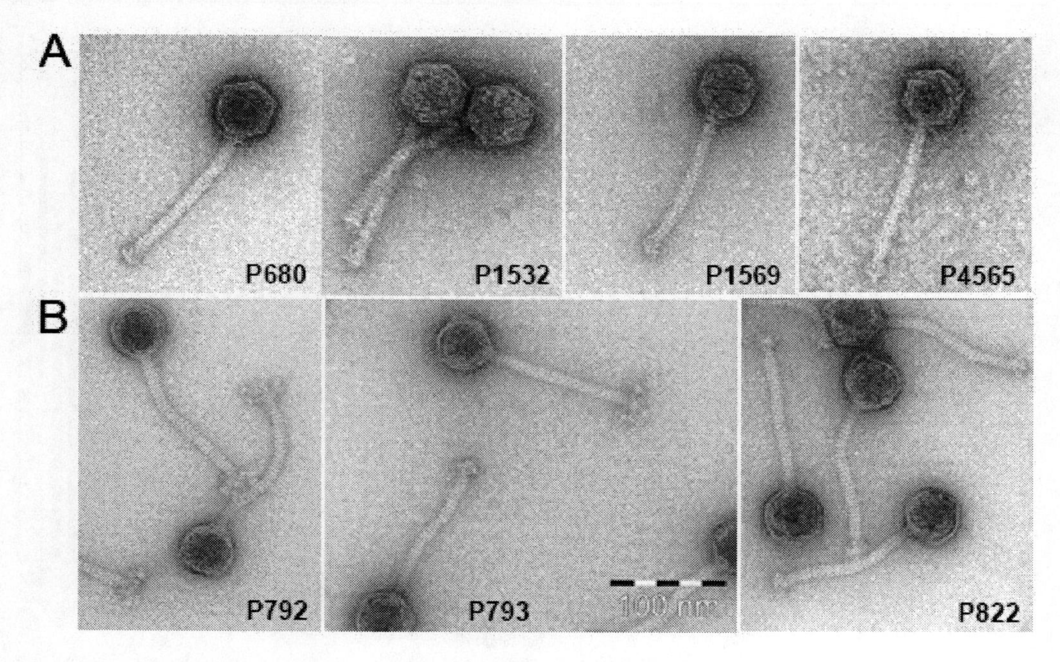

Figure 7. Transmission electron micrographs of the most heat-resistant L. lactis phages (top row A) and of 3 heat-resistant Leuc. pseudomesenteroides phages (bottom row B).

Table 1. Comparison of thermal resistances of bacteriophages infecting lactic acid bacteria. For each study, the phage showing highest heat resistance is given.

Phage	Heating medium	Host species	t_{99}[d] at 72 °C (=2·$D_{72°C}$[e]) [min]	Source
P680	SM[b]	*Lactococcus lactis*	300	Atamer et al. (2009)
CNRZ 832-B1	RSM[c]	*Lactobacillus helveticus*	21	Quiberoni et al. (1999)
001	RSM	*Lactococcus lactis*	20	Suárez and Reinheimer (2002)
0BJ	RSM	*Streptococcus thermophilus*	12	Binetti and Reinheimer (2000)
P008	SM	*Lactococcus lactis*	8.4	Müller-Merbach et al. (2005)
lb3	RSM	*Lactobacillus delbrueckii*	2.9	Quiberoni et al. (2003)
Cb1/204[a]	RSM	*Lactobacillus delbrueckii*	2.4	Ebrecht et al. (2010)
J-1	RSM	*Lactobacillus casei*	~ 2	Capra et al. (2004)
PL-1	RSM	*Lactobacillus paracasei*	~ 2	Capra et al. (2004)

[a]A temperate *Lb. delbrueckii* phage,
[b]SM: Skim milk,
[c]RSM: Reconstituted non fat dry skim milk,
[d]t_{99}: Time to achieve 99% inactivation of phage particles or the time needed to cause 2-log reduction ($t_{99} = 2 \cdot D_{72°C}$),
[e]D-value: Decimal reduction time required to cause 1-log reduction (90% inactivation).

1.4. Comparison of Heat Resistance of Dairy Phages

Studies conducted on heat resistance of bacteriophages revealed that phages of *L. lactis* as well as phages of *Lb. helveticus* showed highest thermal resistance among the studied phages of lactic acid bacteria (Quiberoni et al. 1999, Binetti and Reinheimer 2000, Suárez and Reinheimer 2002, Quiberoni et al. 2003, Capra et al. 2004, Atamer et al. 2009, Ebrecht et al. 2010, Atamer et al. 2011). Table 1 shows the 99% inactivation times (the time needed to cause 2-log reduction) for phages of lactic acid bacteria at 72°C, providing a comparison of the results obtained with those in literature. Recent studies on determination of thermal inactivation kinetics of *L. lactis* phages have shown that the wide-spread reference phage P008 (a 936-type phage) had a t_{99} of 8.4 min in skim milk (kinetic data from Müller-Merbach et al. 2005), whereas the two thermo-resistant *L. lactis* 936-like phages P680 and P1532 had t_{99} of 300 and 219 min in skim milk, respectively (kinetic data from Atamer et al. 2009).

2. Inactivation of Bacteriophages

2.1. Inactivation of Bacteriophages by Thermal Treatment

The presence of phages in the dairy environment cannot be avoided due to their existence in raw milk (Heap et al. 1978, Jarvis 1987, Bruttin et al. 1997) and their resistance to pasteurization conditions (Chopin 1980, Súarez and Reinheimer 2002, Müller-Merbach et al. 2005, Abedon 2008, Atamer et al. 2009). Therefore, processes are necessary for the inactivation of phages by means of heat, high pressure, chemicals or ultraviolet radiation. Thermal inactivation is by far the most common method applied in the elimination of phages. Neither the low temperature long time pasteurization (LTLT, 62-65°C, 30 min) nor the high temperature short time pasteurization (HTST, 72-75°C, 20-40 s) would ensure complete inactivation of all phages (Müller-Merbach et al. 2005, Atamer et al. 2009). However, literature information about the thermal resistance of phages is inconsistent. According to Yakovlev (1941), lactococcal phages suspended in milk resist 10 min heating at 80°C and 5 min heating at 90°C. In dry state they even remained active after heating at 95°C for 2 h. Nichols and Wolf (1945) reported that whey was free of phages after heat treatment at 75°C for 10 to 30 min. D_{70}-values (the time for 1-log inactivation at 70°C) of different *L. lactis* phages in milk determined by de Fabrizio et al. (1999) were in the range of 20 to 45 min. A treatment of at least 1 min at 95°C for inactivation of the most heat resistant phages was recommended by Walstra et al. (1999).

Using kinetic parameters, so-called lines of equal effects were calculated for some phages. Lines of equal effects indicate temperature and time combinations of heating processes leading to an equal effect. The results were presented on a temperature-time diagram for heat treatment of milk for practical applications (Müller-Merbach et al. 2007, Atamer et al. 2009, Atamer et al. 2011). Figure 8 illustrates temperature and time combinations for heat treatment of milk. Conditions for pasteurization (LTLT and HTST), for sterilization, for high temperature heating and for ultra high temperature heating (UHT) of milk are shown, together with lines of equal effects for inactivation of the wide-spread reference phage P008, thermo-resistant lactococcal phages (P680, P1532) and *Leuc. pseudomesenteroides* phage P793 as well as lines of significant thermal effects such as destruction of bacterial endospores.

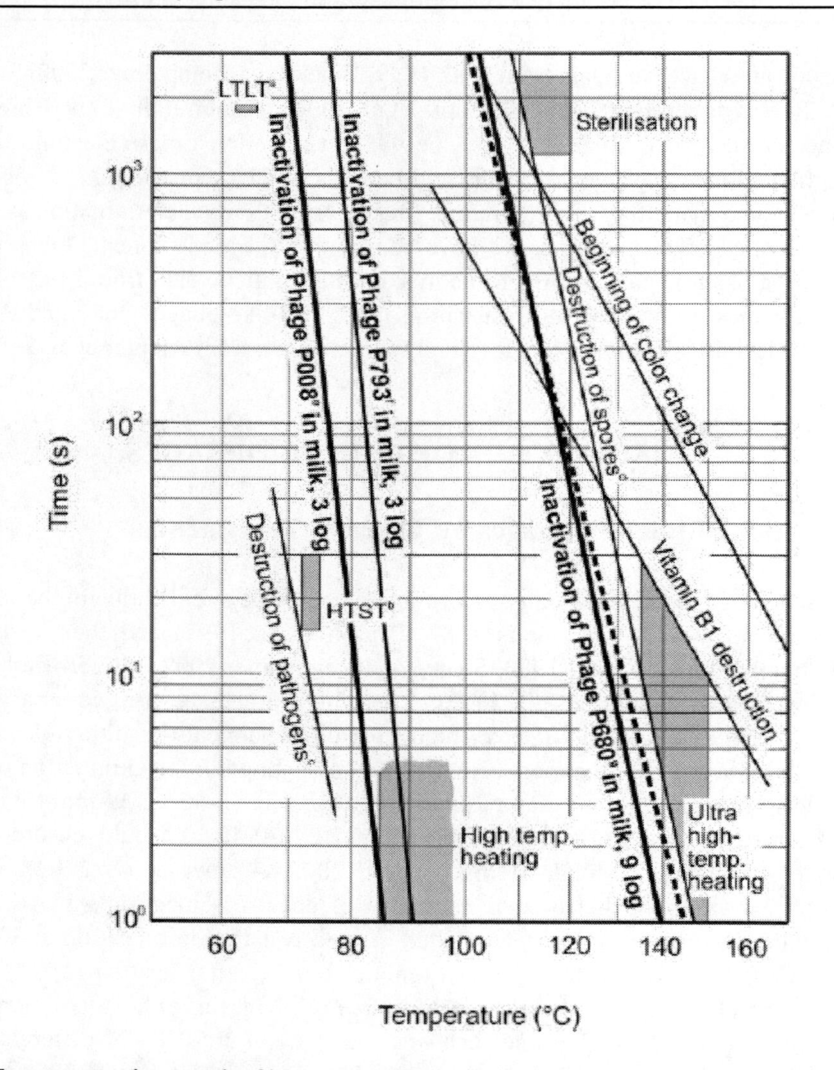

Figure 8. Temperature time graph of heat treatment of milk for inactivation of the wide-spread reference phage P008, of thermo-resistant phages of *Leuc. pseudomesenteroides* (P793) and of *L. lactis* (P680 and P1532) (modified from Kessler, 2002). [a]Low temperature long time pasteurization, [b]high temperature short time pasteurization, [c]destruction of bacterial pathogens, [d]destruction of bacterial endospores, [e]data: Müller-Merbach et al. (2005), [f]data: Atamer et al. (2011), [g]data: Atamer et al. (2009), dashed line: 9-log inactivation of P1532 in milk, data: Atamer et al. (2009).

In Figure 8, 9-log inactivation lines for the thermo-resistant lactococcal phages P1532 and P680 are shown together with the 3-log inactivation lines for the wide-spread reference lactococcal phage P008 and for the thermo-resistant *Leuconostoc* phage P793. The 3-log reduction was proposed for the case in which milk contains up to 10^3 PFU ml^{-1}, whereas the 9-log reduction line was aimed for the case of whey having up to 10^9 PFU ml^{-1}. According to these inactivation lines, pasteurization conditions (either LTLT or HTST) are not sufficient to guarantee even a 3-log reduction of heat-resistant phage populations infecting flavor-producing *Leuconostoc* starter cultures.

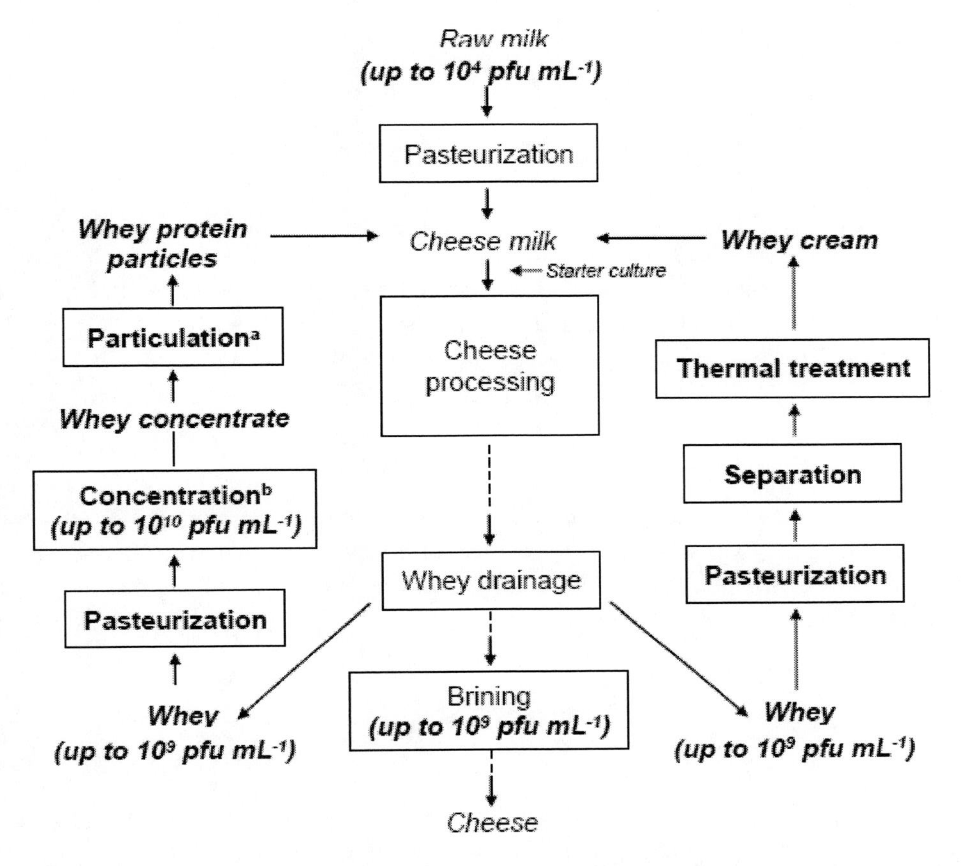

Figure 9. Flow chart of a cheese making process in which concentrated whey proteins and whey cream are being recycled. [a]Combined heat and shear treatment, [b]by means of ultrafiltration (Atamer and Hinrichs, 2010).

Two possible ways of whey recycling are shown in Figure 9, in which a thermal treatment of whey is needed before recycling (Atamer and Hinrichs 2010). The concentration of phages can reach numbers as high as 10^9 phages per ml in whey due to the presence of thermo-resistant phages in milk and their multiplication during fermentation (Neve et al. 1994, Bruttin et al. 1997, Madera et al. 2004). Therefore, a heat treatment providing 9-log reduction of phages should be achieved for the case of whey containing phages up to 10^9 phages per ml. By the concentration of whey, this number can be increased by a factor of 10 and can reach levels up to 10^{10} PFU ml^{-1}, making complete elimination of phages from the process difficult.

2.2. Kinetics of Thermal Inactivation

Microbial inactivation is considered to obey first-order kinetics and it is thus traditionally governed by the loglinear model described in Eq. (1). In this primary model, the rate of reduction, $-dN/dt$, is directly proportional to the number of microorganisms, N, where the proportional factor is the temperature-dependent rate constant, k (s^{-1}).

$$-\frac{dN}{dt} = k \cdot N \text{ for } n = 1 \tag{1}$$

Integration of Eq. (1) leads to

$$N_t = N_0 \cdot \exp(-k.t) \tag{2}$$

where N_t is the number of microorganisms at time t (s), N_0 is the initial number of microorganisms. Eq. (2) can be rearranged into Eq. (3).

$$\log \frac{N_t}{N_0} = -\frac{t}{D} \tag{3}$$

where D is the decimal reduction time ($D = 2.303/k$, units in min or s). The temperature dependence of the D-value is usually expressed as z-value, which is the increase in temperature causing a reduction in D by a factor 10. The z-value can be related to activation energy as follows (van Boekel 2002):

$$z = \frac{2.303 \cdot k \cdot T^2}{E_A} \tag{4}$$

This method assumes that each target organism is equally susceptible to heat (same probability of dying) and inactivation of an individual cell depends on the random chance that a target molecule receives sufficient heat (Peleg 2000). Although this loglinear model is commonly applied, a lot of deviations have been observed indicating that inactivation kinetics do not always follow first-order loglinear relationships (e.g., Suárez and Reinheimer 2002, van Boekel 2002, Quiberoni et al. 2003, Müller-Merbach 2007, Atamer et al. 2009). Figure 10 shows inactivation of lactococcal phage P680 in skim milk at various temperatures. The inactivation curves of the phage differ from the linear course and exhibit a tailing. Survival curves having a downward concavity (presence of a shoulder) and an upward concavity (presence of a tail) are frequently observed. Commonly observed types of survival curves (linear, linear with tailing, sigmoidal-like, linear with a preceding shoulder, biphasic, concave, biphasic with a shoulder, and convex) are shown in Figure 11 (Xiong et al. 1999, Geerhard et al. 2000, Geerhard et al. 2005).

One of the explanations for these deviations was reported as the physical interaction of the population to be inactivated, such as aggregation, adsorption to the walls of the vessel, protecting against inactivating process (mechanistic approach). Another explanation was suggested as the heterogeneous heat stability of the population (vitalistic approach) (Hiatt 1964).

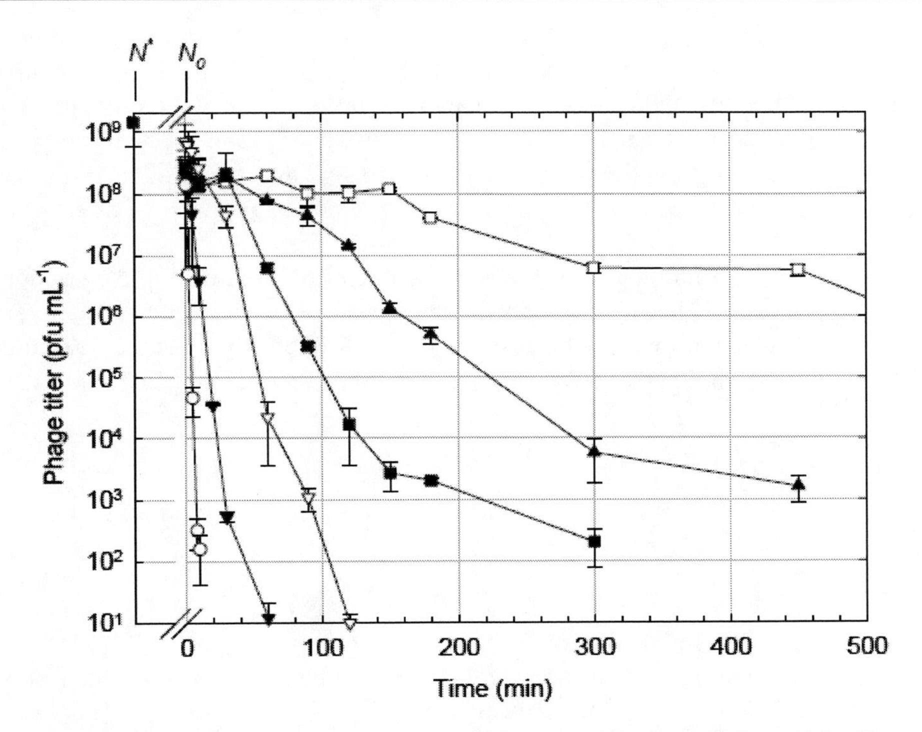

Figure 10. Thermal inactivation curves of thermo-resistant lactococcal phage P680 in skim milk at 70°C (□), 75°C (▲), 80°C (■), 85°C (▽), 90°C (▼) and 95°C (○). N*: titer of phage lysate before any heat treatment. N_0: initial phage titre (PFU ml^{-1}), i.e., after heating-up time (1 min).

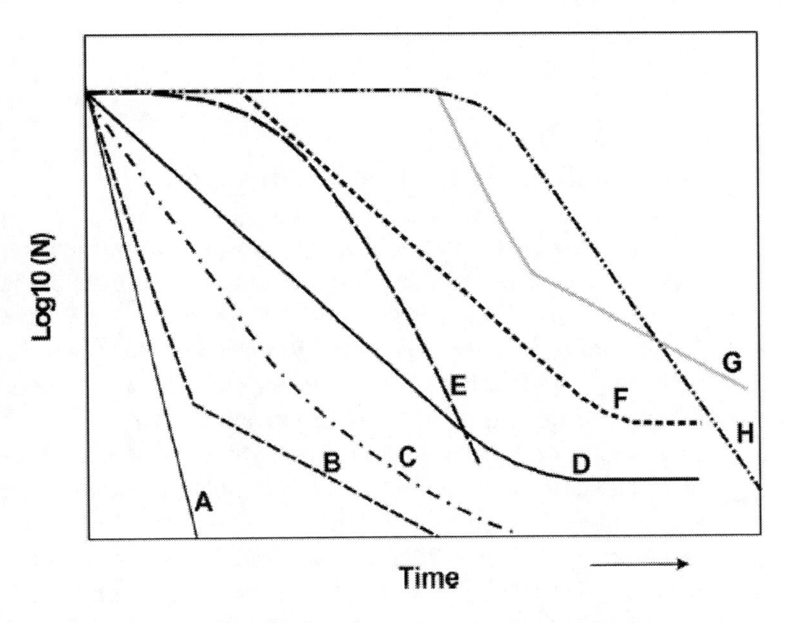

Figure 11. Commonly observed types of survival curves: linear (A), linear with tailing (D), sigmoidal-like (F), linear with a preceding shoulder (H), biphasic (B), concave (C), biphasic with a shoulder (G), and convex (E) (modified from Geerhard et al. 2000, Geerhard et al. 2005, Xiong et al. 1999).

This means that there is a mechanism at molecular level behind inactivation, such as inactivation of certain vital enzymes or DNA, which may vary between individuals (van Boekel 2002). To model these linear and nonlinear survival curves several approaches have been proposed. A detailed analysis of different modeling approaches using a range of experimental data for different species was conducted, and limitations of each approach were discussed by Xiong, Geeraerd and their co-workers (Xiong et al. 1999, Geeraerd et al. 2000, Geerhard et al. 2005).

In some studies, a non-loglinear kinetic model (Eq. 6 based on Eq. 5) was applied for determining the thermal inactivation kinetic parameters of phages (rate constant, activation energy, order of reaction), inactivation curves which differed from linearity (reaction orders of $n \neq 1$) (Müller-Merbach et al. 2005, Atamer et al. 2009, Atamer et al. 2011).

$$-\frac{dN}{dt} = k_n \cdot N^n \quad \text{for } n \neq 1 \tag{5}$$

$$N_t = N_0 \cdot \left[1 + (n-1) \cdot k_n \cdot N_0^{\,n-1} \cdot t\right]^{\frac{1}{1-n}} \quad \text{for } n \neq 1 \tag{6}$$

The effect of the absolute temperature T on the rate constant k_n, which depends on temperature, is formulated by the Arrhenius law (Eq. 7). The pre-exponential factor k_{Tref} refers to the rate constant (s^{-1}) at reference temperature T_{ref}, E_A is the activation energy ($J\ mol^{-1}$), R is the universal gas constant (8.314 $J\ mol^{-1}\ K^{-1}$).

$$k_n = k_{T_{ref}} \cdot \exp\left[-\frac{E_A}{R}\left(\frac{1}{T} - \frac{1}{T_{ref}}\right)\right] \tag{7}$$

2.3. Effect of Suspension Media on Heat Inactivation

The efficiency of heat inactivation in liquid media is influenced by the composition of the suspension media. An effect of various suspension media on thermal resistance has been reported for phages (Hunter and Whitehead 1940, Daoust et al. 1965, Chopin 1980, de Fabrizio et al. 1999, Moroni et al. 2002, Suárez and Reinheimer 2002, Quiberoni et al. 2003, Müller-Merbach et al. 2005, Atamer et al. 2009, Atamer et al. 2011). However, Capra et al. (2006) found no influence of suspension media on inactivation of *Lb. casei* and *Lb. paracasei* phages. Atamer et al. (2009), de Fabrizio et al. (1999), Müller-Merbach (2007) and Dietrich (2008) investigated inactivation of *L. lactis*, *Strep. thermophilus*, *Lb. delbrueckii* and *Leuc. mesenteroides* phages in three different suspension media: broth (either LM17 or MRS), skim milk and water. Quiberoni et al. (1999, 2003) studied inactivation of *Lb. helveticus* and *Lb. delbrueckii* phages, respectively and they used Tris magnesium gelatin (TMG) buffer, MRS broth and reconstituted non fat dry skim milk (RSM) as suspension media in thermal inactivation experiments. The studies showed that milk as a complex suspension medium tends to slow down inactivation of dairy phages. Furthermore, it was also documented for a *Leuc. pseudomesenteroides* phage (i.e., phage P808) that under the same heating conditions, the phage was inactivated approximately 1 log unit in milk after heating at 70 °C for 1 min, whereas it was inactivated around 4 log units in water (Atamer et al. 2011).

Different hypotheses were proposed to explain slower inactivation of phages in milk. Hunter and Whitehead (1940) suggested that milk proteins had a protective effect on inactivation of *L. lactis* phages. Bidawid et al. (2000) demonstrated that heat resistance of hepatitis A virus increased as fat levels in the heating media increased, while Chopin (1980) could not confirm a protective effect of milk fat on thermal inactivation of lactococcal phages. Daoust et al. (1965) reported that thermo-resistance of lactococcal phages increased significantly at higher concentrations of phosphate or whey. The impact of individual milk components (fat, casein, whey proteins, lactose and minerals) was investigated by heating phages in different media to elucidate this protective role of milk (Atamer et al. 2009, Atamer et al. 2011). As a result of heat treatment, loss of phage DNA from capsids, disintegration of phages into head and tail structures, aggregation of either phage tails or heads were observed. Electron micrographs of phages (heat-sensitive *L. lactis* phage P008 and heat-resistant phage P680) inactivated in different heating media (1 M NaCl, 3.5 M NaCl, nanofiltration permeate) are presented in Figure 12. Among milk components, casein was found to be responsible for the protective effect of milk. The buffering capacity of casein may explain this phenomenon (Salaün et al. 2005).

2.4. Survival of Phages in Cleaning Solutions

It was postulated that phages can be present in cleaning solutions, resulting in distribution of phages on the equipment surfaces used in processes or for fermentation. In order to prove this hypothesis, 49 lactococcal phage isolates were tested at different pH levels (Atamer and Hinrichs 2010). After a treatment time of 10 min at pH levels of 2 and 12 at room temperature, all isolated phages, which included thermo-resistant phages surviving a heat treatment at 95°C (5 min), were completely inactivated, whereas at pH levels between 3 and 10 active phages were detected (Figure 13). In industry, sodium hydroxide with a concentration range of 0.2 to 2% (pH 12.5 - 13.5) and nitric acid with a concentration of 0.5% (approximately pH 1) are commonly used as cleaning solutions (Prouty 1950, Kessler 2002, Bogosian 2006). According to the obtained results, a 10-min treatment using these commonly applied cleaning solutions are assumed to be sufficient for complete inactivation of phages (Atmer and Hinrichs 2010). Conducting inactivation studies in cleaning solutions without any impurities as well as with impurities (10%) simulated by skim milk showed that milk had no protection on phages.

Furthermore, Hunter and Whitehead (1940) reported in their study with lactococcal phages that phages lost their infectivity at pH values lower than 2.5 and higher than 11.8. Experiments done by Prouty (1950) with three phages infecting lactic acid streptococci also showed that the phages were completely inactivated after a treatment time of 10 min at pH 2.5. It was shown that in the alkaline range between pH levels of 11.3 to 11.8, inactivation of phages was found to be dependent on the phage type. Capra et al. (2006) studied the effect of pH on viability of the phages of *Lb. casei*/*paracasei* in the pH range from 2 to 11 at 25 °C and 37 °C.

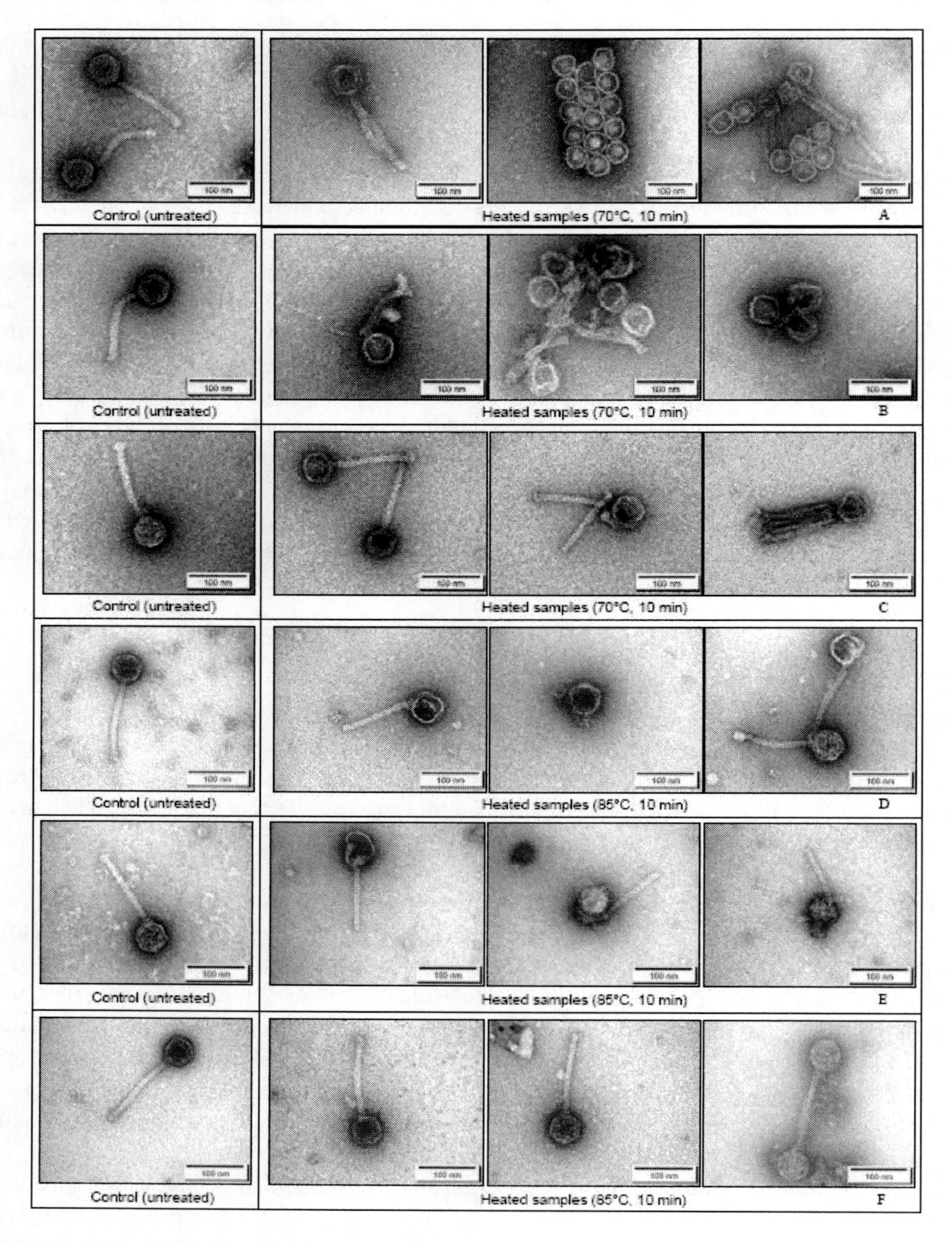

Figure 12. Transmission electron micrographs of the heat-sensitive *L. lactis* phage P008 and of the heat-resistant phage P680 before (untreated) and after a 10-min heat treatment at 70°C (P008) or 85°C (P680) in 1 M NaCl (A and D), in 3.5 M NaCl (B and E), and in nanofiltration permeate (enriched in minerals) derived from low-fat skim milk (C and F) (Atamer et al. 2010).

The phages were completely destroyed after 30 min at pH 2 (25° C), whereas the viability of the phages was high after 30 min in the pH range from 4 to 11. Hence, based on results from literature, the hypothesis that phages may be distributed by an automated cleaning procedure such as CIP (Cleaning In Place) cannot be confirmed. However, if the recommended steps in a CIP operation would be applied improperly, then distribution of the present phages by milk processing equipment could be expected.

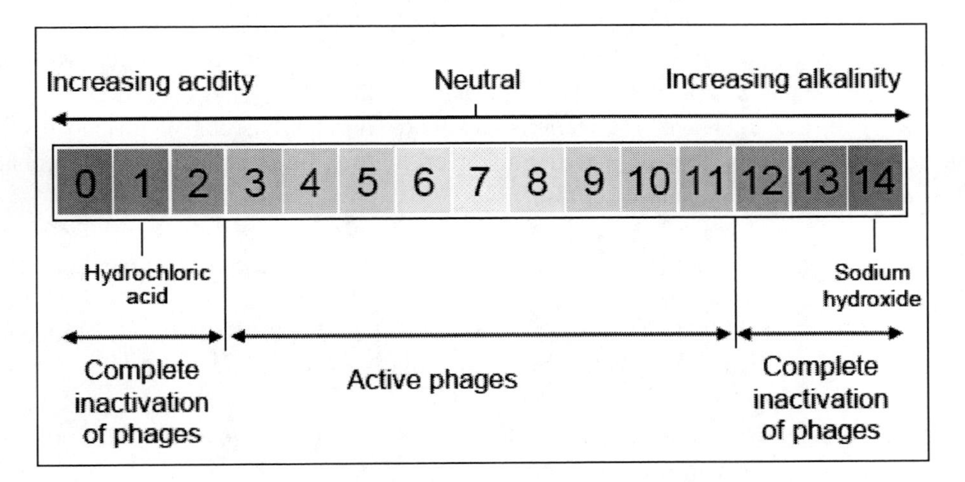

Figure 13. Effect of acid and alkaline cleaning solutions on survival of lactococcal phages, providing information on the effectiveness of the cleaning conditions for industry.

CONCLUSION

Bacteriophage infection of starter cultures during fermentation is a major concern in dairy industry, especially during production of fermented milk products (i.e., yoghurt, all cheese types). The presence of phages cannot be avoided in dairy environment due to their prevalence in raw milk and their noted resistance to pasteurization conditions. Strategies like sanitation, optimization of the factory design, use of air filtration units, regular rotation of starter cultures and use of phage-resistant starter cultures are already applied in dairy industry, in order to prevent propagation of virulent phages. However, new virulent phages emerge regularly, which lead to fermentation failure. Furthermore, processes applied in cheese manufacturing, such as recycling of whey cream and whey protein into cheese milk, carry contamination risk due to the prevalence of thermo-resistant phages in whey. Therefore, fundamental research on occurrence of thermo-resistant bacteriophages is of prime importance for dairy industry to ensure safety of the fermentation process. In this chapter overview of the thermal resistance of dairy phages, their characterization and their occurrence in the dairy environment as well as the kinetics of inactivation are provided. The required heat treatment conditions are presented for thermo-resistant phages. However, considering the fact that heat treatments with such temperature time combinations are restricted due to the effect they might cause on the functional properties of the product, possible solutions are under research. There is still a demand for alternative methods, which can be applied to reduce or eliminate phage concentration in milk, whey and whey product.

REFERENCES

Abedon, T. S. (2008). Bacteriophage ecology: Population growth, evolution, and impact of bacterial viruses. Cambridge, UK: Cambridge University Press. pp. 302–332.

Atamer, Z., Dietrich, J., Müller-Merbach, M., Neve, H., Heller, K. J. and Hinrichs, J. (2009). Screening for and characterization of *Lactococcus lactis* bacteriophages with high thermal resistance. *International Dairy Journal, 19*, 228-235.

Atamer, Z., Dietrich, J., Neve, H., Heller, K. J. and Hinrichs, J. (2010). Influence of the suspension media on the thermal treatment of mesophilic lactococcal bacteriophages. *International Dairy Journal, 20*, 408-414.

Atamer, Z. and Hinrichs, J. (2010). Thermal inactivation of the heat-resistant *Lactococcus lactis* bacteriophage P680 in modern cheese processing. *International Dairy Journal, 20*, 163-168.

Atamer, Z., Ali, Y., Neve, H., Heller, K. J. and Hinrichs, J. (2011) Thermal resistance of bacteriophages attacking flavor-producing dairy *Leuconostoc* starter cultures. *International Dairy Journal, 21*, 327-334.

Bidawid, S., Farber J. M., Sattar, S. A. and Hayward, S. (2000). Heat inactivation of Hepatitis A virus in dairy foods. *Journal of Food Protection, 63*, 522-528.

Binetti, A. G. and Reinheimer, J. A. (2000). Thermal and chemical inactivation of indigenous *Streptococcus thermophilus* bacteriophages isolated from Argentinian dairy plants. *Journal of Food Protection, 63*, 509-515.

Bogosian, G., (2006). In: The bacteriophages, 2nd ed., Calender R. (ed.), Oxford Univ. Press, NY, p. 667-674.

Bruttin, A., Desiere, F., D'Amico, N., Guérin, J.-P., Sidoti, J., Huni, B., Lucchini, S., Brüssow, H. (1997). Molecular ecology of *Streptococcus thermophilus* bacteriophage infections in a cheese factory. *Applied and Environmental Microbiology, 63*, 3144-3150.

Brüssow, H., Fremont, M., Bruttin, A., Sidoti, J., Constable, A. and Fryder V. (1994). Detection and classification of *Streptococcus thermophilus* bacteriophages isolated from industrial milk fermentation. *Applied and Environmental Microbiology, 60*, 4537-4543.

Capra, M. L., Quiberoni, A. and Reinheimer, J. A. (2004). Thermal and chemical resistance of *Lactobacillus casei* and *Lactobacillus paracasei* bacteriophages. *Letters in Applied Microbiology, 38*, 499-504.

Capra, M. L., Quiberoni, A. and Reinheimer, J. (2006). Phages of *Lactobacillus casei/paracasei*: response to environmental factors and interaction with collection and commercial strains. *Journal of Applied Microbiology, 100*, 334-342.

Chopin, M. C. (1980). Resistance of 17 mesophilic lactic *Streptococcus* bacteriophages to pasteurization and spray-drying. *Journal of Dairy Research, 47*, 131-139.

Daoust, D. R., El-Bisi, H. M. and Litsky, W. (1965). Thermal destruction kinetics of a lactic streptococcal bacteriophage. *Applied Microbiology, 13*, 478-485.

de Fabrizio, S. V., Ledford, R. A. and Parada, J. L. (1999). Heat resistance of *Lactococcus lactis* bacteriophages. *Microbiology, Aliments, Nutrition, 17*, 191–198.

Dietrich, J. (2008). Thermische Inaktivierung von Bakteriophagen der mesophilen und thermophilen Milchsäurebakterien (Thermal inactivation of mesophilic and thermophilic bacteriophages). PhD Thesis, Max-Rubner Institut, Kiel, Germany.

Ebrecht, A. C., Guglielmotti, D. M., Tremmel, G., Reinheimer, J. A. and Suárez V. B. (2010). Temperate and virulent *Lactobacillus delbrueckii* bacteriophages: comparison of their thermal and chemical resistance. *Food Microbiology, 27*, 515-520.

Geeraerd, A. H., Herremans, C. H. and Van Impe, J. F. (2000). Structural model requirements to describe microbial inactivation during a mild treatment. *International Journal of Food Microbiology, 59*, 185–209.

Geeraerd, A. H., Valdramidis, V. P. and Van Impe, J. F. (2005). GInaFiT, a freeware tool to assess non-log-linear microbial survivor curves. *International Journal of Food Microbiology, 102*, 95–105.

Heap, H. A., Limsowtin, G. K. Y. and Lawrence, R. C. (1978). Contribution of *Streptococcus lactis* strains in raw milk to phage infection in commercial cheese factories. *New Zealand Journal of Dairy Science and Technology, 13*, 16-22.

Hiatt, C. W. (1964). Kinetics of the inactivation of viruses. *Bacteriological Reviews, 28*, 150-163.

Hunter, G. J. E. and Whitehead, H. R. (1940). The action of chemical disinfectants on bacteriophages for the lactic streptococci. *Journal of Dairy Research, 11*, 62-66.

Jarvis, A. W. (1987). Sources of lactic streptococcal phages in cheese plants. *New Zealand Journal of Dairy Science and Technology, 22*, 93-103.

Kessler, H. G. (2002). Food and Bio Process Engineering – Dairy Technology. Publishing House A. Kessler, Munich, Germany. pp. 471–543.

Madera, C., Monjardín, C. and Suárez, J. E. (2004). Milk contamination and resistance to processing conditions determine the fate of *Lactococcus lactis* bacteriophages in dairies. *Applied and Environmental Microbiology, 70*, 7365-7371.

McIntyre, K., Heap, H. A., Davey, G. P. and Limsowtin, K. Y. (1991). The distribution of lactococcal bacteriophage in the environment of a cheese manufacturing plant. *International Dairy Journal, 1*, 183-197.

Moineau, S. and Levesque, C. (2005). Control of bacteriophages in industrial fermentations. *In*: Bacteriophages: Biology and Applications, E. Kutter and A. Sulakvelidze, *eds*. (Boca Raton: CRC Press), pp. 285-296.

Moroni, O., Jean, J., Autret, J. and Fliss, I. (2002). Inactivation of lactococcal bacteriophages in liquid media using dynamic high pressure. *International Dairy Journal, 12*, 907-913.

Müller-Merbach, M., Neve, H. and Hinrichs, J., (2005). Kinetics of the thermal inactivation of the *Lactococcus lactis* bacteriophage P008. *Journal of Dairy Research, 72*, 281-286.

Müller-Merbach, M. (2007). *Lactococcus lactis* bacteriophages in milk systems: Thermal-hydrostatic inactivation and proliferation by means of propagation and diffusion. PhD Thesis, University of Hohenheim, Germany.

Neve, H., Kemper, U., Geis, A. and Heller, K. J. (1994). Monitoring and characterization of lactococcal bacteriophages in a dairy plant. *Kieler Milchwirtschaftliche Forschungsberichte, 46*, 167-178.

Neve, H., Berger, A. and Heller, K. J. (1995). A method for detecting and enumerating airborne virulent bacteriophage of dairy starter cultures. *Kieler Milchwirtschaftliche Forschungsberichte*, *47*, 193-207.

Neve, H., Dietrich, J. and Heller, K. J. (2005). A short note on long-term stability of *Lactococcus lactis* bacteriophages in cheese brine. *Kieler Milchwirtschaftliche Forschungsberichte, 57,* 191-200.

Neve, H., Dietrich, J., Ali, Y., Atamer, Z., Hinrichs, J. and Heller, K. J. (2008). Phagenkontrolle in Molkereien – Neue Erkenntnisse zur Phageninaktivierung. *Deutsche Milchwirtschaft, 59*, 828-834.

Nichols, M. A. and Wolf, J. Z. (1945). The heat resistance of the bacteriophages of cheese starter with observations on the estimation of phage concentration. *Journal of Dairy Research, 14*, 93-100.

Quiberoni, A., Suárez, V. B. and Reinheimer, J. A. (1999). Inactivation of *Lactobacillus helveticus* bacteriophages by thermal and chemical treatments. *Journal of Food Protection, 62*, 894-898.

Quiberoni, A., Guglielmotti, D. M. and Reinheimer, J. A. (2003). Inactivation of *Lactobacillus delbrueckii* bacteriophages by heat and biocides. *International Journal of Food Microbiology, 84*, 51-62.

Quiberoni, A., Moineau S., Rousseau, G. M., Reinheimer, J. and Ackermann, H.-W. (2010). *Streptococcus thermophilus* bacteriophages. *International Dairy Journal, 20*, 657-664.

Peleg, M. (2000). Microbial survival curves – the reality of flat 'shoulders' and absolute thermal death times. *Food Research International, 33*, 531–538.

Prouty, C. C. (1950). Inactivation of bacteriophage of lactic acid streptococci at high and low pH levels. *Journal of Milk and Food Technology, 13*, 329-331.

Salaün, F., Mietton, B. and Gaucheron, F. (2005). Buffering capacity of dairy products. *International Dairy Journal, 15*, 95-109.

Samson, J. E. and Moineau, S. (2010). Characterization of *Lactococcus lactis* phage 949 and comparison with other lactococcal phages. *Applied and Environmental Microbiology, 76*, 6843–6852.

Suárez, V. and Reinheimer, J.A. (2002). Effectiveness of thermal treatments and biocides in the inactivation of Argentinian *Lactococcus lactis* phages. *Journal of Food Protection, 65*, 1756-1759.

Van Boekel, M. A. J. S. (2002). On the use of the Weibull model to describe thermal inactivation of microbial vegetative cells. *International Journal of Food Microbiology, 74*, 139-159.

Walstra, P., Geurts, T. J., Noomen, A., Jellema, A. and van Boekel, M. A. J. S. (1999). Dairy Technology – Principles of Milk Properties and Processes. New York, Basel: Marcel Dekker.

Xiong, R., Xie, G., Edmondson, A. E. and Sheard, M. A. (1999). A mathematical model for bacterial inactivation. *International Journal of Food Microbiology, 46*, 45–55.

Yakovlev, D. A. (1941). Bacteriophage of lactic acid streptococci. *Chemical Abstracts, 34*, 3674.

In: Bacteriophages in Dairy Processing
Editor: Andrea del Lujan Quiberoni et al.

ISBN 978-1-61324-517-0

Chapter 10

NON-THERMAL TECHNOLOGIES: PULSED ELECTRIC FIELD, HIGH HYDROSTATIC PRESSURE AND HIGH PRESSURE HOMOGENIZATION. APPLICATION ON VIRUS INACTIVATION

María Luján Capra[1*], Francesca Patrignani[2], Maria Elisabetta Guerzoni,[2] and Rosalba Lanciotti[2]

[1]Instituto de Lactología Industrial (UNL-CONICET),
Facultad de Ingeniería Química
Universidad Nacional del Litoral, Santiago del Estero,
Santa Fe, Argentina

[2]Università di Bologna, Dipartimento di Scienze degli Alimenti,
Piazza Goidanich, Cesena, Italy

ABSTRACT

Bacteriophage attacks to starters in the dairy industry cause enormous economic losses. Phage inactivation has always been a matter of study; however, even combining different control strategies, complete removal of phages is not possible and the real aim consists in maintaining their concentration under a critical threshold. This chapter deals with three novel technologies such as pulsed electric field, high hydrostatic pressure and high pressure homogenization. These non-thermal methods are considered the most promising in food processing due to their better preservation of flavor, essential nutrients and vitamins with respect to heat treatments. Aspects such as the principles of the technologies, applications, treatment parameters and factors that influence the inactivation extent, as well as microbial inactivation, are developed below for each technology. For this last aspect and due to the scarce available information regarding phage inactivation, not only the effect of these novel technologies on dairy bacteriophages is referred to, but also on other phages, viruses and even on others microorganisms.

[*] E-mail address: mcapra@fbcb.unl.edu.ar. Tel.: +54 342 4530302; fax: +54 342 4571162

INTRODUCTION

The most commonly used method of food preservation today is thermal processing. In spite of achieving good inactivation in microorganisms and deleterious enzymes, thermal treatment triggers unwanted reactions in foodstuffs, involving the loss of flavor and nutrients (Aronsson et al. 2005). To minimize these reactions, the food industry is developing a wide range of alternative methods that are claimed to be ''quality friendly'' and which correspond to consumer claim for fresh like products (Torregrosa et al. 2006, Huang and Wang 2009). Among the methods alternative to thermal treatment great attention have received pulsed electric field (PEF), high hydrostatic pressure processing (HP) and high pressure homogenization (HPH). Several non-thermal preservation technologies exist but they are not described in this chapter because no data about the efficacy upon the inactivation of phages are yet available.

1. PULSED ELECTRIC FIELD

1.1. General Aspects

Pulsed electric field (PEF) technology has been widely studied and proposed for several purposes in food industry and bioengineering. Well consolidated is the application of PEF for electroporation, cell hybridization and electrofusion purposes in the areas of genetic engineering and biotechnology (De Vos and Simons 1994). PEF applications include also the improvement of mass transfer in plant or animal cells or sludge disintegration. PEF treatments showed a great feasibility and selectivity for fruit juice production, improving tissue softness, texture and the content of valuable components and even replacing the enzymatic maceration (Lebovka et al. 2004a, b, Corrales et al. 2008). Also the microbial inactivation by PEF has been investigated extensively and a wide literature shows that sufficient reduction of pathogenic and spoilage microorganisms can be achieved throughout PEF treatments (Grahl and Märkl 1996, Barsotti and Cheftel 1999, Wouters et al. 1999, Ho and Mittal 2000, Aronsson et al. 2001, Barbosa-Canovas and Sepulveda 2005, Mãnas and Pagãn 2005, Gerlach et al. 2008, Xin et al. 2008). PEF technology is reported to increase liquid food shelf life due to microbial inactivation, while maintaining its sensorial characteristics. In fact, temperature increase due to ohmic heating is typically held low enough (temperature of treatment lower than 60°C) to exploit the advantages of a non-thermal technology better preserving the sensory, nutritional and functional properties of foods (Grahl and Märkl 1996, Mãnas and Pagãn 2005, Toepfl et al. 2007). Potential products for this application are fluid ones that may be pumped such as fruit juices, milk, egg, water, wine and soups. PEF treatment has been applied successfully not only to improve microbial stability but also sensorial properties of high viscosity foods, such as yoghurt and other high viscosity dairy products (Yeom et al. 2004).

PEF treatment has also been investigated in functional food field mainly to preserve immunoglobulins (Igs), vitamins and bioactive peptides, including angiotensin-I-converting enzyme (ACE) inhibitory peptides, generally sensitive (depending on treatment intensity) to thermal treatment applied in dairy products or mixtures of orange juice and milk (Bendicho et al. 2002, Li et al. 2005, Wan et al. 2005, 2009). Moreover, PEF influence on some probiotic features (*i.e.* viability, bile and acid tolerance, and protease activity) of *Lactobacillus acidophilus* LA-K has also been studied. *Lb. acidophilus* LA-K is recognized as a probiotic bacterium able to produce bio-active peptides (Cueva 2009) as well as to enhance the activity of enzymes such as β-D-galactosidase (Ohshima et al. 2007).

1.2. Equipment and Treatment Parameters

Various laboratory and pilot scale PEF equipments have been designed, constructed, manufactured and used for liquid foods. Also some industrial scale PEF treatment systems are available (Buta and Tauscher 2002, Gerlach et al. 2008). A typical PEF food processing unit comprises a high voltage pulse generator, a treatment chamber, a fluid handling system and control and monitoring devices (Min et al. 2007, Wan et al. 2009). PEF treatment process may be either static or continuous. While in the first, discrete portions of fluid foodstuff are treated as a unit by subjecting all the fluid to a PEF treatment chamber, in the continuous processing, the treated foodstuff is flowed into and emitted from the PEF treatment system in a steady stream by a pump (Alkhafaji and Farid 2007). In any case, the design of the treatment chamber is one of the pivotal factors in the development of the PEF treatment (Alkhafaji and Farid 2007), because it should impart uniform electric field to foods with a minimum increase of the temperature and the electrodes should be designed to minimize the effect of electrolysis (Buta and Tauscher 2002, Gerlach et al. 2008). Depending on the system used, typical PEF treatment parameters include pulsed field intensity of 15-50 kV/cm, pulse width of 1-5 μs, and pulses frequency of 200-400 Hz (pulses/s) (Barbosa-Canovas and Supelveda 2005). Maximal PEF treatment intensity is limited by equipment design and food's ability to withstand dielectric breakdown (Lado and Yousef 2002). For food preservation and microbial killing, electric field pulses of short duration (1-100 μs) and high intensity (10-50 kV/cm) are applied to the food placed between electrodes. However, the degree of inactivation strongly depends on the intensity of the pulses in terms of field strength, energy and number of pulses applied as well as on the microbial strain and on the properties of the food matrix under investigation. The critical analysis of literature data indicates that the main factors determining the efficacy of PEF on microbial inactivation include treatment parameters (i.e. electric field strength, treatment time and temperature, pulse width, pulse shape), products features (i.e. electric conductivity, density, pH, water activity) and microbial characteristics (i.e. cell size and morphology, cell wall structure, growth phase and physiological state) (Wan et al. 2009).

1.3. Mechanism of Microbial Inactivation

Microbial cells exposed to PEF can respond by an electrical breakdown and local conformational changes of the phospholipidic bilayer and cell membranes (Neumann et al. 1972, Zimmermann et al. 1973) and, as a consequence of the so named electroporation, membrane permeability dramatically increases (Xin et al. 2008). Alternatively, other researchers suggested different mechanisms for the biocidal action of electricity, considering that the effect of an electric discharge in a liquid media may generate small amounts of microbicidal agents, such as chlorine, free radicals and H_2O_2 that could alter the DNA and cytoplasmic activity during the treatment (Lado and Yousef 2002). Regarding cell electroporation, diverse explanations were formulated (Jayaram 2000). In the classical electromechanical model, Zimmermann (1986) suggested that PEF temporarily increases the transmembrane potential of cells by accumulating compounds of opposite charges in the membrane surroundings. This high trans-membrane potential exerts pressure on the cell membrane decreasing its thickness and ultimately, causing pore formation. On the other hand, Tsong (1991) introduced the poration theory to explain the mechanism of cell death by PEF. Electroporation in protein channels and lipid domains results in osmotic swelling of the cell and thus membrane weakening until the cell bursts. In both hypotheses, excessive pore formation causes irreversible loss of membrane functions. A drastic increase in permeability re-establishes the equilibrium of the electrochemical and electric potential differences of the cell plasma and the extracellular medium forming a Donnan-equilibrium (Glaser 1996). Simultaneously, the neutralization of the trans-membrane gradient irreversibly impairs vital physiological control systems of the cell like osmoregulation and consequently cell death occurs. Depending on the electrical field strength, pulse duration and pulse number, permeabilization can be either reversible or irreversible (Ulmer et al. 2002). In fact, electroporated cells that are permeable to large DNA molecules can recover full viability if incubated in a suitable medium (Mackey 2000). The physical and chemical characteristics of food products are known to strongly influence microbial inactivation and subsequent recovery using PEF and, consequently, safety and shelf-life of PEF treated foods (Wouters et al. 2001, Aronsson et al. 2004).

1.4. Microbial Susceptibility to PEF

Different types of microorganisms vary in their sensitivity to this kind of treatment due to different membrane and cell wall construction. Also size, shape and physiological state of the microorganism play a part as well (Aronsson et al. 2001, Wouters et al. 2001, Heinz et al. 2002).

PEF is widely reported as being ineffective for the inactivation of bacterial spores (Grahl and Märkl 1996, Wan et al. 2009). It is generally reported that yeast cells are more sensitive to PEF treatment than bacterial cells, and that Gram-negative are more sensitive than Gram-positive cells (Toepfl et al. 2007). Also variations of PEF sensitivity among different strains of the same species have been reported (Lado and Yousef 2002). The addition of some antimicrobial agents may improve the efficacy of PEF. Particularly, nisin molecule incorporation through the cell membrane may be improved during the treatment and may cause the formation of permanent pores. Also, PEF could increase the permeability of the cell wall and membrane, enhancing the entry of un-dissociated organic acids into the bacterial cells. Ozone contributes to the breakdown of cells during mild PEF treatments and lysozyme combined with PEF improved the inactivation of spores probably dissolving the spore coat and rendering the cell susceptible to an electric field (Lado and Yousef 2002, Ross et al. 2003).

1.5. Virus and Bacteriophage Inactivation by PEF

Few studies regarding PEF inactivation have been conducted with viruses (Khadre and Yousef 2002, Hirneisen et al. 2010). Viruses are expected to be more resistant to electrochemical inactivation than bacteria and, to date, PEF has shown little effectiveness against viruses (Grove et al. 2006). Some of them (notably bacteriophages) are not enveloped with a membrane and so, they would be immune to inactivation processes involving irreversible membrane permeabilization. Therefore, PEF may be less effective against protein capsids as compared with lipid-rich membranes (Grove et al. 2006). Even enveloped viruses would be more resistant than their cellular counterparts due to their smaller size (Drees et al. 2003) because trans-membrane potential induced by an external electric field depends upon the radius of the cell membrane and thus, from a given electric field, small cells appeared to be less vulnerable to membrane permeabilization than larger cells (Wouters et al. 2001). In 1967, Gilliland and Speck had applied an electrohydraulic shock to inactivate a bacteriophage specific for *Streptococcus cremoris* ML1, finding that it was much more susceptible than other bacteria tested. However, the voltage level applied on the phage (6 kv) was lower than the one applied on the bacteria (10 kv) and also phage initial concentration (5.0×10^5 PFU ml^{-1}) was lower than the bacterial concentration (from 1.0×10^7 to 2.7×10^8 CFU ml^{-1}). PEF inactivation of two eukaryotic viruses was achieved, finding unchanged shell, but damage RNA genome for the non-enveloped Swine Vesicular Disease Virus (SVDV), while both damages to the envelope and to the core of Equine Herpes virus 1 (EHV-1) (double strand DNA virus) (Mizuno et al. 1990). A wide concentration range for human rotavirus (a non-enveloped virus) was treated (from 20 to 29 kv/cm) and the obtained results showed that the virus was almost completely resistant to PEF (Khadre and Yousef 2002). Different population densities of bacteriophages MS2 and PRD1, which are non-enveloped enterobacterial viruses, were exposed to direct current with a variety of magnitudes (25-350 V) and durations of application (Drees et al. 2003). This study demonstrated that bacteriophages were able to tolerate greater magnitudes and greater durations of exposure to direct current than bacteria. It also indicates that the inactivation is primarily due to the disinfectant properties of electrochemically generated oxidants produced at the electrode in the cell. Finally, the authors found a population-dependent effect upon the efficacy of the electrochemical cell in inactivating bacteria and bacteriophage. This last observation could be explained by the fact that a finite amount of oxidants are produced at the electrodes and they can inactivate a greater proportion of the microorganism population when the population is low comparing when it is high. Fang et al. (2006) also investigated phage MS2 inactivation using an electrochemical disinfection system. Better disinfection was achieved increasing sodium chloride content, contact time and applied current.

2. High Hydrostatic Pressure

2.1. General Aspects

High Hydrostatic Pressure (HHP) is a non-thermal method that has the ability to produce high-quality and extended shelf-life with the added benefit of being microbiologically safe

(Patterson 2005). HHP has been proposed since 1899 by Hite and, among food preservation technologies alternative to thermal treatments, it is one of the most scientifically studied (Senorans 2003, Wan et al. 2005, Kelly and Zeece 2008), being its efficacy in inactivating different microbial species well documented (Toepfl et al. 2006, Ananta and Knorr 2009). Foods processed by high pressure have a fresher taste, look better, maintain normal textures, and have no loss of nutrients in comparison to foods treated by thermal methods (Hirneisen et al. 2010). Contrarily to heat-treated foods, all these attributes are retained since HHP does not disrupt covalent bonds thus small molecules, such as vitamins, amino acids, simple sugars and flavor compounds remain unaffected. Also proteins maintain their primary structure. However, usually proteins exhibited changes in the tertiary structure, maintained by hydrofobic and ionic interactions, at pressure values higher than 200 MPa (Baert et al. 2009). Currently, the food industry has used HHP for products such as guacamole, cooked ready-to-eat meats, tomato-based sauce, fruit juices, jams, sauce, rice products and mayonnaise, whole-shell oysters, and other shellfish (Patterson 2005, Aymerich et al. 2008, Hirneisen et al. 2010). HHP treated foods are available in United States, Europe, and Japan (Ramaswamy et al. 2003).

2.2. Physical and Chemical Principles. Process Parameters

Two principles underlie the effect of high pressure. Firstly, the principle of Le Chatelier, according to which any phenomenon (phase transition, chemical reactivity, change in molecular configuration, chemical reaction) accompanied by a decrease in volume is enhanced by pressure. One would expect that temperature would have an antagonistic effect because increasing temperature results in a volume increase. On the other hand, the reaction rate increases with increasing temperature according to Arrhenius' law. The second fundamental principle states that pressure is instantaneously and uniformly transmitted independent of the size and the geometry of the food. This is known as isostatic pressure (Smelt 1998). Moreover, the size and geometry of the package do not affect the effectiveness of the pressure process because, according to the Pascal principle, HHP is applied instantaneously and uniformly throughout the system.

From a process point of view, HHP consists in the application of pressures ranging between 100 and 1000 MPa to packaged liquid and solid foods, with holding times varying from several seconds to minutes (Ross et al. 2003). Also the process temperatures used during the treatment can range from 0 to 100°C with very different exposure time. Microbial inactivation and chemical changes are related to different parameters such as the process temperature and pressure applied, the time required to achieve the treatment pressure, the considered food matrix (in terms of composition, pH and water activity) and the packaging material integrity (Smelt 1998).

2.3. Applications

HHP offers interesting possibilities to restructure food proteins, affecting protein conformation, leading to protein denaturation, gelation or aggregation and, consequently, to create new products with new/improved textures. HHP is also used on dairy proteins for providing low temperature enzyme inactivation and stabilization of fermented dairy products, and also to improve coagulation of milk and to prepare dairy gels and emulsions characterized by novel textures (Senorans et al. 2003). HHP showed high potential also for functional food formulation and most of the literature available deals with the effect of HHP on bioactive milk proteins, i.e. β-lactoglobulin, bovine colostrums (Wan et al. 2005, Chicon et al. 2006, Peñas et al. 2006). These findings have raised interest in pressure processing to produce hypoallergenic foods (Zeece et al. 2008). The use of HHP has increased in this last decade also to produce dairy foods, with particular regard to low-fat yoghurt or fermented milks products (Penna et al. 2006) due to its positive effects on syneresis, texture, shelf life and nutritional and sensory qualities (Udabage et al. 2010). HHP treatment has also been used for the production of fermented probiotic dairy products improving their rheological and textural properties (Penna et al. 2006, 2007).

2.4. HHP Effects on Microorganisms

Generally, Gram-positive bacteria are more resistance to pressure than Gram-negative bacteria (500-600 MPa and 300-400 MPa, respectively, at 25°C during 10 min). Vegetative forms of yeast and moulds are the most pressure sensitive (Trujillo et al. 2002). Gram-positive bacteria are inactivated at higher pressures than Gram-negative ones (Hayakawa et al. 1994), due to the rigidity of the teichoic acids in the peptidoglycan layer of the cell wall (Lado and Yousef 2002). Exponentially growing cells are more sensitive to HHP than cells in the stationary phase. At a given HHP treatment intensity, reduced cell surface area in contact with the environment may limit cell leakage; therefore bacteria of small size and coccoid in shape are generally more resistant to HHP than the large rod-shaped ones. However, other factors than the shape and size of the cell are involved in the cell response to HHP (Lado and Yousef 2002). It was found that microbial HHP inactivation was higher at lower pH. In addition, ethanol, lysozyme, chitosan, sorbic and benzoic acids, and other additives enhanced the effect of pressure on microorganisms, thus permitting lower pressures, temperatures or shorter application times to be used to achieve inactivation (Datta and Deeth 1999). Moreover, there are many reports concerning the application of HHP for bacterial spore inactivation (Wuytack and Michiels 2001), although this approach has still not been solved by the use of pressure because some types of spores are able to survive at the most prohibitive conditions (1000 MPa) (Farkas and Hoover 2001). HHP destruction of bacterial spores is enhanced in the presence of certain antimicrobial compounds (Ross et al. 2003). Furthermore, there is a pressure-related process known as pressure-assisted thermal sterilization (PATS) that can achieve commercial sterilization in a range of foods (Rastogy et al. 2007).

Regarding to the mode of action, pressure induces many changes in the bacterial cell, including inhibition of key enzymes and protein synthesis, alterations in cell morphology and membrane, disturbance of transcription, translation and other cellular functions responsible for survival and reproduction (Huppertz et al. 2006, Abe 2007).

Shah et al. (2008) investigated the potential of HHP treatment at 480 MPa to increase the viable populations of *Lactobacillus delbrueckii* subsp. *bulgaricus*, *Streptococcus thermophilus*, and *Lb. acidophilus* inoculated in reconstituted skim milk during storage. Moreover, some reports indicate the potentiality of mild HHP to increase the heat resistance of some probiotic bacteria, i.e. *Lactobacillus rhamnosus* GG, due to the production of stress metabolites which have a cross-protective action. Analogously to PEF, also HHP is reported to cause some conformational changes, which results in the increasing or decreasing of enzyme activities and/or resistance to adverse conditions (i.e. refrigerated storage; heat treatment). In that sense, it has been reported the promotion while in others the prevention of thermal inactivation of several enzymes (Konisky et al. 1995).

2.5. Virus Inactivation by HHP

The major part of the studies regarding the effect of HHP on viruses deals with the inactivation of foodborne viruses (Grove et al. 2006, Kingsley et al. 2006), or animal viruses (Kingsley et al. 2007, Murchie et al. 2007), and bacteriophages (Guan et al. 2006, 2007, Black et al. 2010) used as their surrogates. Especially in the case of shellfish, HHP is successfully applied (c.a. 250 MPa, 3 min) to produce a safe (with regard to bacteria) and palatable product. However, higher intensities (up to 500 MPa) have been also tested to achieve the inactivation of hepatitis A virus and noroviruses (Black et al. 2010).

Viruses show very different sensitiveness to pressure. While Tobacco mosaic virus was found to be extremely resistant to pressure, needing 920 MPa to have any measurable inactivation; others were found rather sensitive, such as the Feline Calicivirus (FCV) which was completely inactivated at pressures as low as 275 MPa for 5 min. Fortunately, most tested human and animal viruses can be inactivated at pressures < 600 MPa, being poliovirus the most resistant one, capable of surviving an hour at 600 MPa with only modest reductions in infectivity (Grove et al. 2006). Very different responses to pressure treatments were observed even for close related viruses. Such diversity also exists among bacteriophages subjected to pressure, making it difficult to predict inactivation according to viral classification or morphology (Black et al. 2010). Guan et al. (2006) compared the resistances of four coliphages in the range of 350-600 MPa, founding 7.5 $\log_{10}$ PFU ml^{-1} inactivation after a treatment at 400 MPa for one of them (λ imm434) while no loss of infectivity for phage ΦX174 which was treated at 600 MPa. Aertsen et al. (2005) suggested that there was no general correlation between morphology and pressure stability of viruses, but a strong influence due to subtle differences in amino acid sequence of coat proteins. Also other authors (Guan et al., 2006; 2007, Murchie et al., 2007; Baert et al, 2009) agreed that similarity in structure did not guarantee comparable inactivation behaviour, which may result from the different stability of capsid proteins to pressure. On the contrary, Müller-Merbach et al. (2005) suggested that the size and shape of the virus can have an effect on its resistance to HHP. It was found that both P001 and P008 bacteriophages were more resistance than their lactococcal hosts and that P008 was the most resistant of the two phages studied. The resilience of P008 may be attributed to its morphology; P008 is an isometrically headed phage, whereas P001 has a prolate or elongated head (Müller-Merbach et al. 2005).

The mechanisms of viral inactivation by HHP are though to involve the denaturation and/or dissociation of proteins of the virus coat (Silva et al. 1992, Müller-Merbach et al. 2005) or in the case of enveloped viruses, damage to the envelope (Grove et al. 2006) rather than damage to viral nucleic acids. Pressure induced changes to the viral coat can be subtle alterations to capsid proteins or receptor recognition proteins that results in loss of infectivity

(Hirneisen et al. 2010). Above 300 MPa, irreversible denaturation of proteins occurs and thus bacteriophages are inactivated. On the contrary, lipid coated viruses are much more resistant (the Sindbis virus can resist up to 700 MPa) (Lado and Yousef 2002).

Viral inactivation by HHP can be affected by several factors including time, temperature, pressure magnitude and properties of the food matrix. Virus inactivation degree is more dependent on pressure level than on increases on the treatment time (Grove et al. 2006). As pressure magnitudes and/or treatment time increased, virus inactivation also amplified showing nonlinear inactivation curves. Curve shapes were characterized by a rapid initial drop in viral counts follow by tailing caused by a diminishing inactivation rate (Avsaroglu et al. 2006, Kingsley et al. 2007).

Highly variable responses of virus inactivation to different pressure and temperature combinations were observed. Temperature can work either antagonistically or synergistically with pressure (Müller-Merbach et al. 2005, Kingsley et al. 2007, Black et al. 2010). Temperatures higher than 30°C enhanced pressure inactivation of hepatitis A virus, while lower temperatures resulted in less inactivation (Kingsley et al. 2006). Contrarily, treatment of some viruses at low temperature has been found to enhance pressure-induced inactivation probably due to the exposure of nonpolar side chains to water; which is promoted at low temperatures. Water penetration of the viral capsid results in destabilization and these nonpolar interactions are promoted by pressure as they result in a reduction of volume of the system (Hirneisen et al. 2010). It was noted that FCV inactivation increased by as much as 3 $\log_{10}$ when treatments were performed at refrigeration temperatures or above 30°C compared with inactivation performed at room temperature (Kingsley et al. 2007). Working with lactococcal bacteriophages, Müller-Merbach et al. (2005) reported that below 450 MPa and 60°C, pressure and temperature seem to act antagonistically, whereas at higher pressures and temperatures a synergistic effect was observed.

Lethality of pressure treatment is affected by the food matrix. Generally, and as it was previously observed for bacteria, lesser inactivation of viruses by HHP was achieved when they were within foods compared with their suspensions in buffers or media. Sharma et al. (2008) found that some enteric viruses and phages were less sensitive to HHP in a food than in liquid suspension. Some authors (Kingsley et al. 2007, Murchie et al. 2007), studying the HHP susceptibility of surrogates of noroviruses and enteroviruses, demonstrated an increased resistance to inactivation when treated in mussels and oysters, compared to culture medium or seawater. Different properties of foods have been shown to play a protective effect against HHP inactivation of viruses. Fat and proteins present in milk have a protective effect on phage λ cI 857. HHP application in an acidic environment significantly enhanced hepatitis A virus inactivation. The effects of different constituents of foods including sugar, protein and salt content on foodborne virus inactivation by high pressure have been reported; high concentrations of sucrose (40%) and NaCl (6%) provided a baroprotection effect against inactivation of FCV (Hirneisen et al. 2010). Increased salinity has been also found protective to hepatitis A virus, and the effect may be attributed to a higher stabilization of the viral capsid proteins at high pressure (Grove et al. 2006).

2.6. Bacteriophage Inactivation by HHP

HHP has other potential purposes not directly associated with food safety. Though scarce data are available, there are some reports regarding HHP inactivation of dairy phages. Müller-Merbach et al. (2005) demonstrated that HHP could be used for the treatment of milk used in cheese making to inactivate bacteriophages of cheese starter cultures. In addition, high-pressure-treated milk mostly maintains its functionality for cheese making, increasing safety and shelf-life due to microbial inactivation. Inactivation experiments were carried out with the commonly occurring lactococcal phages P001 and P008, in calcium-enriched M17-broth and treated at up to 600 MPa by HHP. The isometric head phage P008 was considerably more resistant than the prolate phage P001, with almost no phage titre reductions after 2h at 450 MPa against 4 $\log_{10}$ cycles after 30 min for the latter. Based on their studies, the authors assumed that phages were inactivated by denaturing essential phage proteins. Avsaroglu et al. (2006), also working with lactococcal phages, demonstrated that phage inactivation by HHP is better described by non-linear semi-logarithmic survival curves than by the first-order model.

Thus, there is a biological variation or heterogeneity between the cells of a population of the same microorganism. HHP inactivation of another lactococcal bacteriophage was also studied (Smiddy et al. 2006).

Although c2 phage does not resemble any known foodborne viruses, it was chosen because of its structural difference to Qβ coliphage (a putative surrogate of human enteric viruses). Samples were treated at 200-800 MPa at 20°C for up to 30 min, both in culture media and in oysters. For both phages and media, almost no inactivation was detected at ≤400 MPa. High reductions (about 7 $\log_{10}$ units) were observed on treatment at 600-800 MPa for 10 min. Pressures necessary for high levels of inactivation of both phages were greater than those required for bacterial inactivation and much greater than those commercially used for HHP treatment of foods.

2.7. Other Applications of HHP

Another interesting use of HHP could be the pasteurization of blood and blood products (Bradley et al. 2000, Silva et al. 2002); inactivating viruses without damaging heat-sensitive plasma proteins. On the other hand, some viruses do not disassociate during pressure treatment or can reassemble afterwards forming non viable viral particles with the potential for use as vaccines.

HHP was successfully applied (260 MPa, 12 h) to produce vaccines against a membrane enveloped virus (vesicular stomatitis virus) to prepare non infectious whole virus particles that are still highly immunogenic (Silva et al. 1992).

One negative aspect on the use of HHP was discovered by Aertsen et al. (2005), who have shown the induction of Stx prophages in some *Escherichia coli* hosts when sublethal HHP treatments were applied. Stx genes are a major virulence factor of Shiga toxin (Stx)-producing *E. coli* strains and their expression was found to be co-regulated with transcription of late phage genes associated with the lytic cycle. The risk associated to the HHP treatment is that HHP-induced production of Stx phages raise the possibility of increased horizontal spread of the virulence trait carried by these phages. Another disadvantage of HHP is that, according to Smiddy et al. (2006), viral strains could develop resistance to this technology. These authors noticed altered plaques of a ss-RNA coliphage after HHP treatment that were able to persist by sub-culturing.

3. High Pressure Homogenization

3.1. General Aspects and Applications

Homogenization was presented for the first time in 1900 by Auguste Gaulin at the World Fair in Paris (French Patent n. 295,596). Since this moment, this technology was introduced at industrial level with many applications, particularly for pharmaceutical, cosmetic and food industries. From a process point of view, it was firstly used to improve the texture, body and flavour of several products such as milk, cream, ice-cream (Dickinson and Stainsby 1988) by using pressure of 20-25 MPa, while in the pharmaceutical, cosmetic and chemical fields it was useful for the preparation of stable emulsions and suspensions. Following, the increasing demand of products with longer shelf life and better microbial stability has stimulated the development of this technology throughout new technological approaches (Diels and Michiels 2006). More specifically, the increase of pressure has allowed using the High Pressure Homogenization (HPH) for the microbial cell disruption and the release and recovery of intracellular metabolites (Paquin 1999). The development, over these last two decades, of a new generation of homogenizers able to work at pressure ranging between 100-400 MPa or more, has allowed to use the HPH approach for microbial inactivation, both vegetative cells (Popper and Knorr 1990, Lanciotti et al. 1994, 1996, Guerzoni et al. 1999, 2002, Vachon et al. 2002) and spores (Chaves-López 2009). Its effectiveness in the inactivation of foodborne pathogens and spoilage microorganisms is well documented both in model systems and real foods including food emulsion, dairy products and fruit juices (Lanciotti et al. 1994, 1996, Vannini et al. 2004, Hayes and Kelly 2003, Diels and Michiels 2006, Burns et al. 2008, Patrignani et al. 2009, Patrignani et al. 2010). Although the reduction of microorganisms is not the primary purpose of the process, it is of great interest because serves to improve the microbiological safety and shelf life of the processed foods. According to the literature available, the HPH has been used, instead of the conventional homogenization, for the modulation of the microstructure and rheology of food emulsions (Guerzoni et al. 1997, Floury et al. 2004), the improvement of the body and texture of yoghurts and cheeses (Lanciotti et al. 2004a, b), the increasing of cheese yield (Lanciotti et al. 2004a Sandra and Dalgleish 2005, Burns et al. 2008) and reduction of cheese ripening time due to the enhanced susceptibility of proteins and triglycerides to proteolysis and lipolysis, respectively (Kheadr et al. 2002, Lanciotti et al. 2004a). Some papers report also the exploitation of HPH for the activation or inactivation of enzymes (Hayes and Kelly 2003, Vannini et al. 2004, Iucci et al. 2006, 2008) and to reduce the biogenic amine content of ripened cheeses (Lanciotti et al. 2007a). HPH has demonstrated great potential in the dairy sector also for the development of new products, differentiated from traditional ones by sensory and structural characteristics or functional properties (Guerzoni et al. 1999, Lanciotti et al. 2004b, Madadlou et al. 2006, Burns et al. 2008). Moreover, Iordache and Jelen (2003) showed HPH as a suitable approach for producing soluble whey protein concentrates/isolates for the production of several dairy products, as well as meat or egg substitutes. HPH showed good potentiality for large scale disruption of *Lb. delbrueckii* subsp. *bulgaricus* 11842 for β-galactosidase recovery (Bury et al. 2001). HPH has been used also for the recovery of several molecules, including (1-6)-β-d-Glucan and endogenous antioxidants from *Saccharomyces cerevisiae* for the improvement of sensorial and functional food properties (Jaehrig et al. 2008, Kogan et al. 2005). Moreover HPH has been proposed for large scale production of enzyme and aroma compounds of yeast and bacteria for functional foods (Bury et al. 2001, Jaehrig et al. 2008). Because HPH has been extensively used to emulsify, disperse, mix and reduce the mean droplet size, simultaneously narrowing the width of the size distribution by reducing the number of microparticles and the polydispersity index, one of the most recent application regards the use of HPH processes to produce nanodispersions and nanoparticles containing bioactive compounds, including functional lipids, with substantial health benefit (Floury et al. 2004, Saito et al. 2005). In the functional dairy-sector HPH has been proposed to produce probiotic fermented milk, bio-yoghurt and probiotic cheeses with improved sensorial or functional properties (Burns et al. 2008, Patrignani et al. 2008). Moreover, because some literature papers proposed HPH to control and enhance the proteolytic and fermentative activities of some *Lactobacillus* species (Lanciotti et al. 2007b), Tabanelli et al. (2008) have adopted this technology to modify/enhance some functional properties of already-known probiotic strains. Additionally, a recent paper (Capela et al. 2007) has investigated on the effect of homogenization on survival and bead-size of encapsulated probiotic bacteria.

3.2. Equipment

From a technological point of view, a homogenizer consists principally of a pump and a homogenizing valve (Figure 1). The pump is used to force the fluid into the valve where the homogenization happens (Diels and Michiels 2006). In the homogenizing valve the fluid is forced under pressure through a small orifice between the valve and the valve seat. The operating pressure is controlled by adjusting the distance between the valve and the seat. Concerning the microbial inactivation, the HPH effectiveness is affected by several parameters such as process and microbial-physiological factors as well as physic-chemical and compositional features of the treated fluid. Among the process parameters, pressure and temperature mainly influence the effectiveness of homogenization for microbial inactivation. The level of microbial inactivation caused by the application of high-pressure homogenization increases with the pressure level (Diels and Michiels 2006). Temperature effects have to be necessarily taken into account in HPH, since, upon homogenization, a rise of the temperature (about 2.5°C per 10 MPa), related to the fluid food employed, is observed in the fluid downstream of the valve. This can be attributed to the viscous stress caused by the high velocity of the fluid flow, which is then impinging on the ceramic valve of the homogenizer, leading to the dissipation of a significant fraction of the mechanical energy as heat in the fluid (Floury et al. 2004). Also, an increase in the number of passes through the homogenizing valve resulted in higher microbial inactivation than a single pass. On the contrary, milk was found more protective than buffers as a bacterial suspension medium. Regarding the microorganisms, it was found that Gram-negative bacteria are less resistant than Gram-positive bacteria and spore destruction is the most difficult, due to their high resistance to both pressure and temperature (Thiebaud et al. 2003).

HHP and HPH are two different processing methods for inactivation although they share some action mechanisms. Several physical phenomena including hydrostatic pressure, shear stress, turbulence, cavitation, impingement, and temperature are imposed on a fluid food subjected to HPH. HPH is a continuous process with relatively lower pressure (400 MPa) and shorter exposure time (seconds) than HHP, which is a batch process requiring longer exposure times (minutes). HPH causes bacterial inactivation and changes to macromolecules probably due to sudden pressure drop, torsion, and shear stresses, but mostly by cavitation shock waves from imploding gas bubbles, turbulence and viscous shears, which seem the most probable mechanisms of action (Diels and Michiels 2006, D'souza et al. 2009).

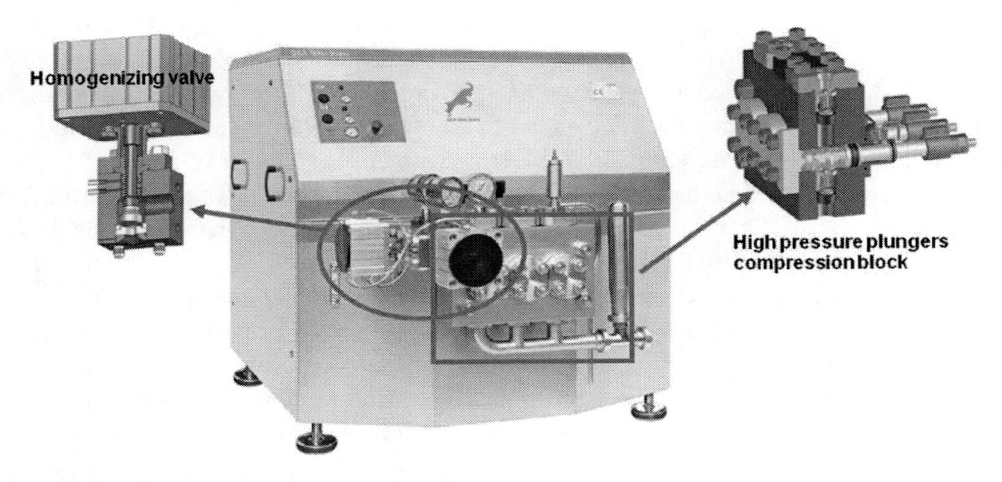

Figure 1. State of the art of homogenizer able to reach 150 MPa (GEA Niro Soavi, Italy).

3.3. Virus Inactivation by HPH

Data about virus inactivation by HPH treatment are even scarcer than by HHP treatment and very little is known about its effect on viruses (Diels and Michiels 2006, Moroni et al. 2002, D'souza et al. 2009). Raw milk is one of the major sources of phages in dairy plants (Neve 1996) and pasteurization is not always useful for its inactivation (Capra et al. 2004, Moineau 1999, Suárez et al. 2007), thus HPH could be used as an alternative treatment. Moroni et al. (2002) used a laboratory device for evaluating the influence of media, pressure, number of passes and initial viral concentration, on the viability of three different species of lactococcal phages. The lactococcal species c2 (prolate-headed), 936 and P335 (isometric-headed) were selected because they are frequently encountered in the dairy industry all around the world. The authors found a statistically significant difference in sensitivity to HPH between the two morphology-types of phages. The long form of the prolate head seemed to be less stable than the isometric structure, being more sensitive to HPH treatment. Influence of the same parameters was also studied for some lactic acid- and probiotic bacteria phages (*Lactobacillus delbrueckii, Lactobacillus helveticus, Streptococcus thermophilus, Lactococcus lactis, Lactobacillus paracasei* and *Lactobacillus plantarum*; Capra et al. 2009).

All phages are *Siphoviridae*, except the *Lb. helveticus* phage, which is *Myoviridae* and showed the highest sensitivity to HPH both in MRS broth and in milk. The inactivation extent achieved by HPH was phage-dependent. Even two related phages MLC-A and MLC-A8, lytic for the same probiotic *Lb. paracasei* strain, showed remarkably different behaviour after HPH treatment. The former was more resistant than the latter and titres of survival phage particles of 1.6 10^2 PFU ml^{-1} (8 passes) and <10 PFU ml^{-1} (5 passes) respectively, were obtained at 100 MPa. Moroni et al. (2002) remarked that the treatment become less effective with greater initial loads. At 200 MPa for 1 pass, while a 1.5 log_{10} decrease was obtained for an initial load of 10^8 PFU ml^{-1}, a 2.6 log_{10} decrease was reached at 10^7 PFU ml^{-1} and no viable particles were detected at 10^4 PFU ml^{-1}. On the contrary, Capra et al. (2009), working with phage MLC-A, have found that the higher the initial load, the higher the reduction achieved. Regarding applied pressure and number of passes, inactivation was proportional to both. The highest inactivation was obtained at 200 MPa after 5 passes, two times higher than at 100 MPa. Using 100 MPa, additional 0.6 and 1.6 log_{10} reductions were observed with 3 and 5 passes respectively compared to 1 pass. When applying 200 MPa for 3 and 5 passes, 1.1 and 2.9 log_{10} reductions, respectively, were added to the inactivation reached for 1 pass (Moroni et al. 2002). Also Capra et al. (2009) observed that with greater number of passes, the rate of inactivation reached was higher. However, even after a treatment of 8 passes at 100 MPa (initial load: c.a. 10^6 PFU ml^{-1}), 50% of the studied phage types remained viable at a concentration of at least 10 PFU ml^{-1}. Concerning the influence of media, some authors (Moroni et al. 2002) reported diminished HPH effectiveness in milk and whey permeate than in buffer, while others (Capra et al. 2009) could not ascertain a protective effect either for MRS broth or milk.

To study phage inactivation mechanism by HPH, Moroni et al. (2002) visualized high pressure treated phages by electron microscopy. Inactivation may result from openings induced in the phage heads, from which genetic material is lost. Some tailess phages or with only a part of the tail were also seen. Therefore, physical effects on phage integrity do not appear to be specific to any part of the phage. The effect of HPH on phages could be due to an alteration of their protein structure (Grove et al. 2006). In fact, it has been reported that HPH could induce a structural rearrangement of proteins, e.g., by increasing the exposure of their hydrophobic regions (Guerzoni et al. 1999). Both the capsid which protects the nucleic acid and tail structures which are essential for the binding of phages to the bacterial cell, are proteins and therefore, they could be the putative targets for high pressure treatments (Guan et al. 2006, Kingsley et al. 2002, Moroni et al. 2002). Furthermore, the sophisticated structure of the tail (base plate, fibres) may increase the pressure sensitivity of the phage (Guan et al. 2006).

Bacteriophages have demonstrated to be more resistant to high pressure than bacteria. Further work is required in order to fully evaluate the efficacy of HPH for the inactivation of bacteriophages, also taken into account the effect of severe treatments on the sensory and nutritional aspects of milk as a raw material. Alternatively, it was proposed that this novel technology may be combined with other conventional preservation processes (Lado and Yousef 2002, Ross et al. 2003) or with other non-thermal process to enhance bacteriophage inactivation. Also the addition of some compounds (lytic enzymes, SDS, EDTA, ions) capable of destabilizing the protein capsid of phages should be tested for enabling the use of reduced HPH treatment intensities (Capra et al. 2009).

Most of the work on human and animal viruses and human enteric virus surrogates has been done using HHP. Recently, a study to evaluate the effect of HPH using pressures of 0, 100, 200, 250, and 300 MPa on the human enteric viral surrogates F-RNA bacteriophage MS2 and murine norovirus (MNV-1) was reported (D'souza et al. 2009). It was found that only homogenization pressures of 300 MPa showed inactivation of approximately 3 $\log_{10}$ PFU ml^{-1} for MS2 and 0.8 $\log_{10}$ PFU ml^{-1} for MNV-1, with initial loads of approximately 6 $\log_{10}$ PFU ml^{-1}. Mechanisms similar to those observed for bacterial inactivation, may be associated with the inactivation of human enteric viruses causing changes or disruption in viral capsid structure. The differences in homogenization pressure resistances of MS2 and MNV-1 might be due to the dissimilarity in their coat proteins as well as morphological differences. Finally, it was concluded that further studies using pressure <300 MPa, depending on the availability of instruments that can reach higher pressures, are warranted to validate this inactivation process, which shows promise for application in industrial environments (D'souza et al. 2009).

CONCLUSION

In answering the demand of consumer claim for fresh like products, pulsed electric field, high hydrostatic pressure and high pressure homogenization have emerged, among others, as preservation technologies designed to produce safe foods, while maintaining its nutritional and sensory qualities. Currently, the most commonly used method of food preservation is thermal processing, even when unwanted reactions in foodstuffs, such as loss of flavour and nutrients occurred.

The analysis of the literature shows that different types of microorganisms considerably vary in their sensitivity to treatments with these novel technologies. The increasing number of outbreaks related to ready to eat minimally processed foods clearly indicate that non-thermal technologies are needed also to target enteric viruses and not only pathogenic and spoilage microorganisms. Taking into accounts that the balance between safety and maintenance of sensory properties is an ongoing challenge of food industry, none of the considered technologies alone can claim the complete elimination of enteric virus from foods.

Concerning bacteriophages, although very few studies have been conducted, they seemed to tolerate greater magnitudes and greater durations of exposure to direct current (PEF) and also to be more resistant to high pressure treatments (both HHP and HPH) than bacteria. However, some interesting results let envisage the potential application of these technologies alone or in combination with other hurdles in dairy industry for bacteriophage inactivation. Since the effects of these processing technologies has been mainly studied in cell culture and synthetic media, there is a defined need for further research in food matrices before their scaling up at industrial level in replacement to thermal treatments.

REFERENCES

Abe, F. (2007). Exploitation of the effects of high hydrostatic pressure on microbial growth, physiology and survival: Perspective from piezophysiology. *Bioscience Biotechnology and Biochemistry, 71,* 2347-2357.

Aertsen, A., Faster, D. and Michiels, C.W. (2005). Induction of shiga toxin-converting prophage in *Escherichia coli* by high hydrostatic pressure. *Applied and Environmental Microbiology, 71,* 1155-1162.

Alkhafaji, S. R. and Farid, M. (2007). An investigation on pulsed electric fields technology using new treatment chamber design. *Innovative Food Science and Emerging Technologies, 8,,* 205-212.

Ananta, E. and Knorr, D. (2009). Comparison of inactivation pathways of thermal or high pressure inactivated *Lactobacillus rhamnosus* ATCC 53103 by flow cytometry analysis. *Food Microbiology, 26*, 542-546.

Aronsson, K., Borch, B., Stenlö, F. B. and Rönner, U. (2004). Growth of pulsed electric field exposed *Escherichia coli* in relation to inactivation and environmental factors. *International Journal of Food Microbiology, 93*, 1-10.

Aronsson, K., Lindgren, M., Johansson, B. R. and Rönner, U. (2001). Inactivation of microorganisms using pulsed electric fields: the influence of process parameters on *Escherichia coli, Listeria innocua, Leuconostoc mesenteroides* and *Saccharomyces cerevisiae*. *Innovative Food Science and Emerging Technologies, 2,* 41-54.

Aronsson, K., Ronner, U. and Borch, E. (2005). Inactivation of *Escherichia coli, Listeria innocua* and *Saccharomyces cerevisiae* in relation to membrane permeabilization and subsequent leakage of intracellular compounds due to pulsed electric field processing. *International Journal of Food Microbiology, 99,* 19-32.

Avsaroglu, M. D., Buzrul, S., Alpas, H., Akcelik, M. and Bozoglu, F. (2006). Use of the Weibull model for lactococcal bacteriophage inactivation by high hydrostatic pressure. *International Journal of Food Microbiology, 108,* 78-83.

Aymerich, T., Picouet, P. A., Monfort, J. M. (2008). Decontamination technologies for meat products. *Meat Science, 78,* 114-129.

Baert, L., Debevere, J. and Uyttendaele, M. (2009). The efficacy of preservation methods to inactivate foodborne viruses. *International Journal of Food Microbiology, 131,* 83-94.

Barbosa-Canovas, G. V. and Sepulveda, D. (2005). Present status and the future of PEF technology. In G. V. Barbosa-Canovas, M. S. Tapia, and M. P. Cano (Eds.), Novel food processing technologies (1st ed., pp. 1-44). Boca Ratón, United States: CRC Press.

Barsotti, L. and Cheftel, J. C. (1999). Food processing by pulsed electric fields II. Biological aspects. *Food Reviews International, 15,* 181-213.

Bendicho, S., Espachs, A., Arántegui, J. and Martín, O. (2002). Effect of high intensity pulsed electric fields and heat treatments on vitamins of milk. *Journal of Dairy Research, 69,* 113-123.

Black, E. P., Cascarino J., Guan, D., Kniel, K. E., Hicks, D. T., Pivarnik, L. F., Hoover, D. G. (2010). Coliphage as pressure surrogates for enteric viruses in foods. *Innovative Food Science and Emerging Technologies, 11,* 239-244.

Bradley, D. W., Hess, R. A., Tao, F., Sciaba-Lentz, L., Remaley, A. T., Laugharn, J. A. Jr. and Manak, M. (2000). Pressure cycling technology: a novel approach to virus inactivation in plasma. *Transfusion, 40,* 193-200.

Burns, P., Patrignani, F., Serrazanetti, D., Vinderola, G., Reinheimer, J., Lanciotti, R. and Guerzoni, M.E. (2008). Probiotic Crescenza cheese containing *Lactobacillus paracasei* and *Lactobacillus acidophilus* manufactured with high pressure-homogeneized milk. *Journal of Dairy Science, 91,* 500-512.

Bury, D., Jelen, P. and Kaláb, M. (2001). Disruption of *Lactobacillus delbrueckii* ssp. *bulgaricus* 11842 cells for lactose hydrolysis in dairy products: A comparison of sonication, high-pressure homogenization and bead milling. *Innovative Food Science and Emerging Technologies*, *2*, 23-29.

Buta, P., Tauscher, B., (2002). Emerging technologies: chemical aspects. *Food Research International, 35,* 279-284.

Capela, P., Hay, T. K. C. and Shah, N. P. (2007). Effect of homogenisation on bead size and survival of encapsulated probiotic bacteria. *Food Research International*, *40*, 1261-1269.

Capra, M. L., Patrignani, F., Quiberoni, A. D. L., Reinheimer, J. A., Lanciotti, R. and Guerzoni M. E. (2009). Effect of high pressure homogenization on lactic acid bacteria phages and probiotic bacteria phages. *International Dairy Journal, 19*, 336-34.

Capra, M. L., Quiberoni, A. and Reinheimer, J. A. (2004). Thermal and chemical resistance of *Lactobacillus casei* and *Lactobacillus paracasei* bacteriophages. *Letters in Applied Microbiology, 38,* 499-504.

Chaves-López, C. , Lanciotti, R., Serio, A., Paparella, A., Guerzoni, M. E. and Suzzi, G. (2009). Effect of high pressure homogenization applied individually or in combination with other mild physical or chemical stresses on *Bacillus cereus* and *Bacillus subtilis* spore viability. *Food Control*, *20*, 691-695.

Chicon, R., Belloque, J., Recio, I. and Lopez-Fandino, R. (2006). Influence of high hydrostatic pressure on the proteolysis of β-lactoglobulin A by trypsin. *Journal of Dairy Research*, *73*, 121-128.

Corrales, M., Toepfl, S., Butz, P. Knorr, D. and Tauscher, B. (2008). Extraction of anthocyanins from grape by-products assisted by ultrasonics, high hydrostatic pressure or pulsed electric fields: A comparison. *Innovative Food Science and Emerging Technologies*, *9*, 85-91.

Cueva, O. A. (2009). Pulsed electric field influences on acid tolerance, bile tolerance, protease activity and growth characteristics of *lactobacillus acidophilus* LA-K. Master's Thesis. http://etd.lsu.edu/docs/available/etd-11192008-161805/

D'souza, D. H., Su, X., Roach, A. and Harte, F. (2009). High-pressure homogenization for the inactivation of human enteric virus surrogates. *Journal of Food Protection, 72,* 2418-2422.

Datta, N. and Deeth, H. C. (1999). High pressure processing of milk and dairy products. *The Australian Journal of Dairy Technology, 54,* 41-48.

De Vos, W. M. and Simons, G. F. M. (1994). Gene cloning and expression systems in lactococci. In M. J. Gasson, and. W. M. De Vos (Eds.), *Genetics and Biotechnology of lactic acid bacteria* (pp. 52-96). London, Great Britain: Blackie Academic and Professional.

Dickinson, E. and Stainby, G. (1988). Emulsion Stability. In E. Dickinson, and G. Stainby (Eds)., *Advances in food Emulsion and Foam* (pages 1-44). London, Great Britain: Elsevier Applied Sciences.

Diels, A. M. J. and Michiels, C. (2006). High-pressure homogenization as a non-thermal treatment for the inactivation of microorganisms. *Critical Reviews in Microbiology*, *32*, 201-216.

Drees, K. P., Abbaszadegan, M. and Maiera, R. M. (2003). Comparative electrochemical inactivation of bacteria and bacteriophage. *Water Research, 37,* 2291-2300.

Fang, Q., Shang, C. and Chen, G. (2006). MS2 inactivation by chloride-assisted electrochemical disinfection. *Journal of Environmental Engineering, 132*, 13-22.

Farkas, D. F. and Hoover, D. G. (2001). High pressure processing. *Journal of Food Science, 66*, 47-63.

Floury, J., Bellettre, J., Legrand, J. and Desrumaux, A. (2004). Analysis of a new type of homogenizer. A study of the flown pattern. *Chemical Engineering and Science, 59*, 843-853.

Gerlach, D., Alleborn, N., Baars, A., Delgado, A., Moritz, J. and Knorr, D. (2008). Numerical simulations of pulsed electric fields for food preservation: A review. *Innovative Food Science and Emerging Technologies, 9*, 408-417.

Gilliland, S. E. and Speck, M. L. (1967). Inactivation of microorganisms by electrohydraulic shock. *Applied Microbiology 15,* 1031-1037.

Glaser, R. (1996). Electric properties of the membrane and the cell surface. In U. Zimmermann, and G. A. Neil (Eds.), *Electromanipulation of Cells,* (pp. 329-364). Boca Raton, United States: CRC Press.

Grahl, T. and Märkl, H. (1996). Killing of microorganisms by pulsed electric fields. *Applied Microbiology and Biotechnologies, 45,* 148-157.

Grove, S. F., Lee, A., Lewis, T., Stewart, C. M., Chen, H. and Hoover, D. G. (2006). Inactivation of foodborne viruses of significance by high pressure and other processes. *Journal of Food Protection 69,* 957-968.

Guan, D., Joerger, R. D., Kniel, K. E., Calci, K. R, Hicks, D. T., Pivarnik, L. F. and Hoover, D. G. (2007). Effect of high hydrostatic pressure on four genotypes of F-specific RNA bacteriophages. *Journal of Applied Microbiology, 102,* 51-56.

Guan, D., Kniel, K., Calci, K. R., Hicks, D. T., Pivarnik, L. F., Hoover, D. G. (2006). Response of four types of coliphages to high hydrostatic pressure. *Food Microbiology, 23,* 546-551.

Guerzoni, M. E., Lanciotti, R., Westall, F. and Pittia, P. (1997). Interrelation between chemicophysical variables, microstructure and growth of *Listeria monocytogenes* and *Yarrowia lipolytica* in food model systems. *Science des les Aliments*, *17*, 507-522.

Guerzoni, M. E., Vannini, L., Chaves-López, C., Lanciotti, R., Suzzi, G. and Gianotti, A. (1999). Effect of high pressure homogenization on microbial and chemico-physical characteristics of goat cheeses. *Journal of Dairy Science, 82*, 851-862.

Guerzoni, M. E., Vannini, L., Lanciotti, R. and Gardini, F. (2002). Optimisation of the formulation and of the technological process of egg-based products for the prevention of *Salmonella enteritidis* survival and growth. *International Journal of Food Microbiology, 73*, 367-374.

Hayakawa, I., Kanno, T., Tomita, M. and Fujio, Y. (1994). Application of high-pressure for spore inactivation and protein denaturation. *Journal of Food Science, 59,*159-163.

Hayes, M. G. and Kelly, A. L. (2003). High pressure homogenisation of raw whole bovine milk (a) effects on fat globules size and other properties. *Journal of Dairy Research*, *70*, 297-305.

Heinz, V., Alvarez, I., Angersbach, A. and Knorr, D. (2002). Preservation of food by high intensity pulsed electric fields-Basic concepts for process design. *Trends in Food Science and Technology, 12,* 103-111.

Hirneisen, K. A., Black, E. P., Cascarino, J. L., Fino, V. R., Hoover, D. G. and Kniel, K. E. (2010). Viral inactivation in foods: a review of traditional and novel food-processing technologies. *Comprehensive Reviews in Food Science and Food Safety, 9,* 3-20.

Ho, S. and Mittal, G. S. (2000). High voltage pulsed electric field for liquid food pasteurization. *Food Reviews International, 16,* 395-434.

Huang, K. and Wang, J. (2009). Designs of pulsed electric fields treatment chambers for liquid foods pasteurization process: A review. *Journal of Food Engineering, 95,* 227-239.

Huppertz, T., Smiddy, M. A., Upadhyay, V. K., Kelly, A. L. (2006). High-pressure-induced changes in bovine milk: a review. *International Journal of Dairy Technology, 59,* 58-66.

Iordache, M. and Jelen, P. (2003). High pressure microfluidization treatment of heat denatured whey proteins for improved functionality. *Innovative Food Science and Emerging Technologies*, *4*, 367-376.

Iucci, L., Lanciotti, R., Kelly, A. and Huppertz, T. (2008). Plasmin activity in high-pressure-homogenised bovine milk. *Milchwissenschaft*, *63*, 68-70.

Iucci, L., Patrignani, F., Vallicelli, M., Guerzoni, M. E. and Lanciotti, R. (2006). Effects of high pressure homogenization on the activity of lysozyme and lactoferrin against *Listeria monocytogenes*. *Food Control*, *18*, 558-565.

Jaehrig, S. C., Rohn, S., Kroh, L. W., Wildenauer, F. X., Lisdat, F., Fleischer, L. -G. and Kurz, T. (2008). Antioxidative activity of (1→3), (1→6)-β-d-glucan from *Saccharomyces cerevisiae* grown on different media. *LWT - Food Science and Technology, 41*, 868-877.

Jayaram, S. H. (2000). Sterilization of liquid foods by pulsed electric fields. *Electrical Insulation Magazine, 16,* 17-25.

Kelly, A. L. and Zeece, M. (2009). Applications of novel technologies in processing of functional foods. *The Australian Journal of Dairy Technology*, *64*, 12-15.

Khadre, M. A. and Yousef, A. E. (2002). Susceptibility of human Rotavirus to ozone, high pressure and pulsed electric field. *Journal of Food Protection 65,* 1441-1446.

Kheadr, E. E., Vachon, J. F., Paquin and P., Fliss, I. (2002). Effect of dynamic pressure on microbiological, rheological and microstructural quality of Cheddar cheese. *International Dairy Journal, 12*, 435-446.

Kingsley, D. H., Guan, D., Hoover, D. G. and Chen H. (2006). Inactivation of hepatitis A virus by high-pressure processing: the role of temperature and pressure oscillation. *Journal of Food Protection, 69,* 2454-2459.

Kingsley, D. H., Holliman, D. R., Calci, K. R., Chen, H. and Flick, G. J. (2007). Inactivation of a norovirus by high-pressure processing. *Applied and Environmental Microbiology, 73,* 581-585.

Kingsley, D. H., Hoover, D. G., Papafragkou, E. and Richards, G. P. (2002). Inactivation of hepatitis A virus and a calicivirus by high hydrostatic pressure. *Journal of Food Protection, 65,* 1605-1609.

Kogan, G., Stasko, A., Bauerová, K., Polovka, M., Soltés, L., Brezová, V., Navarová, J. and Mihalová, D. (2005). Antioxidant properties of yeast (1→3)-β-D-glucan studied by electron paramagnetic resonance spectroscopy and its activity in the adjuvant arthritis. *Carbohydrate Polymers*, *61*, 18-28.

Lado, B. H., Yousef and A. E. (2002). Alternative food-preservation technologies: efficacy and mechanisms. *Microbes and Infection, 4,* 433-440.

Lanciotti, R., Chaves-Lopez, C., Patrignani, F., Papparella, A., Guerzoni, M.E., Serio, A. and Suzzi, G. (2004a). Effects of milk treatment with dynamic high pressure on microbial populations, and lipolytic and proteolytic profiles of Crescenza cheese. *International Journal of Dairy Technology*, *57*, 19-25.

Lanciotti, R., Gardini, F., Sinigaglia M. and Guerzoni M. E. (1996). Effects of growth conditions on the resistance of some pathogenic and spoilage species to high pressure homogenization. *Letters in Applied Microbiology*, *22*, 165-168.

Lanciotti, R., Patrignani, F., Iucci, L., Guerzoni, M. E., Suzzi, G., Belletti, N. and Gardini F. (2007a). Effects of milk high pressure homogenization on biogenic amine accumulation during ripening of ovine and bovine Italian cheese. *Food Chemistry*, *104*, 693-701.

Lanciotti, R., Patrignani, F., Iucci, L., Saracino, P., Guerzoni, M.E. (2007b). Potential of high pressure homogenization in the control and enhancement of proteolytic and fermentative activities of some *Lactobacillus* species. *Food Chemistry*, *102*, 542-550.

Lanciotti, R., Sinigaglia, M., Angelini, P. and Guerzoni, M. E. (1994). Effects of homogenization pressure on the survival and growth of some food spoilage and pathogenic microorganisms. *Letters in Applied Microbiology*, *18*, 319-322.

Lanciotti, R., Vannini, L., Patrignani, F., Iucci, L., Vallicelli, M., Ndagijimana, M. and Guerzoni M. E. (2006). Effect of high pressure homogenisation of milk on cheese yield and microbiology, lipolysis and proteolysis during ripening of Caciotta cheese. *Journal of Dairy Research*, *73*, 1-11.

Lanciotti, R., Vannini, L., Pittia, P. and Guerzoni, M. E. (2004b). Suitability of high dynamic-pressure-treated milk for the production of yogurt. *Food Microbiology*, *21*, 753-760.

Lebovka, N. I., Praporscic, I. and Vorobiev E. (2004a). Effect of moderate thermal and pulsed electric field treatments on textural properties of carrots, potatoes and apples. *Innovative Food Science and Emerging Technologies, 5,* 9-16.

Lebovka, N. I., Praporscic, I., Vorobiev, E. (2004b). Combined treatment of apples by pulsed electric fields and by heating at moderate temperature. *Journal of Food Engineering 65,* 211-217.

Li, S. Q., Boser, J. A. and Zhang, Q. H. (2005). Effect of PEF and heat treatment on stability and secondary structure of bovine immunoglobulin G. *Journal of Agriculture of Food Chemistry*, *53*, 663-670.

Mackey, B. M. (2000). Injured bacteria. In: B. M. Lund, A. C. Baird-Parker, and G.W. Gould (Eds.), *The Microbiological Safety and Quality of Foods* (pp. 315-341). Maryland, USA: Aspen Publishers.

Madadlou, A., Khosroshahi, A., Mousavi, S. M. and Djome, Z. E. (2006). Microstructure and rheological properties of Iranian White cheese coagulated at various temperatures. *Journal of Dairy Science, 89*, 2359-2364.

Mãnas, P. and Pagãn, R. (2005). A review: Microbial inactivation by new technologies of food preservation. *Journal of Applied Microbiology, 98,* 1387-1399.

Min, S., Evrendilek, G. A. and Zhang, H. Q. (2007). Pulsed electric fields: processing system, microbial and enzyme inhibition, and shelf life extension of foods. *IEEE Transactions on Plasma Science 35,* 59-73.

Mizuno, A., Inoue, T., Yamaguchi, S., Sakamoto, K. -I., Saeki, T., Matsumoto, Y. and Minamiyama, K. (1990). Inactivation of viruses using pulsed high electric field. *Industry Applications Society Annual Meeting, Conference Record of the 1990 IEEE, 1,* 713-719.

Moineau, S. (1999). Applications of phage resistance in lactic acid bacteria. *Antoine van Leeuwenhoek, 76,* 377-382.

Moroni, O., Jean, J., Autret, J. and Fliss, I. (2002). Inactivation of lactococcal bacteriophages in liquid media using dynamic high pressure. *International Dairy Journal, 12,* 907-913.

Müller-Merbach, M., Rauscher, T. and Hinrichs, J. (2005). Inactivation of bacteriophages by thermal and high-pressure treatment. *International Dairy Journal, 15,* 777-784.

Murchie, L. W., Kelly, A. L., Wiley, M., Adair, B. M. and Patterson, M. (2007). Inactivation of a calicivirus and enterovirus in shellfish by high pressure. *Innovative Food Science and Emerging Technologies, 8,* 213-217.

Neumann, E. and Rosenheck, K. (1972). Permeability changes induced by electric impulses in vesicular membranes. *Journal of Membrane Biology, 10,* 279-290.

Neve, H. (1996). Bacteriophage. In: T. M. Cogan, and J. P. Accolas (Eds.), *Dairy Starter Cultures* (pp. 157-190). New York, USA: VCH Publishers Inc.

Ohshima, T., Tamura, T. and Sato, M. (2007). Influence of pulsed electric field on various enzyme activities. *Journal of Electrostatics*, *65*, 156-161.

Paquin, P. (1999). Technological properties of high pressure homogenizers: the effect of fat globules, milk proteins and polysaccharides. *International Dairy Journal*, *9*, 329-225.

Patrignani F., Vannini L., Sado Kamdem S. L., Lanciotti R. and Guerzoni M. E. (2009). Effect of high pressure homogenization on *Saccharomyces cerevisiae* inactivation and physico-chemical features in apricot and carrot juices. *International Journal of Food Microbiology, 136,* 26-31.

Patrignani F., Vannini L., Sado Kamdem S. L, Lanciotti R., Guerzoni M. E. (2010). Potentialities of high-pressure homogenization to inactivate *Zygosaccharomyces bailii* in fruit juices. *Journal of Food Science*, *75,* 116-120.

Patrignani, F., Burns, P., Serrazanetti, D., Vinderola, G., Reinheimer, J. A., Lanciotti, R. and Guerzoni, M. E. (2008). Suitability of high pressure-homogenized milk for the production of probiotic fermented milk containing *Lactobacillus paracasei* and *Lactobacillus acidophilus*. *Journal of Dairy Research*, *76*, 74-82.

Patterson, M. F. (2005). Microbiology of pressure-treated foods. *Journal of Applied Microbiology*, *98*, 1400-1409.

Peñas, E., Restani, P., Ballabio, C., Préstamo G., Fiocchi, A. and Gomez, R. (2006). Evaluation of the residual antigenicity of dairy whey hydrolysates obtained by combination of enzymatic hydrolysis and high-pressure treatment. *Journal of Food Protection, 69*, 1707-1712.

Penna, A. L. B., Gurram, S. and Barbosa-Canovas, G. V. (2006). Effect of high hydrostatic pressure processing on rheological and textural properties of probiotic low-fat yogurt fermented by different starter cultures. *Journal of Food Process Engineering, 29,* 447-461.

Penna, A. L. B., Subbarao-Gurram, S. and Barbosa-Cánovas, G. V. (2007). High hydrostatic pressure processing on microstructure of probiotic low-fat yogurt. *Food Research International 40*, 510-519.

Popper, L. and Knorr, D. (1990). Applications of high-pressure homogenization for food preservation. *Food Technology, 44*, 84-89.

Ramaswamy, H. S., Riahi, E., Idziak, E. (2003). High pressure destruction kinetics of *E. coli* (29055) in apple juice. *Journal of Food Science, 68,* 1750-1756.

Rastogy, N. K., Rahavarao, K. S. M. S., Balasubramaniam, V. M., Niranjan K. and Knorr D. (2007). Opportunities and challenges in high pressure processing of foods. *Critical Reviews in Food Science and Nutrition, 47,* 69-112.

Ross, A. I. V., Griffiths, M. W., Mittal, G. S. and Deeth, H. C. (2003). Combining non thermal technologies to control foodborne microorganisms. *International Journal of Food Microbiology, 89,* 125-138.

Saito, M., Yin, L. J., Kobayashi, I. and Nakajima, M. (2005). Preparation characteristics of oil-in-water emulsions stabilized by proteins in straight-through microchannel emulsification. *Food Hydrocolloids, 19*, 745-751.

Sandra, S. and Dalgleish, D. G. (2005). Effects of ultra-high-pressure homogenization and heating on structural properties of casein micelles in reconstituted skim milk powder. *International Dairy Journal, 15*, 1095-1104.

Senorans, F. J., Ibanez, E. and Cifuentes, A. (2003). New trends in food processing. *Critical Reviews in Food Science and Nutrition, 43*, 507-256.

Shah, N. P., Tsangalis, D., Donkor, O. N. and Versteeg, C. (2008). Effect of high pressure treatment on viability of *Lactobacillus delbrueckii* ssp. *bulgaricus*, *Streptococcus thermophilus* and *L. acidophilus* and the pH of fermented milk. *Milchwissenschaft,* 63, 11-14.

Sharma, M., Shearer, A. E. H., Hoover, D. G., Liu, M. N., Solomon, M. B. and Kniel, K. E. (2008). Comparison of hydrostatic and hydrodynamic pressure to inactivate foodborne viruses. *Innovative Food Science and Emerging Technologies, 9,* 418-422.

Silva, J. L., Luan, P., Glaser, M., Voss, E.W. and Weber, G. (1992). Effects of hydrostatic pressure on a membrane-enveloped virus: high immunogenicity of the pressure-inactivated virus. *Journal of Virology, 66,* 2111-2117.

Silva, J. L., Oliveira, A. C., Gomes, A. M. O., Lima, L. M. T. R., Mohana-Borges, R., Pacheco, A. B. F. and Foguel, D. (2002). Pressure induces folding intermediates that are crucial for protein-DNA recognition and virus assembly. *Biochimica et Biophysica Acta, 1595,* 250-265.

Smelt, J. P. P. M. (1998). Recent advances in the microbiology of high pressure processing. *Trends in Food Science and Technology, 9*, 152-158.

Smiddy, M., Kelly, A. L., Patterson, M. F. and Hill, C. (2006). High pressure-induced inactivation of Qβ coliphage and c2 phage in oysters and in culture media. *International Journal of Food Microbiology, 106,* 105-110.

Suárez, V., Binetti, A., Quiberoni, A., Guglielmotti, D. and Capra, M. L. (2007). Tratamientos térmicos y químicos para la inactivación de fagos. In: J. Reinheimer, and C. Zalazar (Eds.), *Avances en microbiología, bioquímica y tecnología de quesos* (pp. 101-115). Santa Fe, Argentina, Facultad de Ingeniería Química, Universidad Nacional del Litoral.

Tabanelli, G., Patrignani, F., Vinderola, G. C., Gardini, F., Guerzoni, M. E. and Lanciotti, R. (2008). Personal communication: High Pressure Homogenization: can it improve Lactobacilli functional properties? Scientific training "How to present your work?" 3-5 September, Berlin.

Thiebaud, M., Dumay, E., Picart, L., Guiraud, J. P., Cheftel, J.C. (2003). High-pressure homogenisation of raw bovine milk. Effects on fat globule size distribution and microbial inactivation. *International Dairy Journal, 13,* 427-439.

Toepfl, S., Heinz, V., and Knorr, D. (2007). High intensity pulsed electric fields applied for food preservation. *Chemical Engineering and Processing, 46,* 537-546.

Toepfl, S., Mathys, A., Heinz, V., Knorr, D. (2006). Review: Potential of high hydrostatic pressure and pulsed electric fields for energy efficient and environmentally friendly food processing. *Food Reviews International, 22*, 405-423.

Torregrosa, F., Cortés, C., Esteve, M. J. and Frígola, A. (2006). Effect of high-intensity pulsed electric fields processing and conventional heat treatment on orange–carrot juice carotenoids. *Journal of Agricultural and Food Chemistry, 53*, 9519-9525.

Trujillo, A. J., Capellas, M., Saldo, J., Gervilla, R. and Guamis, B. (2002). Application of high-hydrostatic pressure on milk and dairy products: a review. *Innovative food Science and Emerging Technologies, 3,* 295-307.

Tsong, T. Y. (1991). Electroporation of cell membranes. *Biophysical Journal, 60,* 297-306.

Udabage, P., Augustin, M. A., Versteeg, C., Puvanenthiran, A., Yoo, J. A., Allen, N., McKinnon, I., Smiddy, M. and Kelly, A. L. (2010). Properties of low-fat stirred yogurts made from high-pressure-processed skim milk. *Innovative Food Science and Emerging Technologies, 11,* 32-38.

Ulmer, H. M., Heinz, V., Gänzle, M. G., Knorr, D., Vogel, R. F. (2002). Effects of pulsed electric fields on inactivation and metabolic activity of *Lactobacillus plantarum* in model beer. *Journal of Applied Microbiology, 93,* 326-335.

Vachon, J. F., Kheadr, E. E., Giasson, J., Paquin, P. and Fliss, I. (2002). Inactivation of foodborne pathogens in milk using dynamic high pressure. *Journal of Food Protection, 65*, 345-352.

Vannini, L., Lanciotti, R., Baldi, D. and Guerzoni, M. E. (2004) Interactions between high pressure homogenization and antimicrobial activity of lysozyme and lactoperoxidase. *Journal of Food Microbiology, 94*, 123-135.

Wan, J., Conventry, J, Sanguansri, P. and Versteeg, C. (2009). Advances in innovative technologies for microbial inactivation and enhancement of food safety-pulsed electric field and low-temperature plasma. *Trends in Food Science and Technology, 20*, 414-424.

Wan, J., Mawson, R., Ashokkumar, M., Ronacher, K., Coventry, M. J., Roginski, H. and Versteeg, C. (2005). Emerging processing technologies for functional foods. *The Australian Journal of Dairy Technology, 60*, 167-169.

Wouters, P. C., Bos, A. P., Ueckert, J. (2001). Membrane Permeabilization in relation to inactivation kinetics of *Lactobacillus* species due to pulsed electric fields. *Applied and Environmental Microbiology,* 67, 3092-3101.

Wouters, P. C., Dutreux, N., Smelt, J. P. P. M. and Lelieveld, H. L. M. (1999). Effects of pulsed electric fields on inactivation kinetics of *Listeria innocua. Applied and Environmental Microbiology 65,* 5364-5371.

Wouters, P., Alvarez, I., Raso, J. (2001). Critical factors determining inactivation kinetics by pulsed electric field food processing. *Trends in Food Science and Technology, 12,* 112-121.

Wuytack, E. Y. and Michiels, C. W. (2001). A study on the effects of high pressure and heat on *Bacillus subtilis* spores at low pH. *International Journal of Food Microbiology,* 64, 333-341.

Xin, Q., Zhang, X., Lei, L. (2008). Inactivation of bacteria in oil-field reinjection water by pulsed electric field (PEF) process. *Industrial and Engineering Chemistry Research, 47,* 9644-9650.

Yeom, H. W., Evrendilek, A., Jin, Z. T. and Zhang, Q. H. (2004). Processing of yogurt-based products with pulsed electric fields: microbial, sensory and physical evaluations. *Journal of Food Processing Preservation, 28*, 161-178.

Zeece, M., Huppertz, T. and Kelly, A. (2008). Effect of high-pressure treatment on in-vitro digestibility of β-lactoglobulin. *Innovative Food Science and Emerging Technologies, 9*, 62-69.

Zimmermann, U. (1986). Electrical breakdown, electropermeabilization and electrofusion. *Reviews of Physiology, Biochemistry and Pharmacology, 105*, 175-256.

Zimmermann, U., Schulz, J. and Pilwat, G. (1973). Transcellular ion flow in *E. coli* and electrical sizing of bacteria. *Biophysical journal, 13*, 1005-1013.

In: Bacteriophages in Dairy Processing
Editor: Andrea del Lujan Quiberoni et al.
ISBN 978-1-61324-517-0

Chapter 11

DETECTION AND QUANTIFICATION OF DAIRY BACTERIOPHAGES

Ana G. Binetti*
Instituto de Lactología Industrial (UNL-CONICET),
Facultad de Ingeniería Química
Universidad Nacional del Litoral, Santiago del Estero,
Santa Fe, Argentina

ABSTRACT

Phage monitoring is critically important for dairy industry because it is a symptom of the degree of phage accumulation in the plant, and allows to take rapid decisions when an infection episode occurs. Standard microbiological methods (double layer agar test, spot test, sensitivity test, etc.), based on the lysis of sensitive strains in agar layers or liquid media are still being recommended in some standardized protocols for bacteriophage detection. However, these methods require the availability of a sensitive strain and are time-consuming (at least, 24 h), which sometimes is crucial for the prevention of the lysis of the whole bacterial culture. Nevertheless, since they are relatively simple to implement and do not require special equipments, they are even nowadays used to detection and quantification of dairy phages for routine controls in dairy plant laboratories. During the last 15 years, advances in DNA technology have allowed the design of an important number of molecular detection methods. In this sense, traditional PCR has been adapted for a relatively rapid (approximately, 4-6 h) and qualitative detection of dairy phages from different industrial samples (milk, cheese whey, fermented milks, etc.) with satisfactory detection limits. The increasing demand for quantitative, more sensitive and faster analysis to detect and enumerate phages from dairy samples coupled with the ever-increasing knowledge on phage genomic sequences has prompted the development of specific real-time qPCR methods to specifically detect dairy phages of industrial relevance, demonstrating a high degree of efficiency, sensitivity and precision. An important benefit of these molecular systems is that neither phage particle concentration

* E-mail address: anabinetti@fiq.unl.edu.ar. Tel.: +54 342 4530302; fax: +54 342 4571162

nor any procedure is required to enrich the samples. Other novel systems, based on different methodologies, have been adapted to indirectly detect dairy phages. Some examples of these methodologies are based on flow citometry and impedimetric measures, which detect phages from morphological modifications of single cells, and from changes of a host biofilm established on a microsensor as a consequence of viral infection, respectively.

INTRODUCTION

Lactic acid bacteria are used extensively in the manufacture of a broad range of fermented dairy products. In many of these applications, infection of the cultures by virulent bacteriophages is a problem leading to either complete loss of the fermentation batch or, if only one out of several strains is affected, an altered flavor or texture (Coffey and Ross 2002). Prevention of phage infections has therefore received great attention over the past decades, and several strategies have been employed as pure cultures and phage-resistant starter strains (Coffey and Ross 2002, Sturino and Klaenhammer 2004, Guglielmotti et al. 2006, Binetti et al. 2007). Detailed programs for rotational changes of phage-resistant starter cultures have been designed (Sing and Klaenhammer 1993) to suppress the propagation of phage mutants that have overcome the resistance mechanism of the bacterial strains. Although these strategies have had success in preventing phage infections, they cannot completely prevent bacteriophage attack and they may limit the number and diversity of dairy starter cultures available for production, which may in turn limit the diversity of dairy products, mainly cheeses. In general, dairy phages are naturally found in raw milk (Bruttin et al. 1997), resist pasteurization (Quiberoni et al. 1999, Binetti and Reinheimer 2000, Suárez and Reinheimer 2002, Capra et al. 2004) and have, in general, a short latent period and a large burst size (Jarvis 1989, Suárez et al. 2002, Quiberoni et al. 2010). Therefore, they can spread rapidly within a fermentation plant and, consequently, constant monitoring is critical. The dairy industry, the main affected, shows a great interest for the development of efficient and fast methods that would detect phage infection at a sufficient early stage for an operator to interfere with the process.

Two types of methods for detecting phages are available: direct and indirect. Direct detection methods put on evidence the presence of lytic phage particles itself or their components (DNA, proteins) in a sample suspected to contain bacteriophages. Indirect methods analyse the effects of a sample potentially infected, on the bacterial metabolism.

In this chapter, the principal methods to detect dairy phages of industrial relevance will be described.

1. DIRECT DETECTION METHODS

1.1. Standard Microbiological Methods

Standard microbiological methods (turbidity test, plaque assays, spot test, limit dilutions test, etc.) are relatively economic, simple and provide information about the sensitivity from a culture, but require the availability of a sensitive strain and are time-consuming. As a consequence, the results may arrive too late to save whole vats of inoculated milk.

Nevertheless, since they are relatively simple to implement and do not require special equipments, these traditional methods are even currently used to detection and quantification of dairy phages for routine controls in dairy plant laboratories, as recommended by the IDF Bulletin 129 (1980).

Turbidity test is based on the clearance of a broth culture caused by the presence of bacteriophages in a sterile filtrate, compared with a control (bacterial culture without filtrate). It must be exclusively used for single strains and lysis can be visually detected or evidenced by absorbance measures (IDF Bulletin 263 1991). In general, three subcultures must be done with the aim to increase the concentration of phages potentially present in the sample which needs several days to complete the analysis.

Methods based on lysis of sensitive cells in agar layers include spot test and double layer plaque titration. In both cases, a small volume of cells is mixed with melted soft-agar and is poured as a thin layer on top of a suitable agar medium, generally containing divalent ions as Ca^{2+} and, eventually, Mg^{2+} (Sozzi et al. 1976, Keogh 1980, Lillehaug 1997, Binetti et al. 2002, Quiberoni et al. 2010). The filtrate, or dilutions thereof, can either be mixed with the indicator bacteria before pouring the soft agar layer (plaque assay) or can be applied as a small drop after the agar has set (spot test). If bacteriophages are present, they will lyse their host cells and, after incubation, lead to a visible clear zone (a lysis plaque or a spot of lysis plaques) on the turbid layer of growth. For double layer plaque assay, by counting the plaques, it is possible to enumerate the phage concentration in the original filtrate. This method gives quantitative information on phage titers in the filtrate and it represents the traditional choice to analyse industrial samples. In case of spot test, a large clear area (a spot) may be seen if the sample has not been diluted (if the phage level in the samples is higher than 10^3-10^4 PFU ml^{-1}), allowing its use as a qualitative test for the presence of phage (non diluted filtrate) or as a semiquantitative method by spotting dilutions of the filtrate (Hull 1977). Both methods can only be used for pure cultures. In case of multiple strain starters, individual strains must be isolated and purified before the analysis, since phage insensitive or resistant strains in the culture will overgrow and mask the lysis plaques. This implies, at least, three extra days for the isolation and purification of singles strains, thus increasing the time required to obtain a final result. Detection of lysis plaques on a semi-solid agar layer may be variable and uncertain, especially for bacteriophages specific for thermophilic lactic acid bacteria as *Streptococcus thermophilus*. Interaction between phage and host cells, phage life clyce (virulent or temperate), growth medium, physical and chemical conditions of the double-agar plaques are the main factors affecting plaque formation and, consequently, false negative results can be obtained from an infected sample. Despite several attempts to optimize these methods (Sozzi et al. 1976, Mullan 1979, Keogh 1980, Lillehaug 1997), the efficiency of phage detection ultimately depends on each phage-host system. To solve the problem concerning the inability of some host-phage systems to form visible lysis plaques, the simultaneous use of another qualitative detection method (generally, the turbidity test) is usually recommended.

Limiting dilution method can be utilized to estimate the concentration of bacteriophages in a sample by using defined and mixed strain cultures. It is a technique more laborious than the previous ones. In this assay, a filtrate from a sample is diluted until the last dilution contains, approximately, one phage particle per ml. In order to detect phage in most diluted samples, a multiplication or propagation step is required. All diluted samples are inoculated in milk (pasteurized or sterile) containing the host culture. The cultures (including a control, without sample) are incubated at the appropriate temperature to allow phage propagation and then, they are filtrated and analyzed for the presence of bacteriophages, using any of the methods previously described (turbidity test, spot test or double layer titration). The number of phages per ml will be lower than the reciprocal of the lowest dilution of the samples with a normal acid production.

1.2. Immunological Methods

Immunological methods represent another way to quantify bacteriophages, by using specific antibodies against phage proteins that bind to viral particles. The most commonly used are founded on the ELISA technique (Enzyme-Linked Immunosorbant Assay), that utilize an enzyme-labelled antibody to detect and quantify phage particles, from the intensity of the colour generated by enzymatic reaction with a substrate. Specific antibodies will exclusively bind to a morphological type of phage, allowing its classification based on a morphological criterion. By using a mixture of antibodies directed against all phage groups, it should be possible to detect the presence of all phage types in one single test. Lembke and Teuber (1979) reported the direct detection of 10^7 lactococcal phage ml^{-1} of whey in 3 h with polyclonal antibodies directed against native phage particles of c2 and 936 species, two of the three most relevant lactococcal phage types. Moineau et al. (1993) and Chibani Azaïez et al. (1998) developed two effective and rapid detection methods for lactococcal phages present in milk or whey samples based on ELISA test but, in these cases, using monoclonal antibodies against the denatured capsid protein of lactococcal phages belonging to the P335 and c2 species, respectively. With these methods, 10^7-10^8 UFC ml^{-1} were detected in 2 h without any steps to enrich or isolate bacteriophages. Evidently, methods with such sensitivity levels will not be adequate for on-line monitoring of phages in dairy plants. To solve this limitation, a concentration step (polyethylene glycol precipitation or immunomagnetic separation) was suggested (Chibani Azaïez et al. 1998), extending consequently the analysis time.

1.3. DNA-Based Methods

a) DNA-DNA Hybridization

DNA-homology studies compare the genome of two bacteriophages (a labelled DNA that acts as probe, and DNA from the sample immobilized in a membrane) and indicate whether or not they are genetically related. This method is based on a dot blot on a nylon membrane, followed by hybridization with DNA probes. In 1984, Jarvis correlated DNA-DNA hybridization with morphological classification of lactococcal (in the past named lactic streptococcal) phages. The lack of homology among the four proposed phage species demonstrated that they do not have a recent common phage ancestor and that one phage type is not derived by mutation from another. Instead this classification was evolved – 10 groups are nowadays recognized for *Lactococcus* phages (Deveau et al. 2006) – this report established an important basis to classify phages of industrial relevance and to utilize this methodology for phage detection from dairy samples. In a similar study from the same author (Jarvis 1984b), DNA-DNA hybridization was used to show homology between temperate and lytic lactococcal bacteriophages. Instead many *Lactococcus* strains have been shown to be lysogenic, the results from this study indicated that virulent phage DNA showed partial similarity with DNA from lysogenic lactococcal strains, suggesting that the release of temperate phages from starter cells was unlikely to be the predominant source of virulent phages in cheese plants.

Moineau et al. (1992) adapted this methodology to design a rapid detection system from milk samples, by using two representative DNAs from c6A (prolate-heated) and 936 (isometric-heated) lactococcal phages as probes. In this work, phages were lysed *in situ* over the nylon membrane with a denaturing solution, prior hybridization. For skim or whole milk and whey samples, a previous clarification step to destabilize casein micelles and solubilize fat globules was included. After this treatment, 10^7 PFU per spot were detected. DNA-DNA hybridization does not require a biological assay with sensitive indicator test and a large number of samples may be processed in 24 h that represent the main advantages of this methodology.

b) Traditional PCR

Microbiological methods provide information about the sensitivity of a culture but do not discriminate phage species. On the other hand, the above described molecular and immunological techniques such as DNA-DNA hybridization and ELISA, respectively, can put on evidence the nature of the phage species but do not detect active lytic cycles. But the principal inconvenience of all these methodologies is that they take many hours, even days, to complete the analysis. Consequently, considering the fast replication cycle of most bacteriophages, the fermentation processes could abort, causing substantial economic losses, before the results of these types of test are ready. This last disadvantage, coupled with the ever-increasing knowledge on phage genomic sequences, has incited the development of molecular methods based on Polymerase Chain Reaction (PCR).

PCR, based on the amplification of DNA molecules, is a reaction extensively used to detect and identify viruses and bacteria from different environments, among numerous applications (Mullis and Faloona 1987). Since the last 15 years, traditional PCR has been adapted for the relatively rapid detection of dairy phages from different industrial samples (milk, cheese whey, fermented milks, etc.) with satisfactory detection limits (Brüssow et al. 1994a, Labrie and Moineau 2000, Binetti et al. 2005, Dupont et al. 2005, Quiberoni et al. 2006, Zago et al. 2006, del Río et al. 2007). For this conventional molecular reaction, a simple agarose gel electrophoresis must be done to verify the result, employing 4-6 h when industrial samples are directly used as templates. An additional benefit is that neither phage particle concentration (purification by precipitation, CsCl gradient, filtration, etc.) nor any procedure is required to enrich the samples. PCR, as well all molecular methods, cannot distinguish viable phage particles from DNA, but the presence of viral DNA could be an indication of the potential presence of infective virions.

Labrie and Moineau (2000) reported the first multiplex PCR system to detect, in one reaction, the presence of the three major lactococcal phage species directly from whey samples. Three sets of primers, one for each species, were designed based on conserved regions of their genomes, belonging to the major capsid protein gene (c2-species phages), the major structural protein gene (936-related phages) and a conserved gene of unknown function located in the replication module for P335-like phages. The detection limit of this method in whey was 10^4-10^7 PFU ml^{-1}, depending on the phage type. When an additional phage concentration step (phage filtration plus polyethylene glycol precipitation) was included, it reached 10^3-10^5 PFU ml^{-1}. The same primer set specific for P335-like phages was then employed to screen prophages in the genomes of lysogenic lactococcal strains isolated from a traditional Spanish cheese, showing that a relatively high proportion (66.6%) of the PCR-positive strain lysed after induction with mitomycin C (Martín et al. 2006). This report evidenced the additional value of PCR techniques to detect prophages or defective phages in bacterial genomes of strains commonly used as starters in industrial processes.

To enhance the sensitivity and specificity of conventional PCR method, Dupont et al. (2005) proposed a system that combines magnetic capture hybridization of phage receptor-binding protein (RBP) genes with a PCR amplification targeting a variable region of these genes. This method was designed based on the RBP sequences from 17 lactococcal bacteriophages belonging to the 936-species, allowing their classification into 6 different groups. Information about RBP type of phages allowed the prediction of their infectivity profiles, since a clear correlation between RBP gene sequences and host range was observed in most of the phages. This represents a great advantage in the selection of nontarget bacterial strains for the implementation of adequate programs of culture rotations in dairy plants. For the magnetic capture hybridization, a biotin-labelled capture probe was attached to streptavidin coated magnetic beads. These hybridized phage DNA beads were directly used as template for PCR analysis, by employing 6 primers set specific for RBP genes of the different lactococcal phage groups. This hybridization step eliminated inhibitors of PCR reaction present in whey and removed irrelevant phage and bacterial DNA, increasing the sensitivity and specificity of the method that achieved a detection limit of 10^2 – 10^3 PFU ml^{-1} of whey.

In case of *Strep. thermophilus* phages, that belong to an unique morphological group (*Siphoviridae*) and share high DNA similarity (Brüssow et al. 1994a), DNA-DNA hybridization is a method that allows bacteriophage detection but only gives information about their presence or absence, since their classification based on any criterion is not possible. In a first report, Brüssow et al. (1994a) designed a PCR detection method for *Strep. thermophilus* bacteriophages from cheese whey samples (detection limit, 10^3 PFU ml^{-1}), amplifying an extremely conserved DNA region, belonging to the replication module (Brüssow et al. 1994b). More recently, another PCR system was developed to detect and classify phages specific of this bacterial species based on the amplification of VR2, a variable region of the antireceptor gene (Binetti et al. 2005). The detection limit was 10^5 PFU ml^{-1} when phage-contaminated milk was used as template without any previous treatment. This method has a significant advantage over other grouping criteria, since the sequence of amplified fragments suggested a correlation between typing profiles and host ranges. This information could be valuable for the dairy industry by allowing a rational starter rotation system to be established (Binetti et al. 2005, Guglielmotti et al. 2009). The primers designed for these authors were then adapted for a multiplex PCR system that, in a single reaction, detects the presence of bacteriophages infecting *Strep. thermophilus* and *Lactobacillus delbrueckii*, plus the three more relevant species of *L. lactis* phages commonly found in dairy plants (P335, 936 and c2). The detection limits ranged from 10^3 to 10^4 PFU ml^{-1}, according to the phage detected (del Río et al. 2007). Another multiplex PCR-based method was developed to classify *Strep. thermophilus* phages into *cos*- or *pac*-types (Quiberoni et al. 2006). In this case, two sets of primers were designed, one for the *cos*-type and another for the *pac*-type group, based on conserved regions in the gene coding for the major structural proteins of specific phages. Although this method was not adapted to detect phages in dairy samples, it allows to clustering *Strep. thermophilus* bacteriophages according to their DNA packaging mechanisms, the most important classification criterion for these dairy phages (Le Marrec et al. 1993).

Lb. delbrueckii is a lactic acid bacteria species of great technological importance. The main subspecies, *Lb. delbrueckii* subsp. *bulgaricus* and *Lb. delbrueckii* subsp. *lactis* are widely used as starters in the fermentation of yoghurt and a great variety of cheeses, respectively (Reinheimer et al. 1996). For that reason, the adequate detection of *Lb. delbrueckii* phages represents another challenge to industry in order to minimize phage dissemination in dairy plants. A PCR-based system was designed and optimised from undefined whey starter cultures of Italian hard cheeses, based on the amplification of a highly conserved fragment of gene *g*17 coding for the major tail protein (MTP) of *Lb. delbrueckii* subsp. *lactis* phages (Zago et al. 2006). In this case, the detection limit was 10^5-10^6 PFU ml^{-1}, descending 3 log orders when a reamplification step was included. Instead the presence of an undefined number of phages coexisting with phage-resistant *Lb. delbrueckii* subsp. *lactis* strains in whey starter cultures for Grana and Padano cheeses are well known (Carminati et al. 1997), the possibility to efficiently detect the incidence of disturbing phages would be useful to prevent potential starter failures.

c) Real-Time Quantitative PCR (qPCR)

In the last years, the Polymerase Chain Reaction has become an essential tool for molecular biologists and its introduction into nucleic acids detection systems has revolutionized the quantitative analysis of DNA and RNA. The technique has rapidly evolved over the last few years and the growing interest in quantitative applications of the PCR has favored the development of real-time quantitative PCR (qPCR), based on four technologies that enable specific sequence detection and quantification (SYBR™ Green I, TaqMan™ probes, Hybridization probes and Molecular Beacon™ probes; Tse and Capeau 2003, Kubista et al. 2006). The field of dairy bacteriophages was not indifferent to this progress due to the increasing demand for quantitative, more sensitive and faster analysis to detect and enumerate phages from dairy samples that has encouraged the design of specific real-time qPCR methods. The first system of this kind of technology was described by del Río et al. (2008), whom developed a fast real-time multiplex PCR to detect and identify *cos*- and *pac*-type *Strep. thermophilus* phages in milk samples. In this case, TaqMan™MGB (Minor Groove Binder) probe technology was applied to allow a rapid (no more than 30 min) detection, without previous or subsequent sample treatment. Taking into account that the monitoring of amplification occurs in real-time, the analysis times could even be shorter in case of positive results. Additionally, this method was able to classify *Strep. thermophilus* bacteriophages in one of the two groups that were established based on the DNA packaging mechanism (Le Marrec et al. 1993). This it was possible by using different fluorescent dyes (VIC and FAM) as fluorophores for each TaqMan probe that let the simultaneous amplification of both types of phages, being possible to detect 10 copies per reaction, with a quantification limit of 10 PFU per reaction (10^4 PFU ml^{-1} of milk). A similar method was designed by the same research group to specifically detect *Lb. delbrueckii* phages, based on the amplification of the highly conserved *mur* gene, encoding for the muramidase of bacteriophage LL-H (Martín et al. 2008). Since the method was mainly assayed on *Lb. delbrueckii* subsp. *bulgaricus* phages, yoghurt process would be the first beneficiary for routine evaluation of milk samples. However, the high similarity of the *mur* (LL-H, *Lb. delbrueckii* subsp. *lactis*) and *lys*A (mv1 and mv4 phages, *Lb. delbrueckii* subsp. *bulgaricus*) genes could also help cheese and dairy starters industries. The detection limit for this assay was 10 copies per reaction, with a quantification limit of 100 PFU per reaction, equivalent to 10^5 PFU ml^{-1} of milk. Since phage concentrations in milk or whey bellow 10^5 PFU ml^{-1} are not considered a threat to industrial fermentations (Suárez et al. 2002), both proposed qPCR methods could be used to guarantee minimum quality requirements of the milk destined to fermentation processes in which *Strep. thermophilus* and/or *Lb. delbrueckii* strains are implicated.

Another application of qPCR technique is the analysis of temperate phage biology to reveal the impact of temperate phages and lysogenic bacteria in diverse ecological fields. Lunde et al. (2005) analyzed the prophage stability in lysogenic *L. lactis* subsp. *cremoris* host strains, when they were growing under different conditions, related to the fermentation processes in dairy processes: temperature, nutrient availability, osmotry and acidity. The frequency of spontaneous ϕLC3 induction was determined by qPCR, by amplifying a fragment of the DNA attachment site (*attB*) on the bacterial chromosomes in a population of lysogenic cells. A variation of prophage stability with all factors estudied was demonstrated but the environmental growth conditions were the decisive factors on induction frequency. In a similar report, Soberón et al. (2007) described a procedure to analyze genotoxicity by the detection of prophage excision from the host chromosome of gram-positive (*Lb. casei*) and gram-negative (*Escherichia coli*) lysogenic bacteria. In this case, two covering oligonucleotides flanking the phage attachment site (*attP*) sequence of a temperate phage were designed to amplify the inverting DNA segment by qPCR. No amplification is expected from an uninduced lysogenic culture, since *attP* half-sites are separated and oppositely orientated.

Also, qPCR may be useful for detection of phages in all the niches of the dairy industry. In this sense, in a very recent and interesting study, qPCR was adapted to compare the efficiency of different aerosol samplers in recovering c2- and 936-like lactococcal phages in a Canadian cheese plant (Verreault et al. 2011) In this assay, SYBR™ Green fluorescence and primers specific to conserved regions were used to amplify and quantify the DNA from the samples, including air samples. The targeted regions belonged to the *orf*6 of phage P008, a gene coding for a highly conserved structural protein of 936-group phages, and the gene coding for the highly conserved major capsid protein for c2-group phages. Instead contamination sources and dissemination routes of viral particles in cheese plants are not easy to identify, the presence of lactococcal phages was demonstrated by detecting, approximately, 10^4 phage genomes per cubic meter of air. These results emphasize the importance of phage dissemination that occurs in dairy plants during the elaboration processes, underlying the requirement of adequate ventilation and airflow control to minimize this problem. The advantage of these kinds of air samplers combined with qPCR phage detection is based on the possibility to collect and detect both entire and deteriorate phage particles, by using a molecular method that is independent of phage infectivity and of bacterial hosts. As occurs for all molecular detection methods, it is inaccurate to assume a strict correlation between phage DNA concentration and PFU m^{-3} of air. Nevertheless, this study clearly demonstrated that phage particles could be found on diverse surfaces (floors, walls, cleaning materials, etc.), since their genomes were detected. Instead phage particles could be inactivated during the sampling/elution procedure, it is safe to presuppose that these DNA molecules were originated from active viral particles, representing a potential serious risk in fermentation processes.

Real-time qPCR assays have the advantage over classical PCR methods of being able to real-time monitoring phage DNA increase, which would be related to bacteriophage replication, ensuring the detection of infectious bacteriophages. However, it represents a laborious and expensive technique that must be conducted by technical expertise, which makes the implementation of the method for routine analysis in industry especially difficult. Moreover, another major inconvenient common to all DNA-based detection methods is that they can only detect bacteriophages whose genome sequences are available.

d) Electron Microscopy- Based Methods

From electron microscopy studies it is possible to classify bacteriophages into different morphological groups, according to Bradley criterion (1967). By this observation it would be possible to assume if the same bacteriophage is the responsible of fermentation failures when repeated infection episodes are registered in a dairy process. The sample suspected to contain phages must be deposit on a carbon-coated Formvar grid for the electron microscope and negative staining must be done, using uranylacetate or phosphotungstate. This method is very simple, when an electron microscopy is available, but it is mainly destined for research purposes (IDF Bulletin 129 1980) (for a review see Chapter 1).

2. Indirect Detection Methods

2.1. Traditional Indirect Methods

One of the methods most commonly implemented for routine analysis in dairy plants is the activity test that evaluates the presence of bacteriophages in a sample as a decrease in acid production (compared with a phage-free control sample) by a starter or strain culture in sterile, steamed or pasteurized milk. If the acid production in the sample is reduced by more than 10 %, or if the pH difference between a sample and a control is higher than 0.2 pH units, the presence of bacteriophage is suspected (IDF Bulletin 129 1980). Important limitations must be considered for this assay, since after 20 h of incubation, the cultures generally show normal acidification levels and it is possible that no phage infection can be detected. This is particularly usual for mixed strain starters where phage-insensitive strains will continue to grow and become dominant.

The Pearce test is a modified activity test where the situation during a cheese (originally Cheddar cheese) manufacture is simulated (Pearce 1969). Filtered samples are submitted to a cooking temperature profile similar to that used during the elaboration process, and the test can predict the probability of phage propagation during a particular cooking procedure.

Other detection methods, named indicator tests, are based on the reduction of an indicator (generally, methylene blue or bromchresol purple) in milk, due to culture acidification in presence and absence of sample filtrate (Steeson and Klaenhammer 1985). If bacteriophages are present in the sample, a time delay or a failure in the colour change is observed, obeying to a delay or a failure in milk acidification, respectively. As in the activity test, mixed cultures may mask the presence of phages, producing false negative results. Moreover, indicator methods are less sensitive than titration and pH measurements and, in consequence, low phage concentrations may be undetected. When a large number or pure strains must be tested against one or several filtrate samples, indicator test may be adapted to microtitre plates, giving rise to the microtitre plate method (IDF Bulletin 263 1991).

2.2. Flow Citometry- Based Methods

The first symptom of a phage infection on a bacterial culture can be evaluated by means of morphological modifications of single cells, as a consequence of the metabolic shutdown of bacterial machinery. This novel methodology to indirectly detect bacteriophages by flow citometry allows monitoring the fermentation process during a lytic cycle. In case of *L. lactis* strains, Michelsen et al. (2007) observed that in the first step of phage attack, the cell growth ceased while cell divisions continued, causing the typical lactococcal chains to be broken up. The lactococcal chains were shorten (2-4 cells) as a consequence of a delay in cell division. For phage P008 and *L. lactis* subsp. *lactis* IL-1403, the phage infection was evidenced 60 min after the viral suspension was added to the culture, which is equivalent to 4×10^5 PFU ml^{-1}. These values vary according to the burst size of a particular phage and, in consequence, the detection limit depends on the phage-host system. Instead, the results are comparable with that obtained by traditional PCR method (Labrie and Moineau 2000), but as the flow citometry analysis can be performed in real time, it allow to take a rapid decision when an infection episode is verified. Moreover, this procedure can be applied from mixed cultures, as the detection is independent of phage and strain. Flow citometry is only based on the detection of cells with low mass that are found late in the lytic cycle. It represents a great advantage when a phage attacks one of the strains integrating a mixed culture, since the reduction in acidification rate may be rarely corroborated as the resistant strains in the culture follow their normal growth. In this report, these cells could be detected when they were present as 0.2 % of the total cells that indicates that a proportion of 1-2 % of the cells are infected, independently of the number of strains integrating the starter and of which phage infected the culture.

2.3. Impedimetric Methods

Another alternative activity test is based on the detection of changes in the electrical impedance or conductance of the milk, due to a decrease in lactic acid production when a phage infection occurs (IDF Bulletin 263 1991). The advances in the last decades in the biosensor technology have improved and simplified the detection of multiple analytes (including bacterial cells) in complex samples by the development of different electrochemical, piezoelectric, magnetic, and optical biosensors. A recent work (García-Aljaro et al. 2009) reports a rapid detection method that analyzes the impedance changes occurred upon infection of a host-biofilm established onto metal (platinum and gold) microelectrodes by non-faradaic impedance spectroscopy. In this case, an *E. coli* phage and its host were chosen as models, but the methodology would be applicable to any dairy bacteriophage of industrial relevance, as long a suitable bacterial host is able to grow on the microelectrode surface. This wide application represents a great benefit when compared with other detection methodologies that were specifically designed to detect a particular type of phage. Impedimetric changes occurring at the biosensor surface, as consequence of phage infection and subsequent cell lysis, were monitored over a 6 h period after the initial inoculation of phage in milk samples. The simplicity and the rapidity of this assay, as well as the possibility of miniaturization increase the utility and versatility of this system.

Conclusion

Usually, milk is examined for phages using standard microbiological methods that allow the detection of active phage particles and the discrimination between phage and nonphage inhibitors, but these assays are time-consuming and require the availability of a sensitive strain. In the last years, the technological advances have improved and simplified the detection of dairy bacteriophages, by developing direct and indirect systems based on different principles (DNA-DNA hybridization, traditional PCR, qPCR, flow citometry, impedance, etc.). Among them, the molecular methods and, particularly, qPCR appears today as the most promising assay for the detection and quantification of viral particles (as DNA molecules) in dairy samples and dairy environment (even air). A limitation for its application, however, is the requirement of special equipment, stringent laboratory precautions and expert personnel. However the increasing availability of dairy phage sequences in database as the growing demand for more sensitive and rapid detection procedures would prompt the routine use of qPCR in dairy industries.

References

Binetti, A. G. and Reinheimer, J. A. (2000). Thermal and chemical inactivation of indigenous *Streptococcus thermophilus* bacteriophages isolated from Argentinian dairy plants. *Journal of Food Protection, 63,* 509–515.

Binetti, A. G., del Río, B., Martín, M. C. and Álvarez, M. A. (2005). Detection and characterization of *Streptococcus thermophilus* bacteriophages based on the antireceptor gene sequence. *Applied and Environmental Microbiology, 71,* 6096–6103.

Binetti, A. G., Suárez, V., Tailliez, P. and Reinheimer, J. (2007). Characterization of spontaneous phage-resistant variants of *Streptococcus thermophilus* by randomly amplified polymorphic DNA analysis and identification of phage-resistance mechanisms. *International Dairy Journal, 17,* 1115–1122.

Binetti, A. G., Quiberoni, A. Reinheimer, J. (2002). Phage adsorption to *Streptococcus thermophilus*. Influence of environmental factors and characterization of cell-receptors. *Food Research International, 35,* 73–83.

Bradley, D. (1967). Ultrastructure of bacteriophages and bacteriocins. *Bacteriological Reviews, 31,* 230–314.

Bruttin, A., Desiere, F., d'Amico, N., Guérin, J.-P., Sidoti, J., Huni, B. and Brüssow. H. (1997). *Applied and Environmental Microbiology, 63,* 3144–3150.

Brüssow, H., Frémont, M., Bruttin, A., Sidoti, J., Constable, A. and Fryder, V. (1994). Detection and classification of *Streptococcus thermophilus* bacteriophages isolated from industrial milk fermentation. *Applied and Environmental Microbiology, 60,* 4537–4543.

Brüssow, H., Probst, A., Fremont, M. and Sidoti, J. (1994). Distinct *Streptococcus thermophilus* bacteriophages share an extremely conserved DNA fragment. *Virology, 200,* 854-857.

Capra, M. L., Quiberoni, A. and Reinheimer, J. A. (2004). Thermal and chemical resistance of *Lactobacillus casei* and *Lactobacillus paracasei* bacteriophages. *Letters in Applied Microbiology, 38,* 499–504.

Carminati, D., Mazzucotelli, L., Giraffa, G. and Neviani, E. (1997). Incidence of inducible bacteriophage in *Lactobacillus helveticus* strains isolated from natural whey starter cultures. *Journal of Dairy Science, 80,* 1505–1511.

Chibani Azaiëz, S. R., Fliss, I., Simard, R. E. and Moineau, S. (1998). Monoclonal antibodies raised against Nnative major capsid proteins of lactococcal c2-like bacteriophages. *Applied and Environmental Microbiology, 64(11),* 4255–4259.

Coffey, A. and Ross, R. P. (2002). Bacteriophage-resistance systems in dairy starter strains: molecular analysis to application. *Antonie van Leeuwenhoek, 82,* 303–321.

del Río, B., Binetti, A. G., Martín, M. C., Fernández, M., H. Magadán, A. and Álvarez, M. A. (2007). Multiplex PCR for the detection and identification of dairy bacteriophages in milk. *Food Microbiology, 24,* 75–81.

del Río, B., Martín, M. C., Martínez, N., H. Magadán, A. and Álvarez, M. A. (2008). Multiplex fast real-time PCR for quantitative detection and identification of *cos-* and *pac-*type *Streptococcus thermophilus* bacteriophages. *Applied and Environmental Microbiology, 74,* 4779–411.

Deveau, H., Labrie, S., Chopin, M.-C. and Moineau, S. (2006). Biodiversity and Classification of Lactococcal Phages. *Applied and Environmental Micro biology,72(6),*4338–4346.

Dupont, K., Vogensen, F.K. and Josephsen, J. (2005). Detection of lactococcal 936-species bacteriophages in whey by magnetic capture hybridization PCR targeting a variable region of receptor-binding protein genes. *Journal of Applied Microbiology, 98,* 1001–1009.

García-Aljaro, C., Muñoz-Berbel, X. and Muñoz, F. (2009). On-chip impedimetric detection of bacteriophages in dairy samples. *Biosensors and Bioelectronics, 24,* 1712–1716.

Guglielmotti, D., Binetti, A. G, Reinheimer, J. A. and Quiberoni, A. (2009). *Streptococcus thermophilus* phage monitoring in a cheese factory: phage characteristics and starter sensitivity. *International Dairy Journal, 19,* 476–480.

Guglielmotti, D., Reinheimer, J., Binetti, A. G., Giraffa, G., Carminati, D. and Quiberoni, A. (2006). Characterization of spontaneous phage-resistant derivatives of *Lactobacillus delbrueckii* commercial strains. *International Journal of Food Microbiology, 111,* 126–133.

Hull, R. (1977). Methods for monitoring bacteriophages in cheese factories. *Australian Journal of Dairy Technology, 32,* 63–64.

IDF Bulletin N° 129. (1980). Starters in the manufacture of cheese, 5–15.

IDF Bulletin N° 263. (1991). Methods for phage monitoring, 29–39.

Jarvis, A. (1984). DNA–DNA homology between lactic streptococci and their temperate and lytic phages. *Applied and Environmental Microbiology, 47(5),* 1031–1038.

Jarvis, A. W. (1984). Differentiation of lactic streptococcal phages into phage species by DNA–DNA homology. *Applied and Environmental Microbiology, 47,* 343–349.

Jarvis, A. W. (1989). Bacteriophages of lactic acid bacteria. *Journal of Dairy Science, 72,* 3406–3428.

Keogh, B. (1980). Appraisal of media and methods for assay of bacteriophages of lactic streptococci. *Applied and Environmental Microbiology, 40(4),* 798–802.

Kubista, M., Andrade, J., Bengtsson, M., Forootan, A., Jonák, J., Lind, K., Sindelka, R., Sjöback, S., Sjögreen, B, Strömbom, L., Stahlberg, A. and Zoric, N. (2006). The real-time polymerase chain reaction. Review. *Molecular Aspects of Medicine, 27,* 95–125.

Labrie, S. and Moineau, S. (2002). Complete genomic sequence of bacteriophage ul36: demonstration of phage heterogeneity within the P335 quasispecies of lactococcal phages. *Virology, 296,* 308–320.

Le Marrec, C., van Sinderen, D., Walsh, L., Stanley, E., Vlegels, E., Moineau, S., Heinze, P., Fitzgerald, G. and Fayard., B. (1997). *Streptococcus thermophilus* bacteriophages can be divided into two distinct groups based on mode of packaging and structural protein composition *Applied and Environmental Microbiology, 63,* 3246–3253.

Lembke, J. and Teuber, M. (1979). Detection of bacteriophages in whey by an enzyme-linked immunosorbent assay (ELISA). *Milchwissenschaft, 34,* 457–458.

Lillehaug, D. (1997). An improved plaque assay for poor plaque-producing temperate lactococcal bacteriophages. *Journal of Applied Microbiology, 83,* 85–90.

Lunde, M., Aastveit, A., Blatny, J. and Nes, I. (2005). Effects of diverse environmental conditions on ϕLC3 prophage stability in *Lactococcus lactis. Applied and Environmental Microbiology, 71(2),* 721–727.

Martín, M. C., del Río, B., Martínez, N., H. Magadán, A. and Álvarez, M. A. (2008). Fast real-time polymerase chain reaction for quantitative detection of *Lactobacillus delbrueckii* bacteriophages in milk. *Food Microbiology, 25(8),* 978–82.

Martín, M. C., Ladero, V. and Álvarez, M. A. (2006). PCR identification of lysogenic *Lactococcus lactis* strains. *Journal für Verbraucherschutz und Lebensmittelsicherheit, 1,* 121–124.

Michelsen, O., Cuesta-Domínguez, A., Albrechtsen, B. and Jensen, A. (2007). Detection of bacteriophage-infected cells of *Lactococcus lactis* by using flow cytometry. *Applied and Environmental Microbiology, 73(23),* 7575–7581.

Moineau, S., Bernier, D., Jobin, M., Hébert, J., Klaenhammer, T. R. and Pandian, S. (1993). Production of monoclonal antibodies against the major capsid protein of the *Lactococcus* bacteriophage ul36 and development of an enzyme-linked immunosorbent assay for direct phage detection in whey and milk. *Applied and Environmental Microbiology, 59,* 2034–2040.

Moineau, S., Fortier, J. and Pandian, S. (1992). Direct detection of lactococcal bacteriophages in cheese whey using DNA probes. *FEMS Microbiology Letters, 92,* 169–174.

Mullan, W. M. (1979). Lactic streptococcal bacteriophage enumeration. A review of factors affecting plaque formation. *Dairy Industries International, 44,* 11–15.

Mullis, K. and Faloona, F. (1987). Specific synthesis of DNA in vitro via a Polymerase-catalyzed Chain Reaction. *Methods in Enzymology, 155,* 335–350.

Pearce, L. (1969). Starters and phage. Activity tests for cheese starter cultures. *New Zealand Journal of Dairy Science and Technology, 4,* 246–247.

Quiberoni, A., Moineau, S., Rousseau, G., Reinheimer, J. and Ackermann, H.-W. (2010). *Streptococcus thermophilus* bacteriophages. Review. *International Dairy Journal, 20,* 657–664.

Quiberoni, A., Suárez, V. B. and Reinheimer, J.A. (1999). Inactivation of *Lactobacillus helveticus* bacteriophages by thermal and chemical treatments. *Journal of Food Protection, 62,* 894–898.

Quiberoni, A., Tremblay, D., Ackermann, H.-W., Moineau, S. and Reinheimer, J. A. (2006). Diversity of *Streptococcus thermophilus* phages in a large production cheese factory in Argentina. *Journal of Dairy Science, 89,* 3791–3799.

Reinheimer, J. A., Quiberoni, A., Tailliez, P., Binetti, A. G. and Suárez, V. B., 1996. The lactic acid microflora of natural whey starters used in Argentina for hard cheese production. *International Dairy Journal, 6*, 869–879.

Sing, W. D. and Klaenhammer, T. R. (1993). A strategy for rotation of different bacteriophage defenses in a lactococcal single-strain starter culture system. *Applied and Environmental Microbiology, 59,* 365–372.

Soberón, N., Martín, R. and Suárez, J. E. (2007). New method for evaluation of genotoxicity, based on the use of real-time PCR and lysogenic gram-positive and gram-negative bacteria. *Applied and Environmental Microbiology, 73(9),* 2815–2819.

Sozzi, T., Maret, R. and Poulin, J. M. (1976). Study of plating efficiency of bacteriophages of thermophilic lactic acid bacteria on different media. *Applied and Environmental Microbiology, 32(1),* 131–13.

Steeson, L. and Klaenhammer, T. (1985). *Streptococcus cremoris* M12R trasnconjugants carrying the conjugal plasmid pTR2030 are insensitive to attack by lytic bacteriophages. *Applied and Environmental Microbiology, 50,* 851-858.

Sturino, J. and Klaenhammer, T. (2004). Bacteriophage defense systems and strategies for lactic acid bacteria. *Advances in Applied Microbiology, 56,* 331–378.

Suárez, V. B. and Reinheimer, J. A. (2002). Effectiveness of thermal treatments and biocides in the inactivation of Argentinian *Lactococcus lactis* phages. *Journal of Food Protection, 65,* 1756–1759.

Suárez, V. B., Quiberoni, A., Binetti, A. G. and Reinheimer, J. A. (2002). Thermophilic lactic acid bacteriophages isolated from Argentinean dairy industries. *Journal of food Protection, 65,* 1597–1604.

Tse, C. and Capeau, J. (2003) Quantification des acides nucléiques par PCR quantitative en temps réel. *Annales de Biologie Clinique, 61,* 279–93.

Verreault, D., Gendron, L., Rousseau, G., Veillette, M., Massé, D., Lindsley, W., Moineau, S. and Duchaine, C. (2011) Detection of airborne lactococcal bacteriophages in cheese plants. *Applied and Environmental Microbiology, 77(2)*, 491–0497.

Zago, M., De Lorentiis, A., Carminati, D., Comaschi, L. and Giraffa, G. (2006). Detection and identification of *Lactobacillus delbrueckii* subsp. *lactis* bacteriophages by PCR. *Journal of Dairy Research, 73,* 1–8.

In: Bacteriophages in Dairy Processing
Editor: Andrea del Lujan Quiberoni et al.
ISBN 978-1-61324-517-0

Chapter 12

PLASMID TRANSDUCTION AND SITE-SPECIFIC SYSTEMS DERIVED FROM LACTIC ACID BACTERIA PHAGES

María J. Olaya Passarell, María C. Aristimuño Ficoseco, Elvira M. Hébert, and Raúl R. Raya*

Centro de Referencia para Lactobacilos (CERELA, CONICET),
Chacabuco, SM Tucumán, Argentina

ABSTRACT

Bacteriophages have always been central for the evolution of bacteria. In this chapter, the transductive capacities of phages to mediate the transfer of either chromosomal or plasmid DNA among strains of lactic acid bacteria, as well as the development of special-purpose genetic tools (i.e., site-specific integration and inducible gene expression systems) derived from either temperate or virulent bacteriophages and their successful exploitation in strain improvement will be presented. The contribution of bacteriophages to the evolution and shaping of lactic acid bacteria will be discussed.

INTRODUCTION

Lactic acid bacteria (LAB) are a diverse and useful group of bacteria that have been used for centuries in the preservation and production of fermented foods of plant and animal origins. One of the most critical problems in these processes is the contamination of LAB starters by bacteriophages (phages) that cause bacterial lysis and significant economic losses. The isolation and deep characterization of many virulent and temperate LAB phages has not only brought information about the origin of lytic phages in the fermentation industry, but also contributed to the identification and characterization of multiple phage defense systems,

* E-mail address: rraya@cerela.org.ar. Tel.: +54 381 431 0465; fax: +54 381 400 5600

particularly of the clustered regularly interspaced short palindromic repeats (CRISPR) system, and significantly enriched comparative phage genome studies, phage taxonomy, and phage-bacteria evolution. Some LAB phage particles have been used as transducing vehicles in the development of gene transfer systems. Furthermore, phage genetic elements have been used in developing invaluable molecular tools (i.e., site-specific integration vectors) to deliver and stabilize genes in the LAB genome (Josephsen and Neve 2004, Brøndsted and Hammer 2006, Brüssow 2006, Brüssow and Desiere 2006, Brüssow and Suárez 2006, Emond and Moineau 2007).

Transduction

The transfer of tryptophan independence in *Streptococcus diacetilactis* 18-16 (Sandine et al. 1962) and streptomycin resistance in *Streptococcus lactis* C2 (Allen et al. 1963) by means of a virulent bacteriophage were the first description of transduction among strains of lactococci. Several mutants unable to ferment maltose, mannose, or lactose were also transduced to the respective carbohydrate-positive phenotypes by temperate phages (McKay et al. 1973, McKay and Baldwin 1974, Molsknes et al. 1974, McKay et al. 1976, McKay and Baldwin 1978, Davies and Gasson 1981, Snook et al. 1981). These experiments showed that transduction of the chromosomally-encoded maltose and mannose genetic markers were greater than that observed with lactose metabolism. However, only the lactose transductants exhibited high-frequency transfer in repeated transduction experiments. It was also shown that proteinase activity could be co-transduced with the lactose by the temperate phage induced from the C2 strain (McKay and Baldwin 1974, Klaenhammer and McKay 1976). The transducing phage discovered by McKay and colleagues was the first reliable genetic transfer system described in dairy lactococci, which made possible to examine the causes for the spontaneous loss of lactose metabolism and proteinase activity observed in these starter cultures and mainly allowed to establish a link between the mentioned phenotypes to extrachromosomal (plasmid) DNA.

In addition to lactococci, transduction has also been demonstrated in lactobacilli (Tohyama et al. 1971, Luchansky et al. 1989, Raya et al. 1989, Ravin et al. 2006) and *Streptococcus thermophilus* (Mercenier et al. 1988, Heller et al. 1996, Ammann et al. 2008). Generalized transduction mediated by the *Lactobacillus salivarius* temperate bacteriophage PLS-1 (Tohyama et al. 1971) and plasmid mediated by *Lactobacillus gasseri* bacteriophage øadh and the lytic *Lactobacillus delbrueckii* phage LL-H are, so far, the three reports on transduction in lactobacilli. Southern blot hybridization of the DNA packaged in the capsids of the temperate *Lactobacillus delbrueckii* subsp. *bulgaricus* phage mv4 provided evidence for the presence of bacterial DNA, which suggests phage mv4 must be capable of generalized transduction (Lahbib-Mansais et al. 1992). Several plasmids were transduced at low frequencies in *Strep. thermophilus* by means of the virulent phages ø56α and ø17α, being the highest frequency observed with plasmid pTG219, a *Strep. thermophilus* plasmid that persists as multimers in its original host (Mercenier et al. 1988). Recently, Ammann et al (2008) provided evidence that plasmid DNA can also be transferred via transduction from *Streptococcus thermophilus* to *Lactococcus lactis*. The theta-replicating plasmid pAG106AE was transduced in *L. lactis* Bu2-60 using three virulent *cos*-site *Strep. thermophilus* phages (P1109, P53 and Pa10/J9). The low transduction frequencies of all these phages, were about 4 orders of magnitude lower in *L. lactis* Bu2-60 than in *Strep. thermophilus* strain a10, and correlated well with the low efficiency of adsorption of these phages to *L. lactis* Bu2-60 (46% adsorption or lower).

Table 1 shows some examples of plasmid DNA transduction systems described in LAB. Although each system has particular characteristics, an analysis of the different transduction systems described so far, both in LAB and other bacterial genera (Vollenweider et al. 1980, Schmidt and Schmieger 1984, Deichelbohrer et al. 1985, Novick et al. 1986, Viret and Alonso 1988, Viret et al. 1991), indicates that plasmid DNA transduction can be enhanced by a factor of 10^2 to 10^6 over the transduction level of the plasmid alone by providing DNA-DNA homology between the plasmid and the phage genomes (in Table 1, see the transduction frequencies of plasmids pGK12 and pTRK170, a pGK12-plasmid derivative containing two *Bgl*II phage øadh-fragments) or by incorporating specific phage packaging signals into the plasmid vector (i.e., *cos-* or *pac*-site, or phage-*ori* functions). Studies of the role of *cos*-site in the transduction mechanism were conducted by Birkeland and Holo (1993) and Chandry et al. (2002): plasmids carrying the cohesive end region from temperate lactococcal bacteriophage φLC3 and sk1, respectively, could be packaged by their respective phages and transduced into their host strains at high frequencies (between 10^{-4} and 10^{-3} transducing particles per PFU; Table 1). Similar results were observed by Damelin et al. (2010) with the *cos*-site of phage øadh; this study also successfully determined the feasibility of utilizing transduction for the expression of two HIV co-receptor antagonists, the CC chemokines CCL5 and CCL3, in *Lactobacillus gasseri*, a predominant vaginal *Lactobacillus* species. In contrast, the high frequency of transduction observed with the LL-H transducing system (Table 1) was attributed to the presence of a *pac*-site or a phage LL-H-homologous replication origin naturally present in the transduced plasmid pX3 (Ravin et al. 2006). Transducing particles were shown to be defective; they encapsidated the transduced plasmid either as monomer (McKay et al. 1976), as high-molecular-weight multimers (Chandry et al. 2002, Ravin et al. 2006, Ammann et al. 2008, Damelin et al. 2010), or as a phage::plasmid co-integrate (Raya et al. 1992). The formation of high-molecular-weight multimers might be induced by the actual insertion of the phage DNA fragment into the plasmid vector (Gruss and Ehrlich 1988); alternatively, upon infection, phage products inhibit RecBC-like DNAases and induce rolling-circle plasmid replication to form a long concatemeric molecule which can be packaged in mature virions (Viret and Alonso 1987). The transduction of "natural" plasmid DNA can also involve the formation of co-integrates mediated by an insertion sequence, homologous recombination through a region of natural homology between the plasmid and phage DNAs, or illegitimate recombination. Co-integrates are then packaged in mature virions and, after transduction, either co-integrates resolve into the monomeric forms or stable co-integrates are found (Shipley and Olsen 1975). In the transduction systems evaluated in LAB, the transduced plasmid DNAs were found as monomers indistinguishable from those of the donor strain. However, this was not the case with pAG106AE; this plasmid, transduced from *Strep. thermophilus* to *L. lactis* Bu2-60 by the virulent *Strep. thermophilus* phage P1109, was isolated from *L. lactis* Bu2-60 transductants as concatemers; it was suggested that the resolvase gene pSt106 of plasmid pAG106AE was poorly expressed in *L. lactis*, since concatemers were also observed when pAG106AE was transferred to Bu2-60 cells via electroporation (Ammann et al. 2008).

Table 1. Examples of plasmid-DNA transduction systems defined in LAB

Phage	Plasmid Transduced	Main host	Frequency[a,b]	Reference
Φadh (temperate; *cos*)	pGK12 (Cm) pTRK170 (Cm; pGK12:: φadh *Bgl*II 1.4 + 0.8-kb fragments) pGK12::φadh *cos* fragment	*Lb. gasseri* ADH	8.7×10^{-9} 1.1×10^{-3} - 4.6×10^{-5} 4.75×10^{-4}	Raya et al. 1989 Raya et al. 1992 Damelin et al. 2010
LL-H (virulent; *pac*)	 pX3 (Em) pJK650 (Em)	*Lb. delbrueckii* LKT ATCC 15808	10^{-2}-1^{b}	Ravin et al. 2006
sk1 (temperate; *cos*)	pMG24 (Km) pSC5029 (pMG24::sk1 *cos* fragment)	*L. lactis* HID113	~ 1×10^{-8} 2×10^{-3}	Chandry et al. 2002
φLC3 (temperate; *cos*)	pSA3 (Em) pSA-COS1 (pSA3:: ØLC3*cos* fragment)	*L. lactis* subsp. *cremoris* NCDO 1201	$<3.0 \times 10^{-7}$ 8×10^{-4}	Birkeland and Holo 1993
φ17α (virulent) φ56α (virulent)	pTG219 (Em) pVA797/pVA838	*Strep.* *thermophilus* AO54	10^{-2} ca. 10^{-6}	Mercenier et al. 1988
P1109 (virulent; *cos*)	pAG106AE (Em)	*Strep.* *thermophilus* a10 *L. lactis* Bu2-60	 1.2×10^{-2} 3.5×10^{-5}	Ammann et al. 2008

[a] Transductants/PFU.
[b] Frequency was defined as number of transductants/total survivors.

Also, a plasmid with a smaller size respect to the original was observed in C2 Lac^{+} transductants due to the fact that the temperate C2 phage is too small to encapsidate the entire original plasmid genome (a process termed "transductional shortening"; McKay et al. 1976).

The plasmid DNA transduction systems described in LAB have been used as an alternative to plaque assays to define the host range of a phage, and as a simple method to transfer plasmid DNA to several LAB cells which, otherwise, did not support lytic growth of the phage and for which an efficient transformation system had not yet been defined (Mercenier et al. 1988, Raya et al. 1992).

Most important, plasmid DNA transduction showed for the first time that phage-mediated horizontal gene transfer occurs in Gram-positive dairy starter cultures (Ammann et al. 2008); this could explain that the significant sequence homology observed among phages of *Strep. thermophilus* and *Lactococcus lactis* have evolved by direct exchange of genetic modules.

SITE-SPECIFIC INTEGRATIVE VECTORS

Integrative recombination of phage lambda provides the classical example of site-specific integration and excision of a temperate bacteriophage (Weisberg et al. 1983). Extensive studies of this system have allowed the characterization of its components in great detail and definition of the mechanism and regulatory circuits involved in the recombination process. This information provides a model which has facilitated the characterization of other genetic elements encoding site-specific recombination. Site-specific recombination is conservative since it does not involve synthesis or loss of DNA.

The reaction, mediated by a recombinase (integrase, resolvase, or invertase), takes place between two attachment sites, *att*P and *att*B, located on the phage and bacterial chromosomes, respectively. The resulting hybrid prophage sites, *att*L and *att*R, are themselves able to recombine to regenerate *att*P and *att*B, during excision of the viral genome. Each attachment site consists of a perfect homologous sequence (or "core" region) where strand exchange occurs, and two unique elements (P "arms") that flank the core region and contains the binding site for the recombinase and other accessory proteins.

The integration system of several temperate LAB phages, including øadh of *Lb. gasseri* (Raya et al. 1992), øLC3 of *L. lactis* (Lillehaug and Birkeland 1993, Lillehaug et al. 1997), Tuc2009 of *L. lactis* (van de Guchte et al. 1994), BK5-T of *L. lactis* (Boyce et al. 1995), mv4 of *Lb. delbrueckii* (Dupont et al. 1995), r1t of *L. lactis* (van Sinderen et al. 1996), TP901-1 of *L. lactis* (Christiansen et al. 1994, 1996), øg1e of *Lb. plantarum* (Kodaira et al. 1997), ø10MC of *Oenococcus oeni* (Gindreau et al. 1997), øO1205 of *Strep. thermophilus* (Stanley et al. 1997), Sfi21 of *Strep. thermophilus* (Bruttin et al. 1997), TP-J34 of *Strep. thermophilus* (Neve et al. 1998), A2 of *Lb. casei* (Alvarez et al. 1998), TPW22 of *L. lactis* (Petersen et al. 1999), øFSW of *Lb. casei* (Shimizu-Kadota et al. 2000), and JCL1032 of *Lb. delbrueckii* (Riipinen et al. 2007) have been functionally characterized. Phage TP901-1 encodes a resolvase-type integrase, while the rest of these characterized systems contain an integrase belonging to the tyrosine integrase family. The attachment site regions of TP901-1 do not show similarity to other members of the Int family, indicating that the integration system of TP901-1 is a unique system for phage integration (Christiansen et al. 1996). Site-specific recombination systems have proved invaluable as vehicles to introduce DNA sequences into the genomes of host cells. Their main use so far has been to study regulation of promoters using reporters genes, where their single copy mimics the chromosomal situation. Table 2 shows the site-specific integration vectors that have been constructed based on the temperate phage integration systems from LAB. These vectors will integrate cloned genes in one copy on a specific location in the chromosome, allowing for stabilization of heterologous or plasmid-encoded genes. The first of these integration vectors was based on the Φadh phage integrase gene and the corresponding *att*P site cloned onto pSA34, a derivative of pSA3 without the Gram replication origin (Raya et al., 1992).

Phages Tuc2009 and ØLC3 are small-isometric-headed, temperate bacteriophages that lysogenize *L. lactis* subsp. *cremoris* strains. Although both phages show considerable differences (in particular, phage ØLC3 has *cos*-site while phage Tuc2009 has a head-full mechanism of DNA-packaging), their *att*L, *att*R, and *att*B sequences are identical (Lillehaug and Birkeland 1993; van de Guchte et al. 1994), van de Guchte et al. (1994) demonstrated that the application of insertional vectors based on bacteriophage integration functions is not limited to the prophage-cured original host strain of the phage. These authors showed the integration of plasmid pIN1 (plasmid carrying the Tuc2009 *int-att*P region) at the *att*B site in the MG1363 chromosome demonstrating that this strain, although resistant to phage Tuc2009, contains the elements (*att*B, host factor) required for integration of the phage in the chromosome.

Christiansen et al. (1994) sequenced of *att*L, *att*R, *att*P, and *att*B regions of phage TP901-1 that infect *Lactococcus lactis* subsp. *cremoris* 3107 strain and identified the short core region 5′-TCAAT-3′ proposed in the recombination between the phage genome and the host chromosome. An integration vector with a fragment harboring a 6.5 Kb fragment *int-att*P of TP901-1 integrated in the chromosome DNA of all recipient strains *L. lactis* subsp. *cremoris* 3107, *L. lactis* MG1363 and LM0230. Brondsted and Hammer (1999) used this system to construct stable chromosomal single-copy transcriptional fusions in *L. lactis.* Two promoter-reporter integration vectors containing the reporter gene *gusA* or *lacLM*, encoding β-glucuronidase or β-galactosidase, respectively, were constructed. By transformation of *L. lactis* subsp. *cremoris* MG1363 containing the integrase gene on a replicating plasmid, the promoter-reporter integration vectors integrated with a high frequency site specifically into the chromosomal attachment site *att*B used by bacteriophage TP901-1 (Christiansen et al. 1994).

Alvarez et al. (1998) showed that the temperate bacteriophage A2 integrates its geneome in the $tRNA^{Leu}$ gene of its host *L. paracasei* ATCC 27092. Further studies using integration functions of bacteriophage A2 (*int-att*P) cloned in plasmid $pG^{+}host9$ (pEM39 plasmid) showed that the highly conserved nucleotide sequence of the $tRNA^{Leu}$ (UUG) gene among Gram positive bacteria could provide integration sites for bacteriophage A2 in other LAB. In contrast, integration of vector pEM39 in the *E. coli* genome took place in *rrnD* operon, located upstream of gene *rrsD* that encodes 16S rRNA but not at the *E. coli* $tRNA^{Leu}$ gene. These studies showed that the A2 derived integrative plasmids can integrate into the Gram positive and Gram negative bacteria, making them useful for heterologous gene transfer (Alvarez et al. 1998). As a general strategy to be used in the protection of LAB against temperate bacteriophages the A2 phage repressor gene (CI) was inserted in the integration vector pEM40 (Alvarez et al. 1998), and inserted in the genome of *L. casei* ATCC393 strain (Alvarez et al. 1999). The strain *L. casei* ATCC393 EM40::CI was completely resistant to bacteriophage A2 attack and able to ferment milk whether phage A2 was present or not. The integration of CI in the genome resulted stable during milk fermentation (selective pressure was not necessary) (Alvarez et al. 1999). Since this system cannot be used in industrial fermentations because pEM40 plasmid contains a resistance gene, the unwanted sequences were removed from the *L. casei* EM76::CI strain with the help of the β-recombinase enzyme (Cruz Martín et al. 2000) that catalyzes DNA resolution between two directly oriented *six* sites to generate a *L. casei* EM77::CI strain that loosed antibiotic resistance gene. Dupont et al. (1995) found that the attachment sites of phage mv4, that lysogenizes the chromosome of Lactobacillus delbrueckii LT4 strain, shared a 17 bp core region (GGACATGA-GAGGAATTA). The integrase protein of mv4 shows the greatest similarity in the C-terminal domain with the integrase of lactococcal temperate phage ØLC3 and with integrases of conjugative transposons

Table 2. Site-specific systems derived from temperate LAB phages

Phage	Integrative vector	Host	Integration site	Reference
ØLC3	pINT2(ts)	*L. lactis* subsp. *cremoris* IMN-C18	3´end of ORF of unknow funtion	Lillehaug and Birkeland 1993, Lillehaug et al. 1997
Tuc2009	pIN1 (Cm)	*L. lactis* subsp. *cremoris* MG1363 (strain resistant to phage Tuc2009 infection)	3´end of ORF of unknow funtion	van de Guchte et al. 1994
TP901-1	Two-plasmid system: pBF20 (*att*P; Cm) plus pLB65 (*int;* Em) also promoter-probe vectors with *gusA* (plasmid pLB85) and *lacLM* (pLB86) as reporter genes	*L. lactis subsp. cremoris MG1363* and LM0230 Also functional in: *E. coli* Mammalian cells	Fimbriae precursor	Christiansen et al. 1994, Brøndsted et al. 1999, Stoll et al. 2002
TPW22	pAP2 (Em) also two-plasmid system: pAP (*att*P; Em) plus pSMAC (*int;* Cm)	*L. lactis subsp. cremoris MG1363*	5´end of putative DNA ligase gene	Petersen et al. 1999
Φadh	pTRK182 (Em)	*Lb. gasseri* NCK102 (Φadh prophage-cured derivative)	n.d.	Raya et al. 1992
A2	pEM40 (Em) Also pEM76 (Em; *att*P + *int* + two directly oriented copies of *six*, the binding site for the ß-recombinase) plus plasmid pEM68 (Cm; ß-recombinase gene)	*Lb. casei* ATCC393 Also functional in: *L. lactis* several lactobacilli; *E. coli*	G⁺: $tRNA^{Leu}$ *E. coli*: *rrnD*	Alvarez et al. 1998, Cruz Martin et al. 2000
ΦFSW	pMSK742 (Em)	*Lb. casei* YIT9029 (ØFSW prophage-cured derivative)	3´end of putative glucose 6-phosphate isomerase gene	Shimizu-Kadota et al. 2000
mv4	pMC1 (Em)	*Lb. plantarum* LP80 Also in other LAB, including *Enteroccus faecalis* and *Streptococcus pneumoniae*	$tRNA^{Ser}$ secondary sites	Dupont et al. 1995, Auvray et al. 1997

n.d., not determined.

Tn*1545* of *Streptococcus pneumoniae* and Tn*916* of *Enterococcus faecalis* (Dupont et al. 1995). The integration of bacteriophages mv4 occurs in the $tRNA^{Ser}$ gene. On the other hand, Auvray et al. (1997) demonstrated that the entire $tRNA^{Ser}$ gene was not required for pMC1 (*att*P, *int* genes) integration since vector pMC1 integrated into *E. faecalis* and *S. pneumoniae* in which the *L. delbrueckii* subsp. *bulgaricus* $tRNA^{Ser}$ gene was not conserved. In these hosts, pMC1 integrated at secondary sites which did not correspond to tRNA genes. pMC1 integration into different sites illustrates the flexibility of the integrase.

BACTERIOPHAGE-BASED GENETIC VECTORS

The discovery of many naturally occurring phage resistance mechanisms in wild-type lactococcal strains and the increased knowledge of LAB phage biology have facilitated the generation of novel resistant starter strains containing artificial phage-resistance systems. Extensive research has been directed toward the development of such artificial phage-resistance systems through food grade techniques.

Table 3. Genetic systems derived from LAB phage genomes

Phage elements	Application
Repressor	controlled- thermoinducible gene expression (Nauta et al. 1996, 1997, Binishofer et al. 2002); block phage propagation (Alvarez et al. 1999, Madsen et al. 1999, Bruttin et al. 2002, Durmaz et al. 2002).
Lysin	cloned and purified enzymes used as cheese- ripening additives (Boizet et al. 1990, Henrich et al. 1995, Vasala et al. 1999) or to generate protoplasts (Watanabe et al. 1990); cloned lysin-holin systems: strains with a leaky phenotype for controlled accelerated cheese ripening (Shearman et al. 1992, de Ruyter et al. 1997, Sanders et al. 1997, Walker and Klaenhammer 2001).
Superinfection exclusion	control phage infections by heterologous phages: i.e., *sie*$_{2009}$ of the temperate *Lactococcus lactis* phage Tuc2009 (McGrath et al. 2002); the *ltp* gene of the temperate *Strep. thermophilus* phage TP-J34 (Sun et al. 2006); *sie* systems found in prophages of MG1363 and IL1403 strains (Mahony et al. 2008).
Origin of replication	construct phage-triggered defense systems ("Per" systems) (Foley et al. 1998, Hill et al. 1990, O´Sullivan et al. 1993, McGrath et al. 1999, Moscoso and Suárez 2000, Stanley et al. 2000).
Promoters	construct phage-triggered suicide systems: expression of the lethal *Lla*IR restriction enzyme under the control of a phage Ø31-inducible middle-expressed promoter (Djordjevic et al. 1997, Djordjevic and Klaenhammer 1997).
Other	antisense RNA strategies: cloning and expression in the antisense orientation of essential phage functions (i.e., the putative replisome organizer; the putative helicase loader gene). This strategy combines the use of a strong phage-inducible promoters and a low copy-number plasmid containing the origin of replication of a phage (Kim and Batt 1991, Kim et al. 1992, Batt et al. 1995, Walker and Klaenhammer 2000, McGrath et al. 2001, Sturino and Klaenhammer 2002).

These include mainly the use of antisense strategies; utilization of origin of replication and phage triggered defense mechanisms; utilization of phage repressors; superinfection exclusion; inducible gene expression systems; and expression of phage encoded lysins or lysine-holin systems (Table 3). Excellent reviews on these systems has been published by, among others, McGrath et al. (2002); Josephsen and Neve (2004); Emond and Moineau (2007); and Sturino and Klaenhammer (2004, 2006).

CONCLUSION

There is no doubt that transduction mediated by LAB phages as well as LAB phage-based genetic tools have contributed in some way to elucidate the genetics of some LAB strains. However, there is growing evidence that LAB phages are important vectors of horizontal gene transfer (HGT) in the natural environment. Horizontal gene transfer has been extensively documented in LAB and appears to play a prominent role in its observed variability. LAB phages not only can transduce chromosomal genes or mobilize plasmid DNA among LAB strains, but can recombine with other phages in mixed infections or with prophage sequences. There are evidences that they can also lead to phage conversion during lysogen formation. Some of the clearest evidence for horizontal transfer has come from recent studies on the genomes of LAB strains. Many genes that are shared by two strains are located adjacent to a prophage suggesting they have been transferred by the phage vehicle. Also, many prophage genes shared by two LAB species probably derived from the same phage integrated into the genome of their common ancestor. The rapid evolutionary changes in the spacer region of CRISPR-related system (CASS) driven by phage exposure is another example of LAB genes disseminated via horizontal gene transfer. Knowledge of more phage genome sequences may provide new insights into phage genomics, phage evolution, and phage-host interactions, which could be further applied in novel ways for LAB strain improvement.

REFERENCES

Allen, L. K., Sandine, W. E. and Elliker, P. R. (1963). Transduction in *Streptococcus lactis*. *Journal of Dairy Research, 30,* 351-357.

Alvarez, M. A., Herrero, M. and Suárez, J. E. (1998). The site-specific recombination system of the *Lactobacillus* spp. bacteriophage A2 integrates in Gram-positive and Gram-negative bacteria. *Virology, 250*, 185-193.

Alvarez, M. A., Rodriguez, A. and Suárez J. E. (1999). Stable expression of the *Lactobacillus casei* bacteriophage A2 repressor blocks phage propagation during milk fermentation. *Journal of Applied Microbiology, 86*, 812-816.

Ammann, A., Neve, H., Geis, A. and Heller, K. J. (2008). Plasmid transfer via transduction from *Streptoccus thermophilus* to *Lactococcus lactis*. *Journal of Bacteriology, 190,* 3083-3087.

Auvray, F., Coddeville, M., Ritzenthaler, P. and Dupont, L. (1997). Plasmid integration in a wide range of bacteria mediated by the integrase of *Lactobacillus delbrueckii* bacteriophage mv4. *Journal of Bacteriology, 179,* 1837-1845.

Batt, C. A., Erlandson, K. and Bsat, N. (1995). Design and implementation of a strategy to reduce bacteriophage infection of dairy starter cultures. *International Dairy Journal, 5,* 949-962.

Binishofer, B., Moll, I., Henrich, B. and Bläsi, U. (2002). Inducible promoter-repressor system from the *Lactobacillus casei* phage ΦFSW. *Applied and Environmental Microbiology, 68,* 4132-4135.

Birkeland, N. K. and Holo, H. (1993). Transduction of a plasmid carrying the cohesive end region from *Lactococcus lactis* bacteriophage ØLC3. *Applied and Environmental Microbiology, 59,* 1966-1968.

Boizet, B., Lahbib-Mansais, Y., Dupont, L., Ritzenthaler, P. and Mata, M. (1990). Cloning, expression and sequence-analysis of an endolysin encoding gene of *Lactobacillus bulgaricus* bacteriophage mvl. *Gene, 94,* 61-67.

Boyce, J. D., Davidson B. E., and Hillier A. J. (1995). Spontaneous deletion mutants of the *Lactococcus lactis* temperate bacteriophage BK5-T and localization of the BK5-T *attP* site. *Applied and Environmental Microbiology, 61, 11,* 4105–4109.

Breuner, A., Brøndsted, L. and Hammer, K. (1999). Novel organization of genes involved in prophage excision identified in the temperate lactococcal bacteriophage TP901-1. *Journal of Bacteriology, 181,* 7291-7297.

Breuner, A., Brøndsted, L. and Hammer, K. (2001). Resolvase-like recombination performed by the TP901-1 integrase. *Microbiology, 147,* 2051-2063.

Brøndsted, L. and Hammer, K. (1999). Use of the integration elements encoded by the temperate lactococcal bacteriophage TP901-1 to obtain chromosomal single-copy transcriptional fusions in *Lactococcus lactis. Applied and Environmental Microbiology, 65, 2,* 752–758.

Brøndsted, L. and Hammer, K. (2006). Phages of *Lactococcus lactis*. In: R. Calendar (Ed.), *The Bacteriophages* (second edition, 572-592), New York: Oxford University Press, Inc.

Brüssow, H. (2006). Prophage genomics. In: R. Calendar (Ed.), *The Bacteriophages* (second edition, 17-25), New York: Oxford University Press, Inc.

Brüssow, H., and Desiere, F. (2006). Evolution of tailed phages: insights from comparative phage genomics. In: R. Calendar (Ed.), *The Bacteriophages* (second edition, 26-36), New York: Oxford University Press, Inc.

Brüssow, H. and Suárez, J. E. (2006). *Lactobacillus* phages. In: R. Calendar (Ed.), *The Bacteriophages* (second edition, 653-666), New York: Oxford University Press, Inc.

Bruttin A., Foley S. and Brüssow H. (2002). DNA-Binding Activity of the *Streptococcus thermophilus* Phage Sfi21 Repressor. *Virology, 303,* 100-109.

Bruttin A., Foley S. and Brüssow H. (1997). The site specific integration system of the temperate *Streptococcus thermophilus* bacteriophage ΦSfi21. *Virology, 237,* 148-158.

Chandry, P. S., Moore, S. C., Davidson, B. E. and Hillier, A. J. (2002). Transduction of concatemeric plasmids containing the *cos* site of *Lactococcus lactis* bacteriophage sk1. *FEMS Microbiology Letters, 216,* 85-90.

Christiansen, B., Johnsen M. G., Stenby E., Vogensen F. K and Hammer K. (1994). Characterization of the lactococcal temperate phage TP901-1 and its site-specific integration. *Journal of Bacteriology, 176,* 1069–1076.

Christiansen, B., Brøndsted, L., Vogensen, F. K. and Hammer, K. (1996). A resolvase-like protein is required for the site-specific integration of the temperate lactococcal bacteriophage TP901-1. *Journal of Bacteriology, 178,* 5164–5173.

Cruz Martín M., Suárez J. E., Alonso J. C. and Alvarez M. A. (2000). Generation of food-grade recombinant lactic acid bacterium strains by site-specific recombination. *Applied and Environmental Microbiology, 66, 6,* 2599-2604.

Damelin, L. H., Mavri-Damelin, D., Klaenhammer, T. R. and Tiemessen, C. T. (2010). Plasmid transduction using bacteriophage Øadh for expression of CC chemokines by *Lactobacillus gasseri* ADH. *Applied and Environmental Microbiology, 76,* 3878–3885.

Davies F. L. and Gasson M. J. (1981). Reviews of the progress of dairy science: genetics of lactic acid bacteria. *Journal of Dairy Research, 48, 2,* 363-376.

Deichelbohrer, I., Alonso, J. C., Luder, G. and Trautner, T. A. (1985). Plasmid transduction by *Bacillus subtilis* bacteriophage SPP1: effects of DNA homology between plasmid and bacteriophage. *Journal of Bacteriology, 162,* 1238-1243.

de Ruyter. P. G., Kuipers, O. P., Meijer, W. C. and De Vos, W. M. (1997). Food-grade controlled lysis of *Lactococcus lactis* for accelerated cheese ripening. *Natural Biotechnology, 15*, 976-979.

Djordjevic, G. M. and Klaenhammer, T. R. (1997). Bacteriophage-triggered defense systems: phage adaptation and design improvement. *Applied and Environmental Microbiology, 63*, 4370-4376.

Djordjevic, G. M., O'Sullivan, D. J., Walker, S. A., Conkling, M. A. and Klaenhammer, T. R. (1997). A triggered-suicide system designed as a defense against bacteriophages. *Journal of Bacteriology, 179*, 6741-6748.

Dupont, L., Boizet-Bonhoure, B., Coddeville, M., Auvray, F. and Ritzenthaler, P. (1995). Characterization of genetic elements required for site-specific integration of *Lactobacillus delbrueckii* subsp. *bulgaricus* bacteriophage mv4 and construction of an integration–proficient vector for *Lactobacillus plantarum*. *Journal of Bacteriology, 177*, 586-595.

Durmaz, E., Madsen, S. A., Israelsen, H. and Klaenhammer, T. R. (2002). *Lactococcus lactis* lytic bacteriophage of the P335 group are inhibited by overexpression of a truncated CI repressor. *Journal of Bacteriology, 184*, 6532-6543.

Emond, E., and Moineau, S. (2007). Bacteriophages in food fermentations. In: S. McGrath and D. van Sinderen (Eds.) *Bacteriophage. Genetics and Molecular Biology* (first edition, 93-124), Norfolk, UK: Caister Academic Press.

Foley, S., Lucchini, S., Zwahlen, M. C. and Brüssow, H. (1998). A short noncoding viral DNA element showing characteristics of a replication origin confers bacteriophage resistance to *Streptococcus thermophilus. Virology, 250*, 377-387.

Gindreau, E., Torlois, S. and Lonvaud-Funel, A. (1997). Identification and sequence analysis of the region encoding the site-specific integration system from *Leuconostoc oenos* (*Oenococcus oeni)* temperate bacteriophage phi 10MC. *FEMS Microbiology Letters, 147, 2,* 279-285.

Gruss, A. and Ehrlich, S. D. (1988). Insertion of foreign DNA into plasmid from gram-positive bacteria induces formation of high-molecular-weight plasmid multimers. *Journal of Bacteriology, 170,* 1183-1190.

Heller, K. J., Geis, A. and Neve, H. (1996). Behaviour of genetically modified microorganisms in yoghurt. *Systematic and Applied Microbiology, 18,* 504-509.

Henrich, B., Binishofer, B. and Bläsi, U. (1995). Primary estructure and functional analysis of the lysis gene of *Lactobacillus gasseri* bacteriophage φadh. *Journal of Bacteriology, 177*, 723-732.

Hill, C., Miller, L. A. and Klaenhammer, T. R. (1990). Cloning, expression, and sequence determination of a bacteriophage fragment encoding bacteriophage resistance in *Lactococcus lactis*. *Journal of Bacteriology, 172*, 6419-6426.

Hols, P., Hancy, F., Fontaine, L., Grossiord, B., Prozzi, D., Leblond-Bourget, N., Decaris, B., Bolotin, A., Delorme, C., Dusko Ehrlich, S., Guédon, E., Monnet, V., Renault, P. and Kleerebezem, M. (2005). New insights in the molecular biology and physiology of *Streptococcus thermophilus* revealed by comparative genomics. *FEMS Microbiology Reviews, 29,* 435–463.

Josephsen, J. and Neve, H. (2004). Bacteriophage and antiphage mechanisms of lactic acid bacteria. In: S. Salminen, A. von wright and A. Ouwehand (eds.) *Lactic acid bacteria. Microbiological and functional Aspects* (third edition, revised and expanded, 295-350). London: CRC Press.

Kim, S. G. and Batt, C. A. (1991). Antisense messenger RNA mediated bacteriophage resistance in *Lactococcus lactis* subsp. *lactis*. *Applied and Environmental Microbiology, 57,* 1109-1113.

Kim S. G., Bor, Y. C. and Batt, C. A. (1992). Bacteriophage resistance in *Lactococcus lactis* ssp *lactis* using antisense ribonucleic acid. *Journal of Dairy Science, 75*, 1761-1767.

Kim. J. H., Kim, S. G., Chung, D. K., Bor, Y. C. and Batt, C. A. (1992). Use of antisense RNA to confer bacteriophage resistance in dairy starter cultures. *Journal of Industrial Microbiology, 10*, 71-78.

Klaenhammer, T. R. and McKay, L. L. (1976). Isolation and examination of the transducing bacteriophage from *Streptococcus lactis* C2. *Journal of Dairy Science, 59,* 396-404.

Koch, B., Christiansen, B., Evison, T., Vogensen, F. K. and Hammer, K. (1997). Construction of specific erythromycin resistance mutations in the temperate lactococcal bacteriophage TP901-1 and their use in studies of phage biology. *Applied Environmental Microbiology, 63,* 2439-2441.

Kodaira K., Oki, M., Watanabe, N., Yasukawa, H., Masamune, Y., Taketo, A. and Kakikawa, M. (1997). Characterization of the genes encoding integrative and excisive functions of *Lactobacillus* phage øg1e: cloning, sequence analysis, and expression in *Escherichia coli*. *Gene, 185, 1*, 119-125.

Lahbib-Mansais, Y., Boizet, B., Dupont, L., Mata, M. and Ritzenthaler, P. (1992). Characterization of a temperate bacteriophage of *Lactobacillus delbrueckii* subsp. *bulgaricus* and its interactions with the host cell chromosome. *Journal of General Microbiology, 138,* 1139-1146.

Lillehaug, D. and Birkeland N. (1993). Characterization of genetic elements required for site-specific integration of the temperate lactococcal bacteriophage ØLC3 and construction of integration-negative ØLC3 mutants. *Journal of Bacteriology, 175,* 1745–1755.

Lillehaug, D., Nes, I. F. and Birkeland, N. K. (1997). A highly efficient and stable system for site-specific integration of genes and plasmids into the phage ØLC3 attachment site (*att*B) of the *Lactococcus lactis* chromosome. *Gene, 188*, 129-13.Luchansky, J. B., Kleeman, E. G., Raya, R. R. and Klaenhammer, T. R. (1989). Genetic transfer systems for delivery of plasmid deoxyribonucleic acid to *Lactobacillus acidophilus* ADH: conjugation, electroporation, and transduction. *Journal of Dairy Science, 72,* 1408-1417.

Madsen, P. L., Johasen, A. H., Hammer, K. and Brondsted, L. (1999). The genetic switch regulating activity of early promoters of the temperate *lactococcal* bacteriophage TP901-1. *Journal of Bacteriology, 181*, 7430-7438.

Mahony, J., McGrath, S., Fitzgerald, G. F. and van Sinderen, D. (2008). Identification and characterization of lactococcal-prophage-carried superinfection exclusion genes. *Applied Environmental Microbiology, 74*, 6206-6215.

McGrath, S., Seegers, J. F., Fitzgerald, G. F. and van Sinderen, D. (1999). Molecular characterization of a phage encoded resistance system in *Lactococcus lactis. Applied and Environmental Microbiology*, *65,* 1891-1899.

McGrath, S., Fitzgerald, G. F. and van Sinderen, D. (2001). Improvement and optimization of two engineered phage resistance mechanism in *Lactococcus lactis*. *Applied and Environmental Microbiology*, *67*, 608-616.

McGrath, S., Fitzgerald, G. F. and van Sinderen, D. (2002). Identification and characterization of phage resistance genes in temperate lactococcal bacteriophages. *Molecular Microbiology, 43*, 509-520.

McKay, L. L., Cords, B. R. and Baldwin, K. A. (1973). Transduction of lactose metabolism in *Streptococcus lactis* C2. *Journal of Bacteriology, 115,* 810-815.

McKay, L. L. and Baldwin, K. A. (1974). Simultaneous loss of proteinase- and lactose-utilizing enzyme activities in *Streptococcus lactis* and reversal of loss by transduction. *Applied Microbiology, 28,* 342-346.

McKay, L. L., Baldwin, K. A., and Efstathiou, J. D. (1976). Transductional evidence for plasmid linkage of lactose metabolism in *Streptococcus lactis* C2. *Applied and Environmental Microbiology, 32*, 45-52.

McKay, L. L. and Baldwin, K. A. (1978). Stabilization of lactose metabolism in *Streptococcus lactis* C2. *Applied and Environmental Microbiology, 36,* 360-367.

Mercenier, A., Slos, P., Faelen, M. and Lecocq, J. P. (1988). Plasmid transduction in *Streptococcus thermophilus. Molecular of General Genetics, 212,* 386-389.

Molsknes, T. A., Sandine, W. E. and Brown, L. R. (1974). Characterization of Lac+ transductants of *Streptococcus lactis*. *Applied Microbiology, 28,* 753-758.

Morelli, L., Vogensen, F. K. and von Wright, A. (2004). Genetics of lactic acid bacteria. In: S. Salminen, A. von Wright and A. Ouwehand (Eds.), *Lactic Acid Bacteria: Microbiological and Functional Aspects* (third edition, 249-293), New York, 270 Madison Avenue, NY10016, USA: Marcel Dekker, Inc.

Moscoso, M. and Suárez, J. E. (2000). Characterization of the DNA replication module of bacteriophage A2 and use of its origin of replication as a defense against infection during milk fermentation by *Lactobacillus casei. Virology, 273*, 101-111.

Nauta, A., van Sinderen, D., Karsens, H., Smit, E., Venema, G. and Kok, J. (1996). Inducible gene expression mediated by a represor-operator system isolated from *Lactococcus lactis* bacteriophage rlt. *Molecular Microbiology, 19*, 1331-1341.

Nauta, A., van Sinderen, B., Karsens, H., Veinema, G. and Kok, J. (1997). Design of thermolabile bacteriophage repressor mutants by comparative molecular modeling. *Nature Biotechnology, 15*, 980-983.

Novick, R.P., Edelman, I. and Lofdahl, S. (1986). Small *Staphylococcus aureus* plasmids are transduced as a linear multimers that are formed and resolved by replicative processes. *Journal of Molecular Biology, 192,* 209–220.

O'Sullivan, D. J., Hill, C. and Klaenhammer, T. R. (1993). Effect of increasing the copy number of bacteriophage origins of replication, in *trans*, on incoming-phage proliferation. *Applied and Environmental Microbiology, 59*, 2449-2456.

Petersen, A., Josephsen, J. and Johnsen, M. G. (1999). TPW22, a lactococcal temperate phage with a site-specific integrase closely related to *Streptococcus thermophilus* phage integrases. *Journal of Bacteriology, 181,* 7034–7042.

Raya, R. R., Kleeman, E. G., Luchansky, J. B. and Klaenhammer, T. R. (1989). Characterization of the temperate bacteriophage Øadh and plasmid transduction in *Lactobacillus acidophilus* ADH. *Applied and Environmental Microbiology, 55,* 2206-2213.

Raya, R. R. and Klaenhammer, T. R. (1992). High-frequency plasmid transduction by *Lactobacillus gasseri* bacteriophage Øadh. *Applied and Environmental Microbiology, 58,* 187-193.

Raya, R. R., Fremaux, C., De Antoni, G. L. and Klaenhammer, T. R. (1992). Site-specific integration of the temperate bacteriophage Φadh into the *Lactobacillus gasseri* chromosome and molecular characterization of the phage (*attP*) and bacterial (*attB*) attachment sites. *Journal of Bacteriology, 174,* 5584-5592.

Ravin, V., Sasaki, T., Räisänen, L., Riipinen, K.-A. and Alatossava, T. (2006). Effective plasmid pX3 transduction in *Lactobacillus delbrueckii* by bacteriophage LL-H. *Plasmid, 55,* 184–193.

Riipinen, K. A., Räisänen, L. and Alatossava, T. (2007). Integration of the group c phage JCL1032 of *Lactobacillus delbrueckii subsp. lactis* and complex phage resistance of the host. *Journal of Applied Microbiology, 103,* 2465–2475.

Sanders, J. W., Venema, G. and Kok, J. (1997). A chloride-inducible gene expression cassette and its use in induced lysis of *Lactoccocus lactis*. *Applied and Environmental Microbiology, 63*, 4877-4882.

Sandine, W. E., Elliker, P. R., Allen, L. K. and Brown, W. C. (1962). Genetic exchange and variability in lactic *Streptococcus* starter organisms. *Journal of Dairy Science, 45,* 1266-1271.

Shimizu-Kadota, M., Kiwaki, M., Sawaki, S., Shirasawa, Y., Shibahara-Sone, H. and Sako, T. (2000). Insertion of bacteriophage ΦFSW into the chromosome of *Lactobacillus casei* strain Shirota (S-1): characterization of the attachment sites and the integrase gene. *Gene, 249*, 127–134.

Schmidt, C. and Schmieger, H. (1984). Selective transduction of recombinant plasmids with cloned *pac* sites by *Salmonella* phage P22. *Molecular of General Genetics, 196,* 123–128.

Shearman, C. A., Jury, K. and Gasson, M. J. (1992). Autolytic *Lactococcus lactis* expressing a *lactococcal* bacteriophage lysine gene. *Bio-Technology, 10*, 196-199.

Shipley, P. L. and Olsen, R. H. (1975). Isolation of a nontransmissible antibiotic resistance plasmid by transductional shortening of R factor RP1. *Journal of Bacteriology, 123*, 20-27.

Snook, R. J., McKay, L. L. and Ahlstrand, G. G. (1981). Transduction of lactose metabolism by *Streptococcus lactis* c3 temperate bacteriophage. *Applied and Environmental Microbiology, 42,* 897-903.

Solem, C., Defoor, E., Jensen, P. R. and Martinussen, J. (2008). Plasmid pCS1966, a new selection/counterselection tool for lactic acid bacterium strain construction based on the *oroP* gene, encoding an orotate transporter from *Lactococcus lactis. Applied and Environmental Microbiology, 74,* 4772–4775.

Stanley, E., Fitzgerald, G. F., Le Marrec, C. Fayard, B. and van Sinderen, D. (1997). Sequence analysis and characterization of ΦO1205, a temperate bacteriophage infecting *Streptococcus thermophilus* CNRZI 205. *Microbiology, 143,* 3417-3429.

Stanley, E., Walsh, L., van der Zwet, A., Fitzgerald, G. F. and van Sinderen, D. (2000). Identification of four loci isolated from two *Streptococcus thermophilus* phage genomes responsible for mediating bacteriophage resistance. *FEMS Microbiology Letters, 182,* 271-277.Stoll, S. M., Ginsburg, D. S. and Calos, M. P. (2002). Phage TP901-1 site-specific integrase funtions in human cells. *Journal of Bacteriology, 184,* 3657-3663.

Sturino, J. M. and Klaenhammer, T. R. (2002). Expression of antisense RNA targeted against *Streptococcus thermophilus* bacteriophages. *Applied and Environmental Microbiology, 68,* 588-596.

Sturino J. M. and Klaenhammer T. R. (2004). Bacteriophage defense systems and strategies for lactic acid bacteria. *Advances in Applied Microbiology, 56,* 331-378.

Sturino J. M. and Klaenhammer T. R. (2006). Engineered bacteriophage-defence systems in bioprocessing. *Nature Review Microbiology. 4,* 395-404.

Sun, X., Göhler, A., Heller, K. J. and Neve, H. (2006). The *ltp* gene of temperate *Streptococcus thermophilus* phage TP-J34 confers superinfection exclusion to *Streptococcus thermophilus* and *Lactococcus lactis*. *Virology, 350,* 146-157.

Tohyama, K., Sakurai, T. and Arai, H. (1971). Transduction by temperate phage PLS-1 in *Lactobacillus salivarius. Japanese Journal of Bacteriology, 26,* 482–487.

van de Guchte, M., Daly, C., Fitzgerald, G. F. and Arendt, E. K. (1994). Identification of *int* and *att*P on the genome of lactococcal bacteriophage Tuc2009 and their use for site-specific plasmid integration in the chromosome of Tuc2009-resistant *Lactococcus lactis* MG1363. *Applied and Environmental Microbiology, 60,* 2324–2329.

van Sinderen D., Karsens H., Kok J., Terpstra P., Ruiters M. H. , Venema G. and Nauta A. (1996). Sequence analysis and molecular characterization of the temperate lactococcal bacteriophage r1t. *Molecular Microbiology, 19, 6,* 1343-1355.

Vasala, A., Isomäki, R., Myllikoski, L. and Alatossava, T. (1999). Thymol-triggered lysis of *Escherichia coli* expressing *Lactobacillus* phage LL-H muramidase. *Journal of Industrial Microbiology and BioTechnology, 22,* 39-43.

Viret, J. F. and Alonso, J. C. (1987). Generation of linear multigenome-length plasmid molecules in *Bacillus subtilis*. *Nucleic Acids Research, 16,* 4389–4406.

Viret, J. F., Bravo, A. and Alonso, J. C. (1991). Recombination dependent concatemeric plasmid replication. *Microbiology Reviews, 55,* 675–683.

Vollenweider, H. J., Fiandt, M., Rosenvold, E. C. and Szybalski, W. (1980). Packaging of plasmid DNA containing the cohesive ends of coliphage lambda. *Gene, 9,* 171-174.

Walker, S. A. and Klaenhammer, T. R. (2000). An explosive antisense RNA strategy for inhibition of a *lactococcal* bacteriophage. *Applied and Environmental Microbiology, 66,* 310-319.

Walker, S. A. and Klaenhammer, T. R. (2001). Leaky *Lactococcus* cultures that externalize enzyme and antigens independently of culture lysis and secretion and export pathways. *Applied and Environmental Microbiology, 67*, 251-259.

Watanabe, K., Kakita, Y., Nakashima, Y. and Sasaki, T. (1990). Protoplast transfection of *Lactobacillus casei* by phage PL-1 DNA. *Agricultural and Biological Chemistry, 54*, 937-941.

Weisberg, R. A, Enquist, L. W, Foeller, C. and Landy, A. (1983) Role for DNA homology in site-specific recombination. The isolation and characterization of a site affinity mutant of coliphage lambda. *Journal of Molecular Biology, 170,* 319-342.

In: Bacteriophages in Dairy Processing
Editor: Andrea del Lujan Quiberoni et al.
ISBN 978-1-61324-517-0

Chapter 13

BACTERIOPHAGES AS BIOCONTROL AGENTS IN DAIRY PROCESSING

Pilar García*, Beatriz Martínez, Lorena Rodríguez, Diana Gutiérrez, and Ana Rodríguez
Instituto de Productos Lácteos de Asturias (IPLA-CSIC),
Villaviciosa, Asturias, Spain

ABSTRACT

Milk and dairy products contain a variety of microorganisms and may be an important source of foodborne pathogens in spite of several advanced preservation technologies and monitoring systems existing in developed countries. However, new food consumer trends, market globalization, societal and demographic changes, and novel foods require new preservation techniques to improve food safety while satisfying new consumer's demands.

Biopreservation of food using bacteriophages or phage biocontrol has been proposed as an innovative strategy to fight undesirable bacteria in food. The specificity of phages for their bacterial host and their harmlessness to eukaryotic cells are the main traits that support the use of phages as novel preservation tools in food production. Moreover, phages lead to microbial destruction without detrimental effects on the organoleptic and nutritional characteristics. In addition to phages, some phage-encoded proteins (endolysins and virion-associated peptidoglycan hydrolases) have also antimicrobial activity with a low probability of developing resistance and a relatively narrow spectrum but active against antibiotic resistant bacteria. Therefore, the dairy industry may benefit from integrating phage-based antimicrobials as part of the current preservation systems within the whole dairy production chain to produce microbially safe minimally processed dairy products with an increased shelf life. However, although this approach is promising, the complex bacteriophage-host interactions in the dairy environment must be further investigated.

---

* E-mail address: pgarcia@ipla.csic.es. Tel.: +34 985 89 21 31; fax: +34 985 89 22 33

At present, several phage products for medical, animal husbandry, food processing and environmental applications have been marketed. In dairy production, the first commercial product, *Listex P100*, has been designed to eradicate *Listeria monocytogenes* in cheese. Other products against *E. coli* O157:H7 and *Staphylococcus aureus* have being developed for veterinary applications in cows.

This chapter summarizes the current status of phages and phage-encoded proteins in biocontrol of foodborne pathogens and its application in dairy processing. We also discuss the criteria to select putative candidates and the factors that might contribute to their performance in dairy products.

INTRODUCTION

Consumption of dairy products as part of a nutrient-rich diet has long been recognized as an important contributor to maintain health and nutrition in people of all ages. However, safety remains as a matter of concern due to the increasing number of outbreaks reported in the last years. A recent report analyzing foodborne outbreaks in USA between 1998 and 2007 indicates that a total of 190 outbreaks and 4,565 illnesses were linked to dairy products (DeWaal et al. 2009). Regarding the European Union, a total of 5,332 foodborne outbreaks were reported in 2008; of these outbreaks, dairy products represented 3.3%. Specifically, dairy products have been involved in 14.3%, 1.4% and 11.2% of *Campylobacter, Salmonella* Typhymurium and *Salmonella* Enteritidis outbreaks, respectively (EFSA 2010). Although the global incidence is difficult to estimate, other pathogens such as *S. aureus, E. coli* O157:H7, *Shigella* spp., *List. monocytogenes* and *Yersinia enterocolitica* are common in milk and fermented milk products (WHO 2007).

Changes in food consumption patterns are likely behind the large numbers of outbreaks linked to dairy products. A greater variety of foods are demanded, thus enlarging both duration and length of distribution chains worldwide. Consumer preferences for trying traditional foods have also increased the consumption of unpasteurized dairy products and, increasingly, meals are eaten outside the home and so the chance of contracting a foodborne illness increases as well. The scenario becomes even more problematic considering demographic changes and the increasing number of elderly and immunocompromised individuals more susceptible to foodborne pathogens (Notermans and Van de Giessen 1993).

Safety of milk and derived products has generally been assured by thermal processing, which destroys vegetative microorganisms by combining temperatures ranging from 60°C to 100°C for few seconds to 30 minutes. However, current nutritional and quality standards (flavor, odor, color, texture, and especially nutrient content) claim for alternative preservation techniques. Several approaches have been proposed such as food biopreservation that makes use of natural preservatives derived from plants, animals, and microorganisms (Tiwari et al. 2009). This concept has been applied for centuries by using, for instance, lactic acid bacteria and/or their antibacterial products such as lactic acid and bacteriocins (Gálvez et al. 2008).

Bacteriophages or phages are viruses that exclusively infect bacteria. They are widespread throughout nature and kill their hosts during their lifecycle. Their antimicrobial activity has been largely exploited in phage therapy for the treatment of infectious diseases (Sulakvelidze and Kutter 2005). After phages were discovered in 1917, phage therapy was applied mainly in the former Soviet Union, while interest in the West declined by the widespread use of antibiotics. Although not always successful, the increase in antibiotic resistance has boosted a renaissance of phage therapy, as exemplified by the marketing of several phage-based products with multiple healthcare applications (Figure 1).

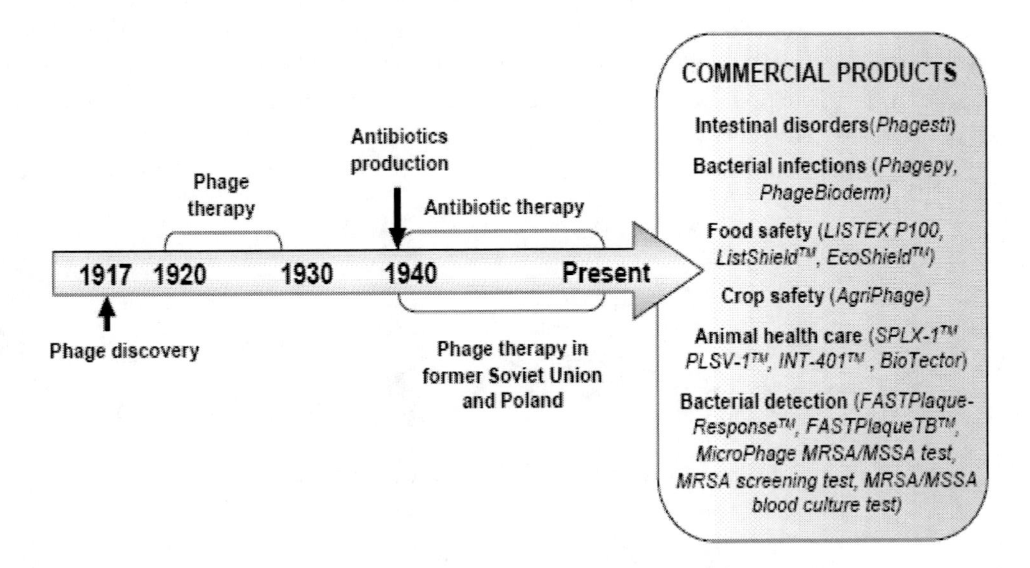

Figure 1. Main events in bacteriophage history and antimicrobial applications. Current commercial phage-based products are also indicated.

In a food context, bacteriophages have been traditionally regarded as leading causes of fermentation failure. Phage infection of dairy starter cultures disrupts normal fermentation cycles inhibiting lactic acid production (Sturino and Klaenhammer 2004). This results in slow milk coagulation, disrupting manufacturing schedules and resulting in low quality end-products. However, the experience gained through the use of phages in phage therapy and the accumulated knowledge in phage biology and phage-host interactions (Hatfull 2008, Hyman and Abedon 2010) have paved the way for novel applications of phages as natural antimicrobials with promising prospects in food biopreservation (García et al. 2010a).

Currently, several reports have explored the application of bacteriophages to reduce pathogenic bacteria during the different stages of food production in order to improve food safety (Hagens and Loessner 2007, Strauch et al. 2007, García et al. 2008). The recent FDA approval of *Listeria* specific bacteriophage preparations for food preservation has set the trigger for a new era in phage biocontrol and new products are being marketed targeting crop safety, animal healthcare and food biopreservation (Figure 1). In this chapter, the potential of phages and their endolysins as antimicrobials in the dairy environment is reviewed. These natural antimicrobials might be used alone or in combination with other thermal and non-thermal technologies. However, several questions may be raised concerning the use of such products regarding efficacy, consumer safety, and application methods. Research priorities and future trends focused on the design of efficient phage biocontrol strategies are also discussed.

1. Microbial Safety Risks and Technological Failures in the Dairy Production Chain

Traditionally, the production of extended shelf-life milk is carried out by HTST or UHT thermal treatments, sometimes coupled with microfiltration or centrifugation. In addition, growth of most foodborne pathogens in dairy products is prevented by refrigeration, post-pasteurization, fermentation by lactic acid bacteria and/or addition of selected antimicrobial agents or biocides. However, milk is a very rich medium that easily supports growth of several spoilage and pathogenic microorganisms. The main routes of pathogen entry into the dairy chain are depicted in Figure 2.

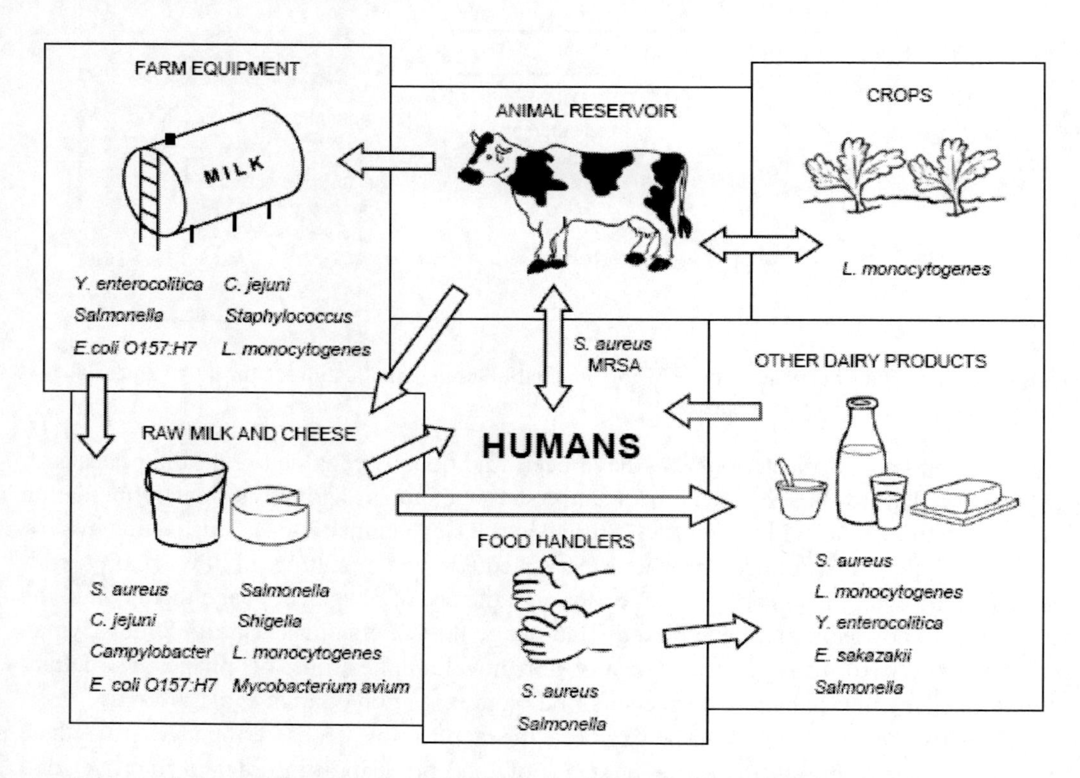

Figure 2. Foodborne pathogens in the dairy farm environment and their transfer to dairy products.

The main reservoir is found in dairy ruminants, but access by foodborne pathogens to milk and dairy products also occurs through direct contact with contaminated sources in the dairy farm environment (bedding, feedstuff, manure), to milking of an infected animal, and poor hygienic processing lines with ineffective cleaning procedures (Oliver et al. 2005). Asymptomatic food handlers may also contaminate the product during manipulation. In dairy plants, an additional source of contamination is found in the formation of biofilms on food contact surfaces, leading to subsequent contamination of dairy products. Dairy biofilms are predominated by bacterial extracellular polymeric substances (EPS) and milk residues. Biofilms in these environments are formed by bacteria belonging to the genus *Enterobacter*, *Lactobacillus*, *Micrococcus*, *Streptococcus*, *Bacillus*, and *Pseudomonas*. They also contain pathogens such *Listeria monocytogenes*, *Yersinia enterocolitica*, *Campylobacter jejuni*, *Salmonella* spp. *Staphylococcus* spp. and *E. coli* O157:H7 (Sharma and Anand 2002).

Milk and dairy products are frequently associated with staphylococcal food poisoning by enterotoxins ingestion (Cremonesi et al. 2007). *S. aureus* has been involved in 15% of recorded foodborne illnesses caused by dairy products in eight developed countries (De Buyser et al. 2001). The most recent large-scale outbreak was caused by consumption of skim milk powder contaminated with *S. aureus* enterotoxin A (Asao et al. 2003). The presence of *S. aureus* in milk has been linked to the presence of clinical or subclinical mastitis in dairy herds, as strains isolated from milk were identical to those from cows, milking units and personnel (Hata et al. 2010). Furthermore, nasal *S. aureus* carriage could be an important risk factor, particularly when methicillin-resistant *S. aureus* (MRSA) are involved (Türkyilmaz et al. 2010)

Fecal contamination is another route of bulk milk contamination by zoonotic inhabitants in dairy ruminants' intestinal tract. *Campylobacter jejuni* is excreted through feces and animal secretions and its presence in bulk tank milk has been reported (Vilar et al. 2010). This bacterium is the most frequently identified cause of acute infectious diarrhea in developed countries. Humans get infected through ingestion of contaminated raw milk or milk not properly pasteurized (Peterson 2003). Recent outbreaks in US due to *Campylobacter* have been associated with drinking raw milk (Babcock 2010).

E. coli O157:H7 is also part of the resident microbiota of the bovine intestinal tract. The presence of *E. coli* O157:H7 in milk is of important economic and public health significance due to the low infectious doses, typically 1-100 colony-forming units. Its prevalence in bulk tank milk and equipment in dairy farms has been reported (Murinda et al. 2002). The principal dairy transmission vectors were shown to be raw milk and cheese (Solomakos et al. 2009). A recent US outbreak was shown to have been caused by raw milk consumption (Marler 2010).

Salmonella spp. is also a common reported cause of food poisoning worldwide. The primary reservoir is the intestinal tract of humans and animals. A wide range of contaminated foods are associated with *Salmonella* food poisoning, including milk and dairy products (Ruzante et al. 2010). Contamination is basically linked to temperature abuse, poor handling practices, and cross-contamination in processed foods.

Shigella is usually transmitted through contaminated raw milk and fresh cheese (García-Fulgueiras et al. 2001). According to surveys by the Center for Disease Control in the US, *Shigella* ranks third among bacterial foodborne pathogens in the number of illness cases. The incidence of infection by this pathogen is estimated at 448000 cases per year, with 20% of these cases being due to foodborne transmission of the pathogen (Mead et al. 1999).

List. monocytogenes has been detected in bulk, pasteurized milk and several milk-derived products (Vilar et al. 2010). Apparently healthy, but nonetheless infected animals shed *Listeria* in the feces. The presence of *List. monocytogenes* biofilms in dairy farm equipment is also an important route of contamination (Latorre et al. 2010). This pathogen can grow in a wide range of pH and adapt easily to grow in refrigerated raw milk and can even survive minimum HTST pasteurization. Several *List. monocytogenes* outbreaks occurred through consumption of raw milk, or products manufactured with raw milk (Lunden et al. 2004).

Other safety concerns in dairying are the spread of antibiotic resistance and the recognition of new microbiological hazards that were not traditionally linked to dairy production but have emerged due to new production technologies and a larger high risk population (Tauxe 2002). There is a notable increase of zoonotic antibiotic resistant bacteria. Modern food production facilitates the emergence and spread of antibiotic resistance through the intensive use of antimicrobial agents and the international trade of both animals and food products. These bacteria can be transmitted to humans through the food chain with serious consequences on public health (Walsh and Fanning 2008). A notable case of an emerging opportunistic pathogen in dairying is exemplified by *Cronobacter* (formerly *Enterobacter sakazakii*), which poses a significant health risk to neonates as a contaminant of powdered infant formula (Craven et al. 2010). *Mycobacterium avium* subsp. *paratuberculosis* is also considered an emerging foodborne pathogen that might be related to the development of Crohn's disease in human patients. Milk from dairy cows infected with this pathogen could be a potential vector in the transmission to humans especially because it may survive pasteurization processes. These bacteria were detected in bulk tank milk samples (Favila-Humara et al. 2010). *Y. enterocolitica* can also be regarded as an emergent pathogen due to its ability to multiply at temperatures close to 0°C. Although the minimum infection dose of *Y. enterocolitica* is high, its multiplication is favoured by longer periods of cold storage and the extensive use of the chilling chain, both in retail handling as well as in households. This bacterium is typically transmitted through consumption of contaminated milk and dairy products (Schiemann 1987). Prevalence of *Y. enterocolitica* in goats and cattle can be an important source of human yersiniosis (Arnold et al. 2006, Bonardi et al. 2007).

Besides safety concerns about the presence of pathogenic bacteria, microbiological alterations of dairy products may also imply important economic losses. Growth of psychotrophic microorganisms during cold storage of fluid milk results in several flavor defects due to the synthesis of proteases and lipases that degrade casein and fat globules, respectively. Moreover, these enzymes may remain active after pasteurization and spoil dairy products made with contaminated milk. *Pseudomonas* spp. is one of the most common milk spoilage bacteria along with other less proteolytic and milk coagulant organisms such as *Achromobacter*, *Aeromonas*, *Flavobacterium*, and *Xanthomonas* spp. (Champagne et al. 1994). *Bacillus* spp. is also very often implicated in food spoilage due to its ubiquity and its highly resistant spores. Strains isolated from raw milk of *Bacillus subtilis*, the *Bacillus cereus* group, and *Bacillus amyloliquefaciens* have been shown to be strongly proteolytic; *Bacillus licheniformis*, and *Bacillus pumilus* exhibit proteolysis to a lesser extent. Lipolytic activity has been demonstrated in strains of *B. subtilis*, *B. pumilus* and *B. amyloliquefaciens*. Qualitative screening for lecithinase activity also revealed that some strains produce this enzyme that is well-known for causing a "bitty cream" defect in pasteurized milk due to lecithinase activity (De Jonghe et al. 2010). Another very serious and economically relevant defect is late blowing or late gas in cheese, namely in brine-salted semi-hard and hard cheeses, e.g. Gouda or Edam cheeses (Bergère and Lenoir 2000). Late blowing is characterized by the formation of large holes in the cheese matrix during ripening. *Clostridium tyrobutyricum* is considered the main agent responsible for fermentation of lactate to butyric acid, acetic acid, carbon dioxide and hydrogen gas. The presence of spores is often linked to the consumption of contaminated silage.

2. Bacteriophages and Bacteriophage-Based Antimicrobials

The antimicrobial activity of bacteriophages resides in their lifecycle. Depending on their life style, bacteriophages are divided into two groups - virulent and temperate phages- (Figure 3). When a phage encounters a putative host bacterium, a specific interaction between phage structures and receptors at the bacterial surface takes place and determines their host range. Some phages are highly specific for single bacterial isolates while others have a broad host range infecting a wide range of bacterial strains within a genus.

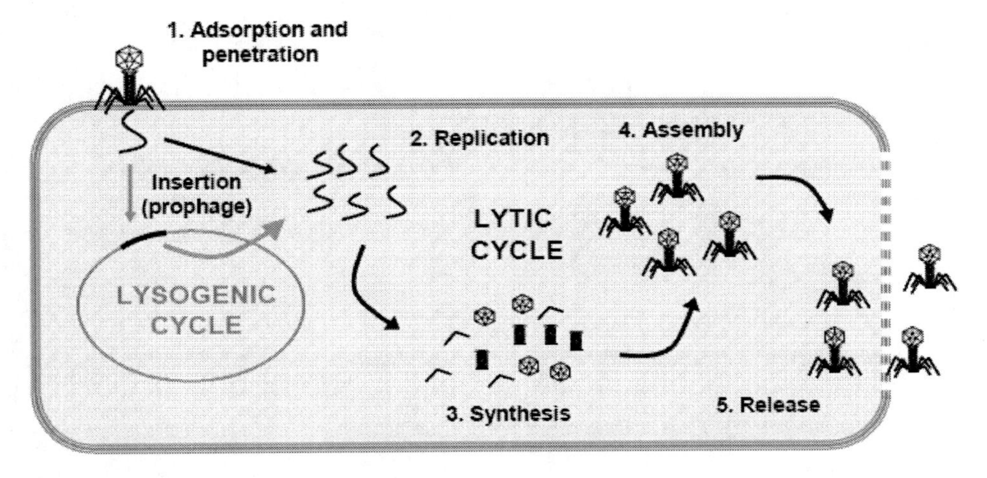

Figure 3. Bacteriophage lytic and lysogenic cycles. The different steps in both life styles are indicated and described in the text.

The next step in the infection process is to introduce the nucleic acid through the host cell wall. Temperate phages can follow a lytic or a lysogenic cycle while virulent phages only follow the lytic cycle. In the lysogenic cycle, the phage DNA is integrated into the bacterial chromosome, where it replicates as part of the host genome (prophage) until it is induced to enter the lytic cycle under adverse environmental conditions. In the lytic cycle, the phage genome is replicated into the cell and phage structural proteins are synthesized and assembled into new virions containing the replicated phage genome. Phage release is accomplished by cell lysis, which kills the host. The new progeny is ready to start a new round of infection, and so their antimicrobial activity increases.

During their lifecycle, bacteriophages display several phage-encoded proteins with antimicrobial activities. The most studied are those with peptidoglycan hydrolytic activities, namely endolysins or lysins, and virion-associated peptidoglycan hydrolases. The endolysins are dsDNA bacteriophage-encoded enzymes produced during the late phase of the lytic cycle and are responsible for the degradation of the bacterial cell wall peptidoglycan (Figure 4). During phage development into the infected bacteria, endolysins accumulate in the cytoplasm until the complete virions are assembled. Timely release of endolysins into the periplasmic space is mediated by holins. These proteins are inserted in the cytoplasmic membrane, thus forming holes through which endolysin molecules translocate to start degrading the peptidoglycan (Wang et al. 2000). In some phages, endolysins are secreted by the general secretion pathway of the host and remain somewhat inactive until the membrane potential is disturbed by the insertion of the holin (Park et al. 2007). The breakages made in the peptidoglycan structure destabilize its structure to such an extent that the osmotic pressure becomes unbearable, cells burst and the new virions are released outside the cytoplasm. Endolysins from phages infecting Gram-positive hosts are able to digest the cell wall when applied exogenously to bacterial cells. It has been precisely this trait that has brought attention to endolysins as potential anti-infectives (Fischetti 2010) and later as potential biopreservatives in food (García et al. 2010a).

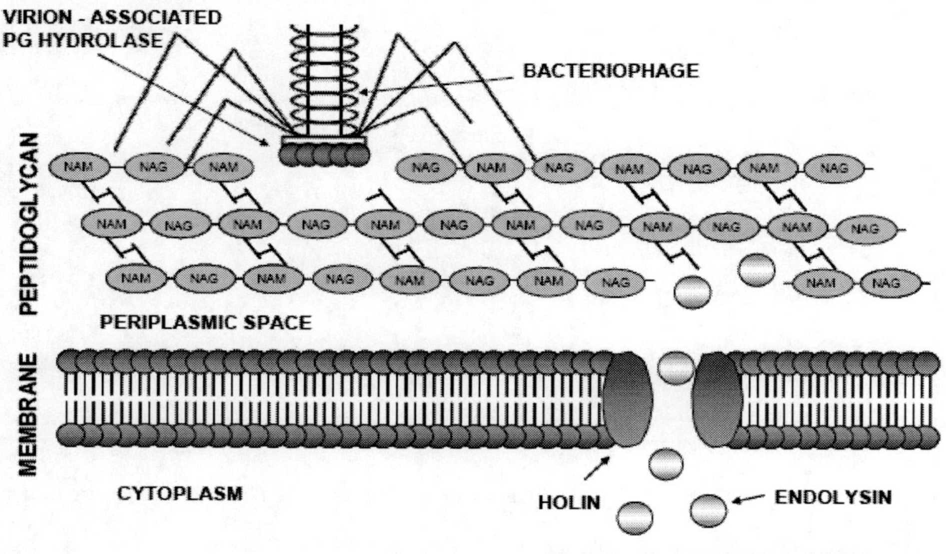

Figure 4. Bacteriophage lytic proteins potentially useful as peptidoglycan degrading antimicrobials. Lysis from without is carried out by virion-associated peptidoglycan hydrolases at the first step of the infection process. The lysis from within is mediated by two proteins: holin and endolysin. (NAG, N-acetyl-D-glucosamine; NAM, N-acetylmuramic acid).

This antimicrobial activity cannot be exploited to lyse Gram-negative pathogens because their peptidoglycan is protected by the outer membrane. However, endolysins able to permeabilize Gram-negative bacteria and induce cell lysis by means of their C-terminal peptide sequence have been described (Orito et al. 2004).

A second type of phage-encoded lytic enzyme -the virion-associated peptidoglycan hydrolases- locally degrades the peptidoglycan of the bacterial wall prior to injecting the genetic material during phage infection (Moak and Molineux 2004). Their action generates a small hole through which the tail crosses the cell envelope. They are responsible for the "lysis from without" caused by the adsorption of a high number of phages to the cell at the initial infection step (Figure 4). Recently, these proteins have been proposed as antimicrobials (Rashel et al. 2008).

Most of the virion-associated peptidoglycan hydrolases and endolysins have a modular structure. Several catalytic domains have been identified with distinct enzymatic activities: N-acetylmuramoyl-L-alanine amidase, interpeptide bridge endopeptidase, N-acetyl-β-D-glucosaminidase, L-alanoyl-D-glutamate endopeptidase, N-acetyl- β-D-muramidase and lytic transglycosylase (Figure 5). Because of this modular structure, it is plausible that different enzyme domains may be swapped, resulting in novel enzymes with different bacterial and catalytic specificities (García et al. 1990). Domain swapping has been performed with endolysins active against *S. aureus* (Manoharadas et al. 2009, Donovan et al. 2006a) and it has been shown that some engineered endolysins display high specificity and a higher lytic potential (Becker et al. 2009). The presence of multiple catalytic domains in the same molecule significantly lowers the probability of resistance development, a very attractive feature in antibiotic design. Indeed, no bacteria resistant to endolysins have been described so far (Fischetti 2005).

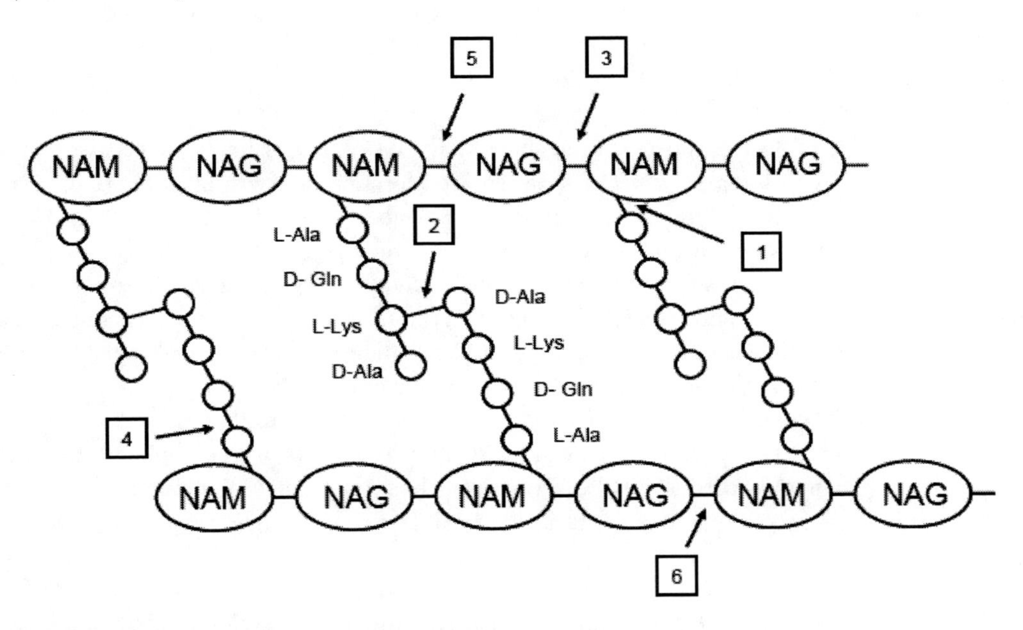

Figure 5. Enzymatic activities in bacteriophage peptidoglycan hydrolases. The structure of peptidoglycan and the chemical bond hydrolyzed by the proteins are indicated. 1: N-acetylmuramoyl-L-alanine amidase; 2: interpeptide bridge endopeptidase; 3: N-acetyl-β-D-glucosaminidase; 4: L-alanoyl-D-glutamate endopeptidase; 5: N-acetyl- β-D-muramidase; 6: Transglycosylase. (NAG, N-acetyl-D-glucosamine; NAM, N-acetylmuramic acid).

Phage-encoded peptidoglycan hydrolytic proteins show a certain degree of substrate specificity, often restricted to the genus the cognate phage infects. Endolysins are endowed with a non-catalytic domain involved in substrate recognition or binding domain. By contrast, virion-associated hydrolases seem to be devoid of such cell binding domain and substrate binding seems to reside in their catalytic domains (Rashel et al. 2008, Rodríguez et al. unpublished) Nevertheless, it is reasonable to think that cell binding domains are not essential since these proteins are part of the virion that already provides for the recognition and binding to the bacterial cell.

Besides peptidoglycan hydrolases, bacteriophages may encode other proteins or peptides involved in arresting cell growth and, thereby, they may be regarded as potential antimicrobials. For example, DNA replication and transcription are often targeted by phage proteins (Mallory et al. 1990, Nechaev and Severinov 1999). Small phages such as fX174 or Qb produce a protein that inhibits synthesis of the murein monomer (Bernhardt et al. 2002). A very clever approach to further exploit the biotechnological potential of phages has been described by Liu et al (2004) in their use of specific phage-inhibition pathways to develop high-throughput settings for antibiotic discovery.

3. Factors Affecting Phage-Based Biocontrol in Dairying

There are several requirements that should be considered prior to designing phage-based biocontrol measurements in food safety. Some are of general concern as they are inherent to phage biology and others have to be considered when a particular food niche is targeted.

Temperate phages are generally less suitable for biocontrol applications because they may lysogenize their hosts. The lysogenic bacteria become immune against superinfection with the same or a closely related phage and survive phage attack. Lytic phages are, therefore, the choice for biocontrol purposes. However, strictly lytic phages may not be available or those available may fail to meet other criteria required for a particular application (e.g. narrow host range). In such cases, lytic variants may be screened by selection of clear lysis plaque phenotype or genetically engineered to delete the lysogenic cassette. Two lytic bacteriophages specific against *S. aureus* were obtained by DNA random deletion from the milk-isolated temperate phages, ΦH5 and ΦA72. The new variants retained the host range and the technological traits (e.g. successful infection in milk) of their temperate counterparts and were more effective in the elimination of the pathogen (García et al. 2007).

The host range is also a criterion to select phages for biocontrol. Phages might only infect a very limited number of strains and even be unable to overcome geographical and niche boundaries (Hyman and Abedon 2010). A restricted host range may be neutralized by using of phage mixtures infecting as many strains as possible to ensure their effectiveness. Phages may also be genetically modified to alter their host binding profile. The long tail fiber genes from the genome of the *E. coli* phage T2 were replaced with those of the phage IP008 by homologous recombination. The host range of the recombinant phages was identical to that of phage IP008 (Mahichi et al. 2009). On the other hand, a restricted host range gives phages an important advantage over other biocontrol approaches in complex microbial ecosystems such as dairy fermentations; they do so by avoiding any interference with starter activity (Modi et al. 2001, García et al. 2007). Moreover, phage specificity guarantees that the gastrointestinal microbiota will remain undisturbed after oral administration of bacteriophages, as shown in human volunteers who ingested up to 10^5 PFU of phage T4 (Bruttin and Brüssow 2005).

Several phages have been shown to carry genes encoding toxins or virulence factors which could confer pathogenic traits onto previously harmless bacteria (Brüssow et al. 2004). Therefore, the full phage genome should be sequenced to discard any phage which may carry any of these traits in order to minimize this risk. In addition, some *S. aureus* phages are able to induce superantigen-carrying pathogenicity islands (SaPIs) leading to high frequencies of virulence transfer (Tormo-Más et al. 2010). In the same line, highly efficient transducers and helpers should be avoided.

Another issue of general concern is the selection of resistant bacteria. A number of phage resistance mechanisms can be acquired by bacteria (Hyman and Abedon 2010). They include loss or mutations in the phage receptor that prevent phage adsorption, restriction-modification systems that digest phage DNA, superinfection immunity, CRISPR, and several abortive infection mechanisms. The use of suitable phage mixtures could minimize the risk of bacteria resistance development. In the case of endolysins, resistant bacteria have not been described so far. Exposure of *Bacillus anthracis* grown on agar plates with low concentrations of endolysin did not lead to the recovery of resistant strains even after 40 cycles (Schuch et al. 2002). This is probably due to the fact that phage endolysins have evolved to target essential molecules for bacterial viability.

Finally, high scale production of the phages and phage-derived proteins has to be accomplished. The specific nature of the bacteriophage infection requires the presence of pathogenic host bacteria to propagate the phage. Therefore, adequate separation and/or sterilization technologies to ensure the complete absence of the pathogen in the final bacteriophage preparation are needed. However, some phages as the *Listeria* phage P100 that infects *List. monocytogenes* could be propagated on an innocuous strain (Carlton et al. 2005). Fast and high yield recombinant production of phage-encoded proteins might be carried out by bioreactor systems coupled with efficient large-scale purification processes to generate safe and reproducible products at reasonable cost.

In addition to the general considerations described above, and due to their intrinsic and extrinsic physical and chemical conditions and processing, there are several specific factors that may have an effect on the antimicrobial activity of phages and phage-encoded proteins in dairy products.

Successful phage infection is determined by the optimal conditions of the growth of target bacteria since cells during the stationary phases or during dormant stage might not be infected by phages (Woody and Cliver 1995). Since many dairy products are stored at refrigeration temperatures, inactivation of undesirable bacteria by phages has to be achieved at a temperature lower than their hosts' optimum growth temperature. Lack of infection in the absence of optimal growth conditions can be overcome by the use of a high phage: host ratio (the multiplicity of infection, or MOI) in which bacterial lysis occurs due to lysis from without (Hudson et al. 2010). Endolysins do not require exponential growth phase cells to be active but the susceptibility to lysis significantly decreases as the culture reaches the late-exponential stage of growth (Celia et al. 2008).

Successful application of phages in dairy products requires a good understanding of the non-linear kinetics of phage-bacteria interactions. Some mathematical models of phage-host interactions suggest that a minimum density of host cells is required in order to support phage replication and significantly reduce the target population of bacteria (Cairns et al. 2009). However, due to the different physico-chemical parameters found in food (contact surface, fluidity, adsorption capacity, proteases, etc) that interfere in the interaction of a phage with its host, the efficacy of phage biocontrol should be determined on a case-by-case basis (Abedon 2009). For example, a probabilistic model was developed to facilitate the successful application of bacteriophages as biocontrol agents against *S. aureus* in pasteurized milk. This model provided the minimum phage concentration (about $2x10^8$ PFU ml^{-1}) required to inactivate *S. aureus* in milk at different temperatures, irrespective of the bacterial contamination level (Obeso et al. 2010).

Another important issue is how a phage-derived agent will survive through processing and intrinsic environmental parameters of dairy products. Phage and phage-derived proteins require specific conditions for its activity, such as pH, temperature, ionic strength, etc., which may not be met in dairy products (Donovan et al. 2006b, Obeso et al. 2008). As such, particular traits have to be selected to warrant viability and antimicrobial activity in a particular application. *S. aureus* phages ΦH5 and ΦA72 were shown to withstand short exposures at high temperatures (72°C, 15 s) but their titre was reduced below the detection limit (<10 PFU ml^{-1}) after 1 min of exposure (García et al. 2009a). Therefore, mixtures of these phages are not inactivated by current pasteurization processes and *S. aureus* might be inhibited during subsequent handling of the pasteurized milk.

Regarding phage lytic proteins, most of the endolysins require divalent cations. The phi11 endolysin has peak activity at 2-3 mM $CaCl_2$, similar to free calcium concentration of bovine milk (Donovan 2007). Endolysins are often inactivated by high temperature and low pH. The endolysin LysH5 from the *S. aureus* bacteriophage ΦH5 was slightly inactivated at pH 6.0 and 48% and 1% residual activity was detected at pHs 5.0 and 4.0, respectively (Obeso et al. 2008). Of note, some virion peptidoglycan hydrolases are quite resistant to high temperatures withstanding even boiling (Rodríguez et al. unpublished). Therefore, they might be used in combination with current food preservation technologies to guarantee the food products safety.

4. Applications of Phage-Derived Antimicrobials in Dairy Processing

There are different approaches that may be adopted within phage-based biocontrol that cover the entire dairy production chain "from farm to fork" (Figure 6). They span from strategies to improve animal health to direct application in the manufactured product to inhibit the development of pathogenic bacteria during storage and also as tools for detecting undesirable bacteria through the different manufacturing steps. Specific applications reported in the literature are summarized in Table 1.

a) Phage therapy and prophylaxis of dairy cattle. For many pathogens, milk contamination occurs either directly or indirectly from sources such the animal's feces or the mammary gland during milking. The use of phages to reduce pathogens within the intestinal tract, udder surface, and the farm environment is therefore seen as a feasible intervention to reduce the number of pathogens in milk. There are some examples of the successful use of phages to lower the carriage of *E. coli* O157:H7 in the gastrointestinal tract. To optimize phage delivery in cattle, phages SH1 and KH1 were applied directly to the rectoanal junction mucosa and were maintained at 10^6 PFU ml^{-1} in the drinking water. This treatment reduced the average number of *E. coli* O157:H7 although it did not eliminate the bacteria (Sheng et al. 2006). The fecal shedding of *E. coli* O157:H7 was also reduced by the oral and rectal administration of four-strain O157-specific bacteriophage cocktail in multiple doses (Rozema et al. 2009).

Phage therapy was also evaluated as an alternative to antibiotics in the prophylaxis and treatment of mastitis and, thereby, reducing the risk of milk contamination during milking. The ability of the lytic *S. aureus* bacteriophage K to eliminate bovine *S. aureus* intramammary infection during lactation was evaluated in cows by intramammary phage infusions. The treatment was far from successful as the cure rate of phage-treated quarters hardly differed from the control quarters (Gill et al. 2006b). The phage concentration in the milk suggested that there was a significant degradation or inactivation of the infused phage within the gland.

Table 1. Reported phage-based biocontrol in dairy

Application	Bacterial target	Phage-derived product	References
Fecal shedding reduction in cattle	*Escherichia coli* O157:H7	Phages SH1 and KH1 Phages rV5, wV7, wV8 and wV11 cocktail	Sheng et al. 2006 Rozema et al. 2009
Mastitis treatment	*Staphylococcus aureus*	Phage K	Gill et al. 2006b
Biofilm elimination	*Listeria monocytogenes*	Phages H387, H387-A and 2671	Roy et al. 1993
Biofilm elimination	*Pseudomonas fluorescens*	Phage ΦIBB-PF7A	Sillankorva et al. 2008
Biofilm elimination	*Staphylococcus aureus*	Phage K endolysin Phage SAP-2 endolysin	Sass and Bierbaum 2007 Son et al. 2010
Milk biopreservation	*Staphylococcus aureus*	Phages ΦH5 and ΦA72 Phages Φ88 and Φ35 and nisin Endolysin LysH5 Endolysin LysH5 and nisin	García et al. 2009a Martinez et al. 2008 Obeso et al. 2008 García et al. 2010b
Curd biopreservation	*Staphylococcus aureus*	Phages Φ88 and Φ35	García et al. 2007
Cheddar cheese biopreservation	*Salmonella* Enteritidis	Phage SJ2	Modi et al. 2001
Red smear soft cheese biopreservation	*Listeria monocytogenes*	Phage P100	Carlton et al. 2005
Chocolate milk and mozzarella cheese brine biopreservation	*Listeria monocytogenes*	Phages A511 and P100	Guenther et al. 2009
Infant formula milk biopreservation	*Enterobacter sakazakii*	Phages ESP 1-3 and ESP 732-1 Phages F, 23, 81,82 and 83 cocktail	Kim et al. 2007 Zuber et al. 2008
Cheese spoilage prevention	*Clostridium tyrobutyricum*	Phage CTP1 endolysin	Mayer et al. 2010
Milk spoilage prevention	*Pseudomonas fluorescens*	Phage ΦIBB-PF7A	Sillankorva et al. 2008
Bacteria detection from dairy products	*Listeria monocytogenes*	*Listeria* phage endolysin–derived cell-wall-binding domain proteins	Kretzer et al. 2007 Schmelcher et al. 2010 Walcher et al. 2010

The ineffectiveness of phage K in the treatment of mastitis could also be due to interference of phage adsorption by whey proteins present in raw milk (O'Flaherty et al. 2005, Gill et al. 2006a). However, phages with wide host ranges and lytic capability in raw milk have been recently described (García et al. 2009a, b, Synnott et al. 2009); these may be active in treatment or prophylaxis of mastitis.

Nevertheless, more research is required to determine the potential of phage therapy including the characterization of new phages and its routes of administration in animals, which requires understanding the ecology and dynamics of the phages and their interactions with pathogen bacteria in the environment (Johnson et al. 2008).

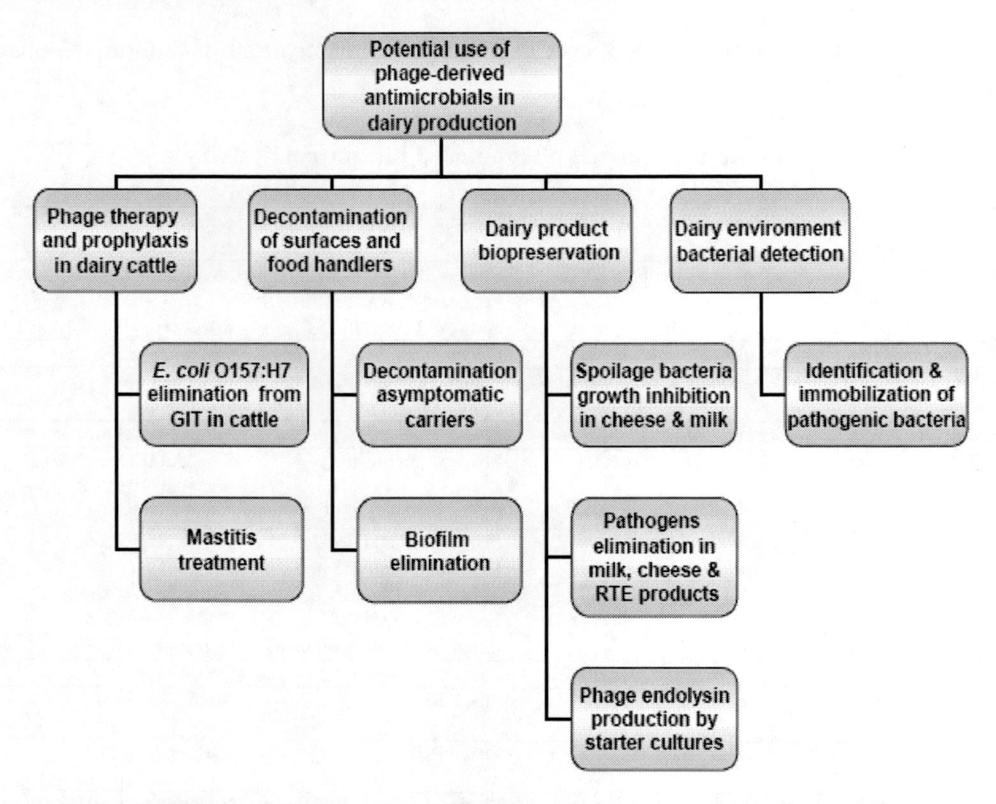

Figure 6. Bacteriophage and phage-derived proteins applications along the dairy production chain.

Another strategy to control *S. aureus* infections involves the use of phage endolysins. It has been shown that phi11, LysK and LysH5 endolysins have high lytic activities against staphylococcal mastitis pathogens (O'Flaherty et al. 2005, Donovan et al. 2006a, Obeso et al. 2008). In addition, the activity of the LysK protein can be enhanced by a synergistic effect in combination with lysostaphin (Becker et al. 2008) and the activity of LysH5 by combination with nisin (García et al. 2010b). Furthermore, new lytic activities can be obtained from virion-associated peptidoglycan hydrolase proteins. *S. aureus* bacteriophages P68 and phiMR11 contains a virion-associated muralytic enzyme that displayed activity against *S. aureus* isolates (Takác and Bläsi 2005, Rashel et al. 2008). Research on staphylococcal lysins as therapeutic agents is still under development and no clinical trials in humans have been performed so far. However, the MV-L lysin from the phage ΦMR11 was shown to reduce MRSA in the nasal cavities of mice and increased survival in a model of systemic MRSA disease (Rashel et al. 2007). More recently, a novel chimeric lysin obtained by fusing the N-terminal catalytic domain of the *S. aureus* Twort phage lysin with the C-terminal cell wall-targeting domain from the *S. aureus* phage phiNM3 lysin showed a synergistic interaction with oxacillin protecting against MRSA septic death in a mouse model (Daniel et al. 2010).

b) Decontamination of food-contact surfaces and food handlers. Biofilms that develop on surfaces in dairy environments (floors, walls, drains, external surfaces of equipment) represent a recurring source of contamination compromising the safety and quality of dairy products. They form a protective environment for undesirable bacteria and serve as reservoirs from which bacteria can potentially come into contact with dairy products (Sharma and Anand 2002). The transmission of pathogens can also result from aerosols during the cleaning procedures of biofilm-contaminated surfaces. Commensal bacteria in food handlers' skin and mucosal surfaces could be also a reservoir of potential pathogens such as *S. aureus* and so strict control in the access of carriers to processing plants is necessary. Regular cleaning and disinfection can reduce the number of bacteria attached to surfaces in biofilms but complete removal is very difficult. *In vitro* experiments have shown that phages are able to infect biofilm cells, although they were killed at a slower rate than planktonic cells (Cerca et al. 2007). Moreover, some phages are able to induce the production of depolymerases that allow the phages penetrating the inner layers of the biofilm by degrading components of the exopolymeric matrix (Azeredo and Sutherland 2008). Recently, phages have been successfully engineered to express an enzyme, DspB, which hydrolyzes the biofilm formed by *Staphylococcus* and *E. coli* (Lu and Collins 2007). In the industrial environment, phage treatment against *List. monocytogenes* on stainless steel and polypropylene surfaces gave a significant reduction in bacteria numbers (Roy et al. 1993). Differences in phage effectiveness to remove *Pseudomonas fluorescens* biofilms were observed depending on the biofilm age (Sillankorva et al. 2008). Additionally, staphylococcal endolysins of phage K and phage SAP-2 have been also shown to be effective in disrupting *S. aureus* biofilms (Sass and Bierbaum 2007, Son et al. 2010).

c) Dairy products biopreservation. The main goal of the addition of phages and endolysins as biopreservatives in dairy products is to inhibit pathogen and spoilage bacteria growth and, thus, increase the shelf life of these products. There are several reports in the literature that study the efficacy and bottlenecks of phage biopreservation in milk and dairy products. Phages ΦH5 and ΦA72 inhibited *S. aureus* growth in UHT and pasteurized whole-fat milk. However, the phages were less active in semi-skimmed raw milk and little inhibition was achieved in whole raw milk. Killing of *S. aureus* was observed at room temperature and at 37°C, but not at refrigeration temperature. However, phage titres were stable during cold storage and counteracted *S. aureus* multiplication in a simulated temperature breakdown (García et al. 2009a). Therefore, the presence of phages may preclude undesirable bacteria from reaching higher numbers in case of unexpected temperature abuse during milk processing or storage. This is of practical relevance in the case of *S. aureus* contamination to avoid toxin accumulation beyond poisoning levels. Lytic phages added to milk were also able to decrease rapidly the viable counts (3×10^6 CFU ml^{-1}) of *S. aureus* during curd manufacture. In acid curd, the pathogen was not detected after 4 h of incubation at 25°C, whereas pathogen clearance was achieved within 1 h of incubation at 30°C in renneted curd (García et al. 2007). The ability of *Salmonella* Enteritidis (10^4 CFU ml^{-1}) to survive in the presence of phage SJ2 (10^8 PFU ml^{-1}) during manufacture, ripening, and storage of cheddar cheese produced from raw and pasteurized milk was investigated. Counts of *Salmonella* Enteritidis decreased by 1 to 2 log cycles in raw and pasteurized milk cheeses made from milk containing the phage. *Salmonella* did not survive in pasteurized milk cheese after 89 days of storage in the presence of phage SJ2. However, *Salmonella* counts of approximately 50 CFU g^{-1} were detected in raw milk cheese containing the phage after 99 days of storage

(Modi et al. 2001). *List. monocytogenes* contamination was also controlled on surface-ripened red smear soft cheese with phage P100. Cheeses were surface-contaminated with low concentrations of *List. monocytogenes* at the beginning of the ripening period, and P100 was surface-applied during the rind washings. Depending on the frequency and dose of phage applications, phage P100 was able to reduce at least 3.5 unit-log or to eliminate *Listeria* viable counts (Carlton et al. 2005). Control of *List. monocytogenes* growth by the virulent and broad-host-range phages A511 and P100 in different ready-to-eat (RTE) foods was also evaluated. Bacterial counts rapidly dropped below the level of direct detection in chocolate milk and mozzarella cheese brine contaminated with bacteria ($1x10^3$ CFU g^{-1}) and treated with the phage ($3x10^6$ to $3x10^8$ PFU g^{-1}) and stored at 6°C for 6 days (Guenther et al. 2009). Six new *Enterobacter sakazakii* phages isolated from sewage were used to control bacterial growth in reconstituted infant formula milk. Phages effectively prevented development of *Enterobacter sakazakii* in formula at various temperatures (12, 24 and 37°C) (Kim et al. 2007). In another report, a total of 67 phages isolated from environmental samples were tested against a representative collection of 40 *Enterobacter sakazakii* strains. A cocktail of five phages prevented the outgrowth of 35 out of 40 test strains in artificially contaminated infant formula (Zuber et al. 2008).

The lytic activity of endolysins can also be used to prevent pathogen growth in milk. The purified protein LysH5 from the *S. aureus* phage ΦH5 was able to rapidly kill 10^6 CFU ml^{-1} of *S. aureus* growing in pasteurized milk and the pathogen was not detected after 4 h of incubation at 37°C. However, the amount of lysin required in milk was 10 fold higher than in buffer to obtain the same antimicrobial effect (Obeso et al. 2008).

The *in situ* production of lysins during fermentation by genetically modified starter cultures able to synthesize and secrete them has also been explored as an alternative to the exogenous addition of endolysins. Synthesis of the Ply511 endolysin active against *List. monocytogenes* was achieved in *Lactococcus lactis* MG1363. High levels of secreted active enzyme were produced by fusion to the *Lactobacillus brevis* S-layer protein signal peptide and caused rapid lysis of *List. monocytogenes* cells (Gaeng et al. 2000). In a similar approach, the expression and secretion signals of the Sep protein from *Lactobacillus fermentum* BR11 were used to export Ply511 by *Lb. fermentum* BR11, *Lactobacillus rhamnosus* GG, *Lactobacillus plantarum* ATCC 14917 and *L. lactis* MG1363. However, supernatants containing Ply511 were unable to inhibit *List. monocytogenes* growth (Turner et al. 2007).

In addition, the antimicrobial potential of phages and phage-derived proteins can be applied in hurdle technology in combination with current preservation methods used in the dairy industry. A synergistic effect of nisin and two lytic phages against *S. aureus* was observed in short-time challenge experiments performed in pasteurized milk. However, nisin-adapted cells developed after nisin exposure and became partially resistant to both phages (Martínez et al. 2008). Studies reporting the effective use of nisin against *S. aureus* in milk in combination with endolysins as preservatives have been also reported. When LysH5 endolysin was combined with nisin in pasteurized milk contaminated with *S. aureus*, strong synergy was observed. Clearance of the pathogen could be achieved by the combined activity of both antimicrobials at lower concentrations than when used alone (García et al. 2010b).

Finally, control of spoilage bacteria by phages and endolysins has been also approached. The partially purified endolysin from the virulent bacteriophage CTP1 induced lysis of *Clostridium tyrobutyricum* cells and reduced viable counts in milk without affecting the cheese microbiota (Mayer et al. 2010). The T7-like phage, phage ΦIBB-PF7A, is a lytic phage capable of infecting a variety of *Pseudomonas fluorescens* strains, a spoilage microorganism in milk. Based on its fast and efficient bacterial lysis, this phage has been proposed as a good candidate to be used as a sanitation agent to control the prevalence of spoilage strains in dairy environments (Sillankorva et al. 2008).

d) Bacterial detection in dairy production. Phages encode proteins which have evolved to specifically interact with molecules on the surface of bacterial cell walls. Specific recognition of the host cell is mostly carried out by tail fiber proteins. This property has been exploited to develop means for detection of pathogenic bacteria in foods or in the food manufacturing environment. Several methods have been reported so far to detect bacteria in foods using phages (Rees and Dodd 2006). A recombinant phage specific for *E. coli* 0157:H7 that carries the *lux*I gene, coding for the synthesis of the quorum-sensing signaling molecule N-3-(oxohexanoyl)-L-homoserine lactone was constructed. In the presence of 10^4 CFU ml^{-1} of *E. coli* 0157:H7, the signaling molecule accumulates and induces light in bioreporter cells that contain the luxR and luxCDABE genes under the control of the lux promoter. This system was able to detect 10 CFU ml^{-1} in pure culture (Brigati et al. 2007). Bacteriophages can also be used as a platform to present the green fluorescent protein (GFP) on the phage capsid. Adsorption of the GFP-labeled PP01 phages to the *E. coli* cell surface enabled visualization and detection of cells under a fluorescence microscope (Oda et al. 2004). More recently, bacteriophage immobilization has been proposed to detect, concentrate, and identify target bacteria (Tolba et al. 2010). T4 recombinant bacteriophages displaying the biotin binding peptide (BCCP) and the cellulose binding module (CBM) on their heads were immobilized on nano-aluminum fiber-based filter and coupled with a bioluminescent ATP assay. These phages allowed simultaneous concentration and detection of as low as 6×10^3 CFU ml^{-1} of *E. coli* in the samples (Minikh et al. 2010). Detection of *Salmonella typhimurium* on fresh tomato surfaces (5×10^2 CFU ml^{-1}) was achieved using a phage-based magnetoelastic (ME) biosensor composed of a ME resonator platform coated with filamentous E2 phage (Li et al. 2010).

In addition, the high affinity of the cell binding domain (CBD) of many endolysins has been exploited for immobilization and concentration of pathogenic bacteria such *List. monocytogenes* on magnetic beads. The CBD-based magnetic separation (CBD-MS) procedure was shown to be more sensitive that the established standard procedures requiring less time (48 h versus 96 h) for capture and detection of *List. monocytogenes* from artificially and naturally contaminated Camembert soft cheese, red smear cheese and pasteurized milk samples (Kretzer et al. 2007). Rapid multiplexed detection and differentiation of *Listeria* strains in artificially contaminated milk or cheese was possible by combining CBDs of different binding specificities with fluorescent markers of various colors (Schmelcher et al. 2010). A similar approach was followed by combination of the bead-based method with subsequent detection and quantification by the traditional plate-count technique and real-time polymerase chain reaction (PCR). The detection limit in raw milk ranged from 10^2 to 10^3 CFU ml^{-1} (Walcher et al. 2010). The specific binding of the lysin PlyG from *Bacillus anthracis* gamma phage has been used for the reliable identification of this pathogen (Schuch et al. 2002). Recently, synthetic peptides based in PlyG, coupled with Qdot-nanocrystals were developed as high-sensitivity bio-probes for detecting *B. anthracis* (Sainathrao et al. 2009).

CONCLUSION

The application of phages or phage-encoded proteins as biocontrol agents in the dairy industry is a relatively new and challenging concept which still requires further research. Their possibilities in food safety have to be assessed in a case-by-case basis to provide enough evidence of their efficacy required for their approval by regulatory bodies and, more importantly, for consumer acceptance.

REFERENCES

Abedon, S. T. (2009). Kinetics of phage-mediated biocontrol of bacteria. *Foodborne Pathogens and Disease, 6,* 807-815.

Arnold, T., Neubauer, H., Ganter, M., Nikolaou, K., Roesler, U., Truyen, U. and Hensel, A. (2006). Prevalence of *Yersinia enterocolitica* in goat herds from Northern Germany. *Journal of Veterinary Medicine B, Infectious Diseases and Veterinary Public Health, 53,* 382-386.

Asao, T., Kumeda, Y., Kawai, T., Shibata, T., Oda, H., Haruki, K., Nakazawa, H. and Kozaki, S. (2003). An extensive outbreak of staphylococcal food poisoning due to low-fat milk in Japan: estimation of enterotoxin A in the incriminated milk and powdered skim milk. *Epidemiology and infection, 130,* 33-40.

Azeredo, J. and Sutherland, I. (2008). The use of phages for the removal of infectious biofilms. *Current Pharmaceutical Biotechnology, 9,* 261-266.

Babcock, D. (2010). 2010 Raw milk scoreboard - *E. coli*, *Salmonella*, and *Campylobacter*. *Food Poison Journal* [http://www.foodpoisonjournal.com/admin/trackback/204252].

Becker, S. C., Foster-Frey, J., Donovan, D. M. (2008). The phage K lytic enzyme LysK and lysostaphin act synergistically to kill MRSA. *FEMS Microbiology Letters, 87,* 185-91.

Becker, S. C., Foster-Frey, J., Stodola, A. J., Anacker, D. and Donovan, D. M. (2009). Differentially conserved staphylococcal SH3b_5 cell wall binding domains confer increased staphylolytic and streptolytic activity to a streptococcal prophage endolysin domain. *Gene, 443,* 32-41.

Bergère, J. L. and Lenoir, J. (2000). Cheese manufacturing accidents and cheese defects. In: Eck, A. and Gillis, J. C. (Eds.). *Cheesemaking: From science to quality assurance.* (2nd edition, pp. 447-508). Paris, France: Lavoisier Publishing.

Bernhardt, T. G., Wang, I. N., Struck, D. K. and Young, R. (2002). Breaking free: "protein antibiotics" and phage lysis. *Research in Microbiology, 153,* 493–501.

Bonardi, S., Paris, A., Bacci, C., D'Incau, M., Ferroni, L. and Brindani, F. (2007). Detection and characterization of *Yersinia enterocolitica* from pigs and cattle. *Veterinary Research, 1,* 347-350.

Brigati, J. R., Ripp, S. A., Johnson, C. M., Iakova, P. A., Jegier, P. and Sayler, G. S. (2007). Bacteriophage-based bioluminescent bioreporter for the detection of *Escherichia coli* 0157:H7. *Journal of Food Protection, 70,* 1386-1392.

Brüssow, H., Canchaya C. and Hardt, W. D. (2004). Phages and the evolution of bacterial pathogens: from genomic rearrangements to lysogenic conversion. *Microbiology and Molecular Biology Reviews, 68,* 560-602.

Bruttin, A. and Brüssow, H. (2005). Human volunteers receiving *Escherichia coli* phage T4 orally: a safety test of phage therapy. *Antimicrobial Agents and Chemotherapy, 49,* 2874-2878.

Cairns, B. J., Timms, A. R., Jansen, V. A., Connerton, I. F. and Payne, R. J. (2009). Quantitative models of in vitro bacteriophage-host dynamics and their application to phage therapy. *PLoS Pathogen, 5,* e1000253.

Carlton, R. M., Noordman, W. H., Biswas, B., de Meester, E. D. and Loessner, M. J. (2005). Bacteriophage P100 for control of *Listeria monocytogenes* in foods: Genome sequence, bioinformatic analyses, oral toxicity study, and application. *Regulatory Toxicology and Pharmacology, 43,* 301-312.

Celia, L. K., Nelson, D. and Kerr, D. E. (2008). Characterization of a bacteriophage lysin (Ply700) from *Streptococcus uberis*. *Veterinary Microbiology, 130,* 107-117.

Cerca, N., Oliveira, R. and Azeredo, J. (2007). Susceptibility of *Staphylococcus epidermidis* planktonic cells and biofilms to the lytic action of staphylococcus bacteriophage K. *Letters in Applied Microbiology, 45,* 313-317.

Champagne, C. P., Laing, R. R., Roy, D. and Mafu, A. A. (1994). Psychrotrophs in dairy products: their effects and their control. *Critical Reviews in Food Science and Nutrition, 34,* 1-30.

Craven, H. M., McAuley, C. M., Duffy, L. L. and Fegan, N. (2010). Distribution, prevalence and persistence of *Cronobacter* (*Enterobacter sakazakii*) in the nonprocessing and processing environments of five milk powder factories. *Journal of Applied Microbiology, 109,* 1044-1052.

Cremonesi, P., Pérez, G., Pisoni, G., Moroni, P., Morandi, S., Luzzana, M., Brasca, M. and Castiglioni, B. (2007). Detection of enterotoxigenic *Staphylococcus aureus* isolates in raw milk cheese. *Letters in Applied Microbiology, 45,* 586-591.

Daniel, A., Euler, C., Collin, M., Chahales, P., Gorelick, K. J. and Fischetti, V. A. (2010). Synergism between a novel chimeric lysin and oxacillin protects against infection by methicillin-resistant *Staphylococcus aureus*. *Antimicrobial Agents and Chemotherapy, 54,* 1603-1612.

De Buyser, M. L., Dufour, B., Maire, M. and Lafarge, V. (2001). Implication of milk and milk products in food-borne diseases in France and in different industrialised countries. *International Journal of Food Microbiology, 67,* 1-17.

De Jonghe, V., Coorevits, A., De Block, J., Van Coillie, E., Grijspeerdt, K., Herman, L., De Vos, P. and Heyndrickx, M. (2010). Toxinogenic and spoilage potential of aerobic spore-formers isolated from raw milk. *International Journal of Food Microbiology, 136,* 318-325.

DeWaal C. S., Tian, X. A. and Plunkett, D. (2009). Outbreak alert report. analyzing foodborne outbreaks 1998 to 2007. Center for Science in the Public Interest. www.cspinet.org/new/pdf/outbreak_report_2009.pdf.

Donovan, D. M., Lardeo, M. and Foster-Frey, J. (2006a). Lysis of staphylococcal mastitis pathogens by bacteriophage phi11 endolysin. *FEMS Microbiology Letters, 265,* 133-139.

Donovan, D. M., Dong, S., Garrett, W., Rousseau, G. M., Moineau, S. and Pritchard, D. G. (2006b). Peptidoglycan hydrolase fusions maintain their parental specificities. *Applied and Environmental Microbiology, 72,* 2988-2996.

Donovan, D. M. (2007). Bacteriophage and peptidoglycan degrading enzymes with antimicrobial applications. *Recent Patents on Biotechnology, 1,* 113-122.

EFSA (European Food Safety Authority). (2010). The Community summary report on trends and sources of zoonoses, zoonotic agents and food-borne outbreaks in the EU. *EFSA Journal, 8,* 1496.

Favila-Humara, L. C., Chávez-Gris, G. G., Carrillo-Casas, E. M. and Hernández-Castro, R. (2010). *Mycobacterium avium* subsp. *paratuberculosis* detection in individual and bulk tank milk samples from bovine herds and caprine flocks. *Foodborne Pathogens and Disease, 7,* 351-355.

Fischetti, V. A. (2005). Bacteriophages lytic enzymes: novel anti-infectives. *Trends in Microbiology, 13,* 491-496.

Fischetti, V. A. (2010). Bacteriophage endolysins: a novel anti-infective to control Gram-positive pathogens. *International Journal of Medical Microbiology, 300,* 357-362.

Gaeng, S., Scherer, S., Neve, H. and Loessner, M. J. (2000). Gene cloning and expression and secretion of *Listeria monocytogenes* bacteriophage-lytic enzymes in *Lactococcus lactis. Applied and Environmental Microbiology, 66,* 2951-2958.

Gálvez, A., López, R. L., Abriouel, H., Valdivia, E. and Omar, N. B. (2008). Application of bacteriocins in the control of foodborne pathogenic and spoilage bacteria. *Critical Reviews in Biotechnology, 28,* 125-152.

García, P., García, J. L., García, E., Sánchez-Puelles, J. M. and López, R. (1990). Modular organization of the lytic enzymes of *Streptococcus pneumoniae* and its bacteriophages. *Gene, 31,* 81-88.

García, P., Madera, C., Martínez, B. and Rodríguez, A. (2007). Biocontrol of *Staphylococcus aureus* in curd manufacturing processes using bacteriophages. *International Dairy Journal, 17,* 1232-1239.

García, P., Martínez B., Obeso J. M. and Rodríguez, A. (2008). Bacteriophages and their application in food safety. *Letters in Applied Microbiology, 47,* 479-485.

García, P., Madera, C., Martínez, B., Rodríguez, A. and Suárez, J. E. (2009a). Prevalence of bacteriophages infecting *Staphylococcus aureus* in dairy samples and their potential as biocontrol agents. *Journal of Dairy Science, 92,* 3019-3026.

García, P., Obeso, J. M., Martinez, B., Lavigne, R. and Rodríguez, A. (2009b). Functional genomic analysis of *Staphylococcus aureus* phages isolated from the dairy environment. *Applied and Environmental Microbiology, 75,* 7663-7673.

García, P., Rodríguez, L., Rodríguez, A. and Martínez, B. (2010a). Food biopreservation: Promising strategies using bacteriocins, bacteriophages and endolysins. *Trends in Food Science and Technology, 21,* 373-382.

García, P., Martínez, B., Rodríguez, L. and Rodríguez, A. (2010b). Synergistic effect of the endolysin *LysH5* and nisin against *Staphylococcus aureus* in milk. *International Journal of Food Microbiology, 141,* 151-155.

García-Fulgueiras, A., Sánchez, S., Guillén, J. J., Marsilla, B., Aladueña, A. and Navarro, C. (2001). A large outbreak of *Shigella sonnei* gastroenteritis associated with consumption of fresh pasteurised milk cheese. *European Journal of Epidemiology, 17,* 533-538.

Gill, J. J., Sabour, P. M., Leslie, K. E. and Griffiths, M. W. (2006a). Bovine whey proteins inhibit the interaction of *Staphylococcus aureus* and bacteriophage K. *Journal of Applied Microbiology, 101,* 377-386.

Gill, J. J., Pacan, J. C., Carson, M. E., Leslie, K. E., Griffiths, M. W. and Sabour, P. M. (2006b). Efficacy and pharmacokinetics of bacteriophage therapy in treatment of subclinical *Staphylococcus aureus* mastitis in lactating dairy cattle. *Antimicrobial Agents and Chemotherapy, 50,* 2912-2918.

Guenther, S., Huwyler, D., Richard, S. and Loessner, M. J. (2009). Virulent bacteriophage for efficient biocontrol of *Listeria monocytogenes* in ready-to-eat foods. *Applied and Environmental Microbiology, 75,* 93-100.

Hagens, S. and Loessner, M. J. (2007). Application of bacteriophages for detection and control of foodborne pathogens. *Applied Microbiology and Biotechnology, 76,* 513-519.

Hata, E., Kobayashi, H., Nakajima, H., Shimizu, Y. and Eguchi, M. (2010). Epidemiological analysis of *Staphylococcus aureus* isolated from cows and the environment of a dairy farm in Japan. *The Journal of Veterinary Medical Science, 72,* 647-652.

Hatfull, G. F. (2008). Bacteriophage genomics. *Current Opinion in Microbiology, 11,* 447-453.

Hudson, J. A., Billington, C., Carey-Smith, C. and Greening, G. (2005). Bacteriophages as biocontrol agents in food. *Journal of Food Protection, 68,* 426-437.

Hyman, P. and Abedon, S. T. (2010). Bacteriophage host range and bacterial resistance. *Advances in Applied Microbiology, 70,* 217-248.

Johnson, R. P., Gyles, C. L., Huff, W. E., Ojha, S., Huff, G. R., Rath, N. C. and Donoghue, A. M. (2008). Bacteriophages for prophylaxis and therapy in cattle, poultry and pigs. *Animal Health Research Reviews, 9,* 201-215.

Kim, K. P., Klumpp, J. and Loessner, M. J. (2007). *Enterobacter sakazakii* bacteriophages can prevent bacterial growth in reconstituted infant formula. *International Journal of Food Microbiology, 115,*195-203.

Kretzer, J. W., Lehmann, R., Schmelcher, M., Banz, M., Kim, K., Korn, C. and Lossner, M. J. (2007). Use of high-affinity cell wall-binding domains of bacteriophage endolysins for immobilization and separation of bacterial cells. *Applied and Environmental Microbiology, 73,* 1992-2000.

Latorre, A. A., Van Kessel, J. S., Karns, J. S., Zurakowski, M. J., Pradhan, A. K., Boor, K. J., Jayarao, B. M., Houser, B. A., Daugherty, C. S. and Schukken, Y. H. (2010). Biofilm in milking equipment on a dairy farm as a potential source of bulk tank milk contamination with *Listeria monocytogenes*. *Journal of Dairy Science, 93,* 2792-2802.

Li, S., Li, Y., Chen, H., Horikawa, S., Shen, W., Simonian, A. and Chin, B. A. (2010). Direct detection of *Salmonella typhimurium* on fresh produce using phage-based magnetoelastic biosensors. *Biosensors and Bioelectronics*, doi:10.1016/j.bios.2010.07.029

Liu, J., Dehbi, M., Moeck, G., Arhin, F., Bauda, P., Bergeron, D., Callejo, M., Ferretti, V., Ha, N., Kwan, T., McCarty, J., Srikumar, R., Williams, D., Wu, J. J., Gros, P., Pelletier,

J. and DuBow, M. (2004). Antimicrobial drug discovery through bacteriophage genomics. *Nature Biotechnology, 22*, 185-191.

Lu, T. and Collins, J. (2007). Dispersing biofilms with engineered enzymatic bacteriophage. *Proceedings of the National Academy of Sciences of the United States of America, 104,* 11197-11202.

Lunden, J., Tolvanen, R. and Korkeala, H. (2004). Human Listeriosis outbreaks linked to dairy products in Europe. *Journal of Dairy Science, 87,* 6-11.

Mahichi, F., Synnott, A. J., Yamamichi, K., Osada, T. and Tanji, Y. (2009). Site-specific recombination of T2 phage using IP008 long tail fiber genes provides a targeted method for expanding host range while retaining lytic activity. *FEMS Microbiology Letters, 295,* 211-217.

Mallory, J. B., Alfano, C. and Mc Macken, R. (1990). Host virus interactions in the initiation of bacteriophage lambda DNA replication. Recruitment of *Escherichia coli* DnaB helicase by lambda P replication protein. *Journal of Biological Chemistry, 265*, 13297-13307.

Manoharadas, S., White, A. and Bläsi, U. (2009). Antimicrobial activity of a chimeric enzybiotic towards *Staphylococcus aureus. Journal of Biotechnology, 139,* 118-123.

Marler, C. (2010). Hartmann farm dairy raw milk *E. coli.* About E. coli, featured Outbreak [Outbreakhttp://www.about-ecoli.com/ecoli_outbreaks/view/e.-coli-o157h7-outbreak-tied-to-raw-milk-from-hartmann-farm-dairy-in-minnes/].

Martínez, B., Obeso, J. M., Rodríguez, A. and García, P. (2008). Nisin-bacteriophage crossresistance in *Staphylococcus aureus. International Journal of Food Microbiology, 122,* 253-258.

Mayer, M. J., Payne, J., Gasson, M. J. and Narbad, A. (2010). Genomic sequence and characterisation of the virulent bacteriophage phiCTP1 from *Clostridium tyrobutyricum* and heterologous expression of its endolysin. *Applied and Environmental Microbiology, 76,* 5415-5422.

Mead, P. S., Slutsker, L., Dietz, V., Mc Caig, L. F., Bresee, J. S., Shapiro, C., Griffin, P. M. and Tauxe, R. V. (1999). Food-Related illness and death in the United States. *Emerging Infectious Diseases, 5,* 607-625.

Minikh, O., Tolba, M., Brovko, L. Y. and Griffiths, M. W. (2010). Bacteriophage-based biosorbents coupled with bioluminescent ATP assay for rapid concentration and detection of *Escherichia coli. Journal of Microbiological Methods, 82*, 177-183.

Moak, M. and Molineux, I. J. (2004). Peptidoglycan hydrolytic activities associated with bacteriophage virions. *Molecular Microbiology, 51,* 1169–1183.

Modi, R., Hirvi, Y., Hill, A. and Griffiths, M. W. (2001). Effect of phage on survival of *Salmonella enteritidis* during manufacture and storage of Cheddar cheese made from raw and pasteurized milk. *Journal of Food Protection, 64,* 927-933.

Murinda, S. E., Nguyen, L. T., Ivey, S. J., Gillespie, B. E., Almeida, R. A., Draughon, F. A. and Oliver, S. P. (2002). Prevalence and molecular characterization of *Escherichia coli* O157:H7 in bulk tank milk and fecal samples from cull cows: a 12-month survey of dairy farms in east Tennessee. *Journal of Food Protection, 65,* 752-759.

Nechaev, S. and Severinov, K. (1999). Inhibition of *Escherichia coli* RNA polymerase by bacteriophage T7 gene 2 protein. *Journal of Molecular Biology, 289*, 815-826.

Notermans, S. and Van de Giessen, A. (1993). Foodborne diseases in the 1980s and 1990s. *Food Control, 4,* 122-124.

O'Flaherty, S., Coffey, A., Meaney, W. J., Fitzgerald, G. F. and Ross, R. P. (2005). Inhibition of bacteriophage K proliferation on *Staphylococcus aureus* in raw bovine milk. *Letters in Applied Microbiology, 41,* 274-279.

Obeso, J. M., Martínez, B., Rodríguez, A. and García, P. (2008). Lytic activity of the recombinant staphylococcal bacteriophage PhiH5 endolysin active against *Staphylococcus aureus* in milk. *International Journal of Food Microbiology, 128,* 212-218.

Obeso, J. M., García, P., Martínez, B., Arroyo-López, F. N., Garrido-Fernández, A. and Rodríguez, A. (2010). Use of logistic regression for prediction of the fate of *Staphylococcus aureus* in pasteurized milk in the presence of two lytic phages. *Applied and Environmental Microbiology, 76,* 6038-6046.

Oda, M., Morita, M., Unno, H. and Tanji, Y. (2004). Rapid detection of *Escherichia coli* O157:H7 by using green fluorescent protein-labeled PP01 bacteriophage. *Applied and Environmental Microbiology, 70,* 527-534.

Oliver, S. P., Jayarao, B. M. and Almeida, R. A. (2005). Foodborne pathogens in milk and the dairy farm environment: food safety and public health implications. *Foodborne Pathogens and Disease, 2,* 115-129.

Orito, Y., Morita, M., Hori, K., Unno, H. and Tanji, Y. (2004). *Bacillus amyloliquefaciens* phage endolysin can enhance permeability of *Pseudomonas aeruginosa* outer membrane and induce cell lysis. *Applied Microbiology and Biotechnology, 65,* 105-109.

Park, T., Struck, D. K., Dankenbring, C. A. and Young, R. (2007). The pinholin of lambdoid phage 21: control of lysis by membrane depolarization. *Journal of Bacteriology, 189,* 9135-9139.

Peterson, M. C. (2003). *Campylobacter jejuni* enteritis associated with consumption of raw milk. *Journal Environmental Health, 65,* 20-21.

Rashel, M., Uchiyama, J., Takemura, I., Hoshiba, H., Ujihara, T., Takatsuji, H., Honke, K. and Matsuzaki, S. (2008). Tail-associated structural protein gp61 of *Staphylococcus aureus* phage ΦMR11 has bifunctional lytic activity. *FEMS Microbiology Letters, 284,* 9-16.

Rashel, M., Uchiyama, J., Ujihara, T., Uehara, Y., Kuramoto, S., Sugihara, S., Yagyu, K., Muraoka, A., Sugai, M., Hiramatsu, K., Honke, K. and Matsuzaki, S. (2007). Efficient elimination of multidrug-resistant *Staphylococcus aureus* by cloned lysin derived from bacteriophage phi MR11. *Journal of Infection Disease, 15,* 1237-1247.

Rees, E. D. and Dodd, C. E. R. (2006). Phage for rapid detection and control of bacterial pathogens in food. *Advances in Applied Microbiology, 59,* 159-186.

Rodríguez, L., Martínez, B., Zhou, Y., Rodríguez, A., Donovan, D. and García, P. Thermostable lytic activity of the virion-associated peptidoglycan hydrolase HydH5 of *Staphylococcus aureus* bacteriophage vB_SauS-phiIPLA88. (unpublished).

Roy, B., Ackermann, H. W., Pandian, S., Picard, G. and Goulet, J. (1993). Biological inactivation of adhering *Listeria monocytogenes* by listeria phages and a quaternary ammonium compound. *Applied and Environmental Microbiology, 59,* 2914-2917.

Rozema, E. A., Stephens, T. P., Bach, S. J., Okine, E. K., Johnson, R. P., Stanford, K. and McAllister, T. A. (2009). Oral and rectal administration of bacteriophages for control of *Escherichia coli* O157:H7 in feedlot cattle. *Journal of Food Protection, 72,* 241-250.

Ruzante, J. M., Lombard, J. E., Wagner, B., Fossler, C. P., Karns, J. S., Van Kessel, J. A. and Gardnerm, I. A. (2010). Factors associated with *Salmonella* presence in environmental samples and bulk tank milk from US dairies. *Zoonoses and Public Health,* DOI: 10.1111/j.1863-2378.2010.01333.x

Sainathrao, S., Mohan, K. V. and Atreya, C. (2009). Gamma-phage lysin PlyG sequence-based synthetic peptides coupled with Qdot-nanocrystals are useful for developing detection methods for *Bacillus anthracis* by using its surrogates, *B. anthracis*-Sterne and *B. cereus*-4342. *BMC Biotechnology, 22,* 67.

Sass, P. and Bierbaum, G. (2007). Lytic activity of recombinant bacteriophage F11 and F12 endolysins on whole cells and biofilms of *Staphylococcus aureus*. *Applied and Environmental Microbiology, 73,* 347-352.

Schiemann, D. A. (1987). *Yersinia enterocolitica* in milk and dairy products. *Journal of Dairy Science, 70,* 383-391.

Schmelcher, M., Shabarova, T., Eugster, M. R., Eichenseher, F., Tchang, V. S., Banz, M. and Loessner, M. J. (2010). Rapid multiplex detection and differentiation of *Listeria* cells by use of fluorescent phage endolysin cell wall binding domains. *Applied and Environmental Microbiology, 76,* 5745-5756.

Schuch, R., Nelson, D. and Fischetti, V. A. (2002) A bacteriolytic enzyme that detects and kills *Bacillus anthracis*. *Nature, 418,* 884-889.

Sharma, M. and Anand, S. K. (2002). Characterization of constitutive microflora of biofilms in dairy processing lines. *Food Microbiology, 19,* 627-636.

Sheng, H., Knecht, H. J., Kudva, I. T. and Hovde, C. J. (2006). Application of bacteriophages to control intestinal *Escherichia coli* O157:H7 levels in ruminants. *Applied Environmental Microbiology, 72*, 5359-5366.

Sillankorva, S., Neubauer, P. and Azeredo, J. (2008). *Pseudomonas fluorescens* biofilms subjected to phage phiIBB-PF7A. *BMC Biotechnology, 27,* 8-79.

Solomakos, N., Govaris, A., Angelidis, A. S., Pournaras, S., Burriel, A. R., Kritas, S. K. and Papageorgiou, D. K. (2009). Occurrence, virulence genes and antibiotic resistance of *Escherichia coli* O157 isolated from raw bovine, caprine and ovine milk in Greece. *Food Microbiology, 26,* 865-871.

Son, J. S., Lee, S. J., Jun, S. Y., Yoon, S. J., Kang, S. H., Paik, H. R., Kang, J. O. and Choi, Y. J. (2010). Antibacterial and biofilm removal activity of a podoviridae *Staphylococcus aureus* bacteriophage SAP-2 and a derived recombinant cell-wall-degrading enzyme. *Applied Microbiology and Biotechnology, 86,* 1439-1449.

Strauch, E., Hammerl, J. A. and Hertwig, S. (2007). Bacteriophages: new tools for safer food?. *Journal für Verbraucherschutz und Lebensmittelsicherheit*, *2,* 138-143.

Sturino, J. M. and Klaenhammer, T. R. (2004). Bacteriophage defense systems and strategies for lactic acid bacteria. *Advances in Applied Microbiology, 56,* 331-378.

Sulakvelidze, A. and Kutter, E. (2005). Bacteriophage therapy in humans. In E. Kutter and A. Sulakvelidze (Eds.), *Bacteriophages: Biology and Application* (pp. 381–436). Boca Raton, FL: CRC Press.

Synnott, A. J., Kuang, Y., Kurimoto, M., Yamamichi, K., Iwano, H. and Tanji, Y. (2009). Isolation from sewage influent and characterization of novel *Staphylococcus aureus* bacteriophages with wide host ranges and potent lytic capabilities. *Applied and Environmental Microbiology, 75,* 4483-4490.

Takác, M. and Bläsi, U. (2005). Phage P68 virion-associated protein 17 displays activity against clinical isolates of *Staphylococcus aureus*. *Antimicrobial Agents and Chemotherapy, 49,* 2934-2940.

Tauxe, R. V. (2002). Emerging foodborne pathogens. *International Journal of Microbiology, 78,* 31-41.

Tiwari, B. K., Valdramidis, V. P., O'Donnell, C. P., Muthukumarappan, K., Bourke, P. and Cullen, P.J. (2009). Application of natural antimicrobials for food preservation. *Journal of Agricultural and Food Chemistry, 57,* 5987-6000.

Tolba, M., Minikh, O., Brovko, L. Y., Evoy, S. and Griffiths, M. W. (2010). Oriented immobilization of bacteriophages for biosensor applications. *Applied and Environmental Microbiology, 76,* 528-535.

Tormo-Más, M. A., Mir, I., Shrestha, A., Tallent, S. M., Campoy, S., Lasa, I., Barbé, J., Novick, R. P., Christie, G. E. and Penadés, J. R. (2010). Moonlighting bacteriophage proteins derepress staphylococcal pathogenicity islands. *Nature*, *10*, 779-782.

Türkyilmaz, S., Tekbiyik, S., Oryasin, E. and Bozdogan, B. (2010). Molecular epidemiology and antimicrobial resistance mechanisms of methicillin-resistant *Staphylococcus aureus* isolated from bovine milk. *Zoonoses Public Health*, *57*, 197-203.

Turner, M. S., Waldherr, F., Loessner, M. J. and Giffard, P. M. (2007). Antimicrobial activity of lysostaphin and a *Listeria monocytogenes* bacteriophage endolysin produced and secreted by lactic acid bacteria. *Systematic and Applied Microbiology, 30,* 58-67.

Vilar, M. J., Peña, F. J., Pérez, I., Diéguez, F. J., Sanjuán, M. L., Rodríguez-Otero, J. L. and Yus, E. (2010). Presence of *Listeria, Arcobacter,* and *Campylobacter* spp. in dairy farms in Spain. *Berliner und Munchener Tierarztliche Wochenschrift, 123,* 58-62.

Walcher, G., Stessl, B., Wagner, M., Eichenseher, F., Loessner, M. J. and Hein, I. (2010). Evaluation of paramagnetic beads coated with recombinant *Listeria* phage endolysin-derived cell-wall-binding domain proteins for separation of *Listeria monocytogenes* from raw milk in combination with culture-based and real-time polymerase chain reaction-based quantification. *Foodborne Pathogen Diseases, 7,* 1019-1024.

Walsh, C. and Fanning, S. (2008). Antimicrobial resistance in foodborne pathogens--a cause for concern?. *Current Drug Targets, 9,* 808-815.

Wang, I. N., Smith, D. L. and Young, R. (2000). Holins: the protein clocks of bacteriophage infections. *Annual Review of Microbiology, 54,* 799-825.

WHO (World Health Organization). (2007). Food safety and foodborne illness. Fact sheet N°237. WHO Media Centre [http://www.who.int/mediacentre/factsheets/fs237/en/].

Woody, M. A. and Cliver, D. O. (1995). Effects of temperature and host cell growth phase on replication of F-specific RNA coliphage Q beta. *Applied and Environmental Microbiology, 61*, 1520-1526.

Zuber, S., Boissin-Delaporte, C., Michot, L., Iversen, C., Diep, B., Brüssow, H. and Breeuwer, P. (2008). Decreasing *Enterobacter sakazakii* (*Cronobacter* spp.) food contamination level with bacteriophages: Prospects and problems. *Microbial Biotechnology, 1*, 532-543.

INDEX

A

B

C

D

E

F

G

H

I

K

L

M

N

O

P

Q

R

S

T

U

V

W

Y